Social and Affective Neuroscience of Everyday Human Interaction

Paulo Sérgio Boggio • Tanja S. H. Wingenbach
Marília Lira da Silveira Coêlho
William Edgar Comfort
Lucas Murrins Marques
Marcus Vinicius C. Alves

Editors

Social and Affective Neuroscience of Everyday Human Interaction

From Theory to Methodology

 Springer

Editors
Paulo Sérgio Boggio (iD)
Social and Cognitive Neuroscience
Laboratory, Developmental Disorders
Program, Center for Health and Biological
Sciences
Mackenzie Presbyterian University
São Paulo, Brazil

Marília Lira da Silveira Coêlho (iD)
Social and Cognitive Neuroscience
Laboratory, Developmental Disorders
Program, Center for Health and Biological
Sciences
Mackenzie Presbyterian University
São Paulo, Brazil

Lucas Murrins Marques (iD)
Instituto de Medicina Fisica e Reabilitacao,
Hospital das Clinicas HCFMUSP
Faculdade de Medicina, Universidade
de Sao Paulo
São Paulo, Brazil

Tanja S. H. Wingenbach (iD)
School of Human Sciences
Faculty of Education, Health, and
Human Sciences
University of Greenwich
Greenwich, London, UK

William Edgar Comfort (iD)
Social and Cognitive Neuroscience
Laboratory, Developmental Disorders
Program, Center for Health and Biological
Sciences
Mackenzie Presbyterian University
São Paulo, Brazil

Marcus Vinicius C. Alves (iD)
Faculty of Health Sciences of Trairi
Universidade Federal do Rio Grande
do Norte
Santa Cruz, Brazil

ISBN 978-3-031-08653-3   ISBN 978-3-031-08651-9   (eBook)
https://doi.org/10.1007/978-3-031-08651-9

This Springer imprint is published by the registered company Springer Nature Switzerland AG
The registered company address is: Gewerbestrasse 11, 6330 Cham, Switzerland

Foreword: "Social and Affective Neuroscience: Open Questions Worth Trying to Answer"

Social and affective neuroscience has exploded in visibility and popularity over the past decade, with its own meetings, journals, and even graduate programs. And rightly so: the topic is not only inherently interesting, but of very high clinical relevance, strongly interdisciplinary, and, perhaps most important for the young scientist, full of open unanswered questions. In short, it offers something for everyone. So does this collection of chapters, which weaves a comprehensive path through the field, surveying theories, basic and clinical research, and methods. Along the way, it raises perhaps the most interesting—and most difficult questions facing the field—questions that should be required material to ponder for every student and faculty. The very first set of Chapters already introduces us to these.

A perennial question begins with the words we use. What are the categories or dimensions that can help organize our understanding of social behavior? Many schemes are possible. Perhaps the most general is a dimension of approach or avoidance, but this is too broad since it is not specific to the social domain. Emotions (the topic of Chap. 1) are more finely differentiated, but also not uniquely social. Modules for mating or aggression have been well-studied in animals and seem both social and categorical. Chapter 7 tells us about categories of moral judgment. How do we make sense of this variety?

One nagging worry is that the schemes we currently have available to make sense of it are all made up in the minds of scientists. That is, they are derivative to the words and concepts we happen to have for understanding the social world and, as such, are no better than folk psychology. In support of this view are constructivist theories of emotion, which point to the failure of cognitive neuroscience to find any specific dimensions or categories in the brain. Several chapters, especially the methodological ones in the latter part of the book, provide an antidote to this view: one just needs the right measures. Just like studies in animals have certainly found quite specific circuits for different types of social behavior, so too can we find systems for emotions in humans if only we develop the right tools. Nonetheless, it is clear that our current schemes will require revision—a requirement that is important not only theoretically but also for the practical diagnosis of disorders.

The second chapter introduces another long-standing question: To what extent is our social and emotional behavior rapid, modular, and automatic, and to what extent is it under deliberative control? This question is of course a species of the long-standing dual processing view in cognitive psychology. Once again complementing the methods Chapters at the end, this question is addressed with a multimodal approach that amasses evidence from many sources, and that also leaves us with one of the most popular anatomical answers: the amygdala subserves rapid, non-conscious processing. The view is not without its distractors (myself amongst them), but it remains a viable hypothesis and one with patent clinical implications.

While the dual processing view originally focused on controlled versus automatic processing, it has since become associated with a myriad of attributes—cognitive/emotional, slow/fast, conscious/nonconscious, and—the topic of Chap. 3—brain/body. The role of the body, or neural representations of it, has been emphasized at both sensory and motor ends of social cognition, and there is now substantial evidence that embodiment matters. But this finding raises what is perhaps one of the deepest questions: are some psychological variables literally in the body? What exactly are the commitments to constitutive and causal relations in social and affective neuroscience? Is an emotion or a personality trait in the behavior, in the body, in the brain, or in all of these? My colleagues and I have recently argued that these should be thought of as in the brain, albeit with of course potent causal connections to body, genes, and environment.

This first batch of most difficult questions is followed by a set of no less interesting but perhaps more practically oriented questions that motivate several of the chapters in the middle of the book. Mirror neurons illustrate the power of a cross-species approach, and an analysis of sex differences raises the importance of considering development: clearly, both comparative and developmental approaches need to be well represented in a mature social affective neuroscience. We are also confronted with vexing challenges whose solution may not be an empirical finding, but rather a practical shift in the field itself. One ubiquitous challenge, not only in this discipline but in many, is simply imprecise language. Findings from the brain are "involved in," "related to," "underpin," or perhaps even are "important to" various presumed functions. But what exactly does this mean? That they play a causal role? Usually this is not tested. That they are the only mechanisms causing a certain function? Even less so. In the end, understanding human social and affective behavior requires not only multiple methods, comparisons with animal studies, and new theories, but it also requires more rigor in our terminology.

The book concludes with a nice overview of the many methods used in this field, ranging from ERPs to facial EMG, TMS, fMRI hyperscanning, and others. In many cases, these are illustrated with respect to specific applications and research questions, ranging from examples of studies in language to tutorials on machine learning. In the end, the reader will have traveled through theory, clinical application, numerous methods, and a lot of specific case studies of topics, well illustrating the richness of this vibrant field.

Two chapters on mirror neurons and on the brain's default-mode network tie together the need for strong methods development in tandem with tackling deep conceptual questions and raise perhaps the biggest question for the field: what exactly is social and affective neuroscience? This question brings us back in a way to the first question that I raised. Is social affective neuroscience merely neuroscience of a particular topic, just like chair perception neuroscience, or cake tasting neuroscience, or any other arbitrary topic? Or are there systems, specializations of some sort in the brain that show us that this discipline really carves nature at its joints in some way, and is not merely the invention of its practitioners? Quick answers have not stood up to scrutiny. True, the default-mode network is engaged during mind wandering and social cognition—but it is also there to some degree in monkeys and under anesthesia, examples where mind wandering seems unlikely. Specific structures like the amygdala and ventromedial prefrontal cortex are often brought up—but these are involved in many functions. The way forward in both cases may be to forge methods that can discover more subtle functional network signatures or specific neuronal subpopulations.

By way of closing on a personal note, this last question, related to the first, has also been a topic I have debated. Whereas my colleague Lisa Feldman Barrett thinks that emotion categories are constructed in some way, I think they are objectively to be discovered in Nature. Neither view is straightforward to unpack, but our debates have certainly helped me identify some of the difficulties with my own view. I also think this kind of back-and-forth characterizes the atmosphere of social affective neuroscience. We realize that the questions are fun but difficult, and we acknowledge that any findings are always preliminary, and often turn out to be wrong. Like much of neuroscience and psychology, the field has weathered the replication crisis and has put in place more robust practices and analyses. Yet the very complexity and interdisciplinary nature of social and affective neuroscience defy any simple formula for how to do research in it. In the end, understanding human social and affective behavior requires not only multiple methods, comparisons with animal studies, and new theories; it requires a lot of ingenuity and detective work to tell a plausible story. Chapters collected here will give readers a good sense of these issues, and I hope they will motivate them not only to read more but also to question and debate.

California Institute of Technology   Ralph Adolphs
Pasadena, CA, USA

Preface

This book on social and affective neuroscience was inspired by an event, the Sao Paulo School of Advanced Science on Social and Affective Neuroscience (SPSAN), which took place in August 2018 over a period of 10 days. The SPSAN was organised by the Social and Cognitive Neuroscience Laboratory (SCNL) of the Mackenzie Presbyterian University in Sao Paulo, Brazil, and funded by the Sao Paulo Research Foundation (FAPESP). The aim of the SPSAN was to deepen the understanding of social and affective neuroscience and learn more about how the current state of research can explain phenomena we experience in our daily lives as social beings. The event constituted a unique opportunity for 100 competitively selected undergraduate and postgraduate students as well as early career researchers (at post-doc level) from all over the world to be part of ten knowledge-enriching days of theoretical and practical learning. National and international leading scientists from the field of social and affective neuroscience were recruited as speakers and flown in.

The editors of this book would like to thank all speakers that contributed to the SPSAN and all participants who engaged in the event and together made it the success it was. We are grateful to FAPESP for their generous funding, which allowed us to bring together scientists from all over the world. We also thank the Brazilian Academy of Science and the companies 'Rogue', 'Natura', 'Proibras' and 'Ergoneers' for their support. We would like to express thanks to Fanny Lachat who was part of the core team developing the theme that became the SPSAN. The event would not have been possible without the help and hard work of all the volunteers from the SCNL, especially Ruth Lyra Espinosa, Carol Nakao, Beatriz Ribeiro, Leticia Yumi, Patricia Cabral, Graziela Bonato, Fernanda Pantaleão, Carolina Gudin and Camila Valim. Last but not least, we are grateful to the Mackenzie Presbyterian University who provided us with the needed physical space and technical support to host the event.

The SPSAN was an enriching and inspiring event. With the intent to present a state-of-the-art overview of current topics within social and affective neuroscience, the idea of this anthology was born. Speakers from the SPSAN alongside other leading researchers from social and affective neuroscience contributed to the book. With slightly varying areas of expertise, the book provides some answers to the question of how social and affective neuroscience can explain various aspects of human everyday interaction. The book entails current state methods and theory of social and affective neuroscience applying an evidence-based approach. Combining the knowledge on social and affective neuroscience with the methodology to conduct social-affective neuroscientific research, the book is likewise of interest to researchers, university teachers and laymen with interest in the topics.

São Paulo, Brazil   Paulo Sérgio Boggio
London, UK   Tanja S. H. Wingenbach
São Paulo, Brazil   Marília Lira da Silveira Coêlho
São Paulo, Brazil   William Edgar Comfort
São Paulo, Brazil   Lucas Murrins Marques
Santa Cruz, Brazil   Marcus Vinicius C. Alves

About the Book

This book seeks to address central aspects for the scientific understanding of social and affective neuroscience as a whole. The book contains four parts: (I) Affective Neuroscience; (II) Social Neuroscience and Moral Emotions; (III) Clinical Neuroscience; and (IV) Methods Used in Social and Affective Neuroscience.

The first part, *Affective Neuroscience*, presents the current state of affective neuroscience research. The term 'affective' relates to moods and emotions and their processing, which plays a crucial role in human social interactions. We are constantly presented with our own emotions and moods and those of others. In social interactions, perceived affect processing and the processing of one's own affect constitute ongoing necessities. Affective states guide our attention as well as motivation and thus have an effect on social interactions. The chapters in this part investigate psychological, neural and molecular aspects of affective neuroscience.

The *Social Neuroscience and Moral Emotions* part covers phenomena present in society. Social and moral emotions guide our behaviour towards others, but the magnitude with which individuals experience these emotions varies greatly. The chapters in this part present neurobiological and behavioural processes in relevance to social interactions covering topics such mirror neurons and sex differences in social cognition as well as the development of morality and trust in the realm of social interaction.

The *Clinical Neuroscience* part focuses on disorders/conditions that affect social cognition. As much as neuroscience can be used to explain everyday phenomena in social interactions, neuroscience can explain disorders/conditions of clinical relevance. The investigation of brains of healthy individuals compared to those with clinical diagnoses provides invaluable information on the disorders and the associated symptoms. With some conditions affecting social functioning, atypical brain processes can explain abnormalities in regard to social skills. Neuroscience can further explain emotion regulation and deficiency thereof. The chapters in this part stretch from clinical neuroscience in childhood and adolescence to adulthood

The last part covers *Methods Used in Social and Affective Neuroscience*. Experts present state-of-the-art methods to investigate social and affective neuroscience with the typical currently widely applied equipment. That is, brain imaging (MRI,

NIRS), electrophysiology (EEG, facial EMG), brain stimulation (TMS AND tDCS) and eye-tracking. The chapters in this part familiarise the reader with the listed methods by providing the necessary basic information but also to deepen the understanding and usability of these methods in social and affective neuroscience for more experienced readers.

Contents

Part I   Affective Neuroscience

1   **Molecular Imaging of the Human Emotion Circuit**   3
    Lauri Nummenmaa, Kerttu Seppälä, and Vesa Putkinen

2   **The Neurocognitive Mechanisms of Unconscious Emotional
    Responses**   23
    Wataru Sato

3   **Social and Affective Neuroscience of Embodiment**   37
    Marília Lira da Silveira Coêlho, Tanja S. H. Wingenbach,
    and Paulo Sérgio Boggio

4   **The Neuroscience of Beauty**   53
    William Edgar Comfort and Ana Luísa Freitas

Part II   Social Neuroscience and Moral Emotions

5   **Mirror Neurons in Action: ERPs and Neuroimaging Evidence**   65
    Alice Mado Proverbio and Alberto Zani

6   **Sex Differences in Social Cognition**   85
    Alice Mado Proverbio

7   **Development of Morality and Emotional Processing**   107
    Lucas Murrins Marques, Patrícia Cabral, William Edgar Comfort,
    and Paulo Sérgio Boggio

8   **Trust in Social Interaction: From Dyads to Civilizations**   119
    Leonardo Christov-Moore, Dimitris Bolis, Jonas Kaplan,
    Leonhard Schilbach, and Marco Iacoboni

Part III   Clinical Neuroscience

9   **The Time Has Come to Be Mindwanderful: Mind Wandering
    and the Intuitive Psychology Mode** 145
    Óscar F. Gonçalves and Mariana Rachel Dias da Silva

10 **Social Cognition Development and Socioaffective Dysfunction
    in Childhood and Adolescence** 161
    Claudia Berlim de Mello, Thiago da Silva Gusmão Cardoso,
    and Marcus Vinicius C. Alves

11 **Clinical Neuroscience Meets Second-Person Neuropsychiatry** 177
    Leonhard Schilbach and Juha M. Lahnakoski

Part IV   Methods Used in Social and Affective Neuroscience

12 **EEG and ERPs in the Study of Language and Social
    Knowledge** 195
    Alice Mado Proverbio

13 **Brain Imaging Methods in Social and Affective Neuroscience:
    A Machine Learning Perspective** 213
    Lucas R. Trambaiolli, Claudinei E. Biazoli Jr, and João R. Sato

14 **fMRI and fNIRS Methods for Social Brain Studies:
    Hyperscanning Possibilities** 231
    Paulo Rodrigo Bazán and Edson Amaro Jr

15 **Modulating the Social and Affective Brain with Transcranial
    Stimulation Techniques** 255
    Gabriel Rego, Lucas Murrins Marques, Marília Lira
    da Silveira Coêlho, and Paulo Sérgio Boggio

16 **What Our Eyes Can Tell Us About Our Social and Affective
    Brain?** 271
    Paulo Guirro Laurence, Katerina Lukasova,
    Marcus Vinicius C. Alves, and Elizeu Coutinho de Macedo

17 **Facial EMG – Investigating the Interplay of Facial Muscles
    and Emotions** 283
    Tanja S. H. Wingenbach

Index 301

Contributors

Marcus Vinicius C. Alves Faculty of Health Sciences of Trairi, Universidade Federal do Rio Grande do Norte, Santa Cruz, Brazil

Edson Amaro Jr. LIM-44, Departamento de Radiologia, Hospital das Clínicas da Faculdade de Medicina da Universidade de São Paulo, São Paulo, Brazil

Paulo Rodrigo Bazán LIM-44, Departamento de Radiologia, Hospital das Clínicas da Faculdade de Medicina da Universidade de São Paulo, São Paulo, Brazil

Hospital Israelita, Albert Einstein, São Paulo, Brazil

Claudinei E. Biazoli Jr. Center for Mathematics, Computing, and Cognition, Federal University of ABC, São Bernardo do Campo, Brazil

Paulo Sérgio Boggio Social and Cognitive Neuroscience Laboratory, Developmental Disorders Program, Center for Health and Biological Sciences, Mackenzie Presbyterian University, São Paulo, Brazil

Dimitris Bolis Independent Max Planck Research Group for Social Neuroscience, Max Planck Institute of Psychiatry, Munich-Schwabing, Germany

International Max Planck Research School for Translational Psychiatry (IMPRS-TP), Munich, Germany

Munich Medical Research School (MMRS), Dekanat der Medizinischen Fakultat, Ludwig-Maximilians- Universitat Munchen, Munich, Germany

Patrícia Cabral Social and Cognitive Neuroscience Laboratory, Developmental Disorders Program, Center for Health and Biological Sciences, Mackenzie Presbyterian University, São Paulo, Brazil

Leonardo Christov-Moore Brain and Creativity Institute, University of Southern California, Los Angeles, CA, USA

William Edgar Comfort Social and Cognitive Neuroscience Laboratory, Developmental Disorders Program, Center for Health and Biological Sciences, Mackenzie Presbyterian University, São Paulo, Brazil

Thiago da Silva Gusmão Cardoso Centro Adventista Universitário de São Paulo, São Paulo, Brazil

Claudia Berlim de Mello Department of Psychobiology, Universidade Federal de São Paulo, São Paulo, Brazil

Mariana Rachel Dias da Silva Tilburg University Cognitive Science and Artificial Intelligence Department, Tilburg, The Netherlands

Ana Luísa Freitas Social and Cognitive Neuroscience Laboratory, Developmental Disorders Program, Center for Health and Biological Sciences, Mackenzie Presbyterian University, São Paulo, Brazil

Óscar F. Gonçalves Proaction Lab, CINEICC – Faculty of Psychology and Educational Sciences, University of Coimbra, Coimbra, Portugal

Marco Iacoboni Department of Psychiatry and Biobehavioral Sciences, Ahmanson-Lovelace Brain Mapping Center, Brain Research Institute, David Geffen School of Medicine at UCLA, Los Angeles, CA, USA

Jonas Kaplan Brain and Creativity Institute, University of Southern California, Los Angeles, CA, USA

Juha M. Lahnakoski Forschungszentrum Jülich, Institute of Neurosciences and Medicine (INM), Jülich, Germany

Paulo Guirro Laurence Social and Cognitive Neuroscience Laboratory and Developmental Disorders Program, Center for Health and Biological Sciences, Mackenzie Presbyterian University, São Paulo, Brazil

Katerina Lukasova Postgraduate Program in Neuroscience and Cognition – PPGNC, Federal University of ABC – UFABC, São Bernardo, Brazil

Elizeu Coutinho de Macedo Social and Cognitive Neuroscience Laboratory and Developmental Disorders Program, Center for Health and Biological Sciences, Mackenzie Presbyterian University, São Paulo, Brazil

Lucas Murrins Marques Instituto de Medicina Fisica e Reabilitacao, Hospital das Clinicas HCFMUSP, Faculdade de Medicina, Universidade de Sao Paulo, Sao Paulo, Brazil

Lauri Nummenmaa Turku PET Centre and Turku University Hospital, Turku, Finland
Department of Psychology, University of Turku, Turku, Finland

Alice Mado Proverbio Department of Psychology, University of Milano-Bicocca, Milan, Italy

Vesa Putkinen Turku PET Centre and Turku University Hospital, Turku, Finland

Gabriel Rego Social and Cognitive Neuroscience Laboratory, Developmental Disorders Program, Center for Health and Biological Sciences, Mackenzie Presbyterian University, São Paulo, Brazil

João R. Sato Center for Mathematics, Computing, and Cognition, Federal University of ABC, São Bernardo do Campo, Brazil

Wataru Sato Psychological Process Research Team, Guardian Robot Project, RIKEN, Kyoto, Japan

Leonhard Schilbach Independent Max Planck Research Group for Social Neuroscience, Max Planck Institute of Psychiatry, Munich-Schwabing, Germany

LVR Klinikum Dusseldorf/Kliniken der Heinrich-Heine-Universitat Dusseldorf, Düsseldorf, Germany

Ludwig-Maximilians-Universitat, Medical Faculty, Munich, Germany

Kerttu Seppälä Turku PET Centre and Turku University Hospital, Turku, Finland

Marilia Lira da Silveira Coelho Social and Cognitive Neuroscience Laboratory, Developmental Disorders Program, Center for Health and Biological Sciences, Mackenzie Presbyterian University, São Paulo, Brazil

Lucas R. Trambaiolli Basic Neuroscience Division, Mclean Hospital – Harvard Medical School, Belmont, MA, USA

Tanja S. H. Wingenbach School of Human Sciences, Faculty of Education, Health, and Human Sciences, University of Greenwich, Greenwich, London, UK

Alberto Zani School of Psychology, Vita-Salute San Raffaele University, Milan, Italy

About the Editors

Paulo Sérgio Boggio holds a Bachelor's degree in Psychology from the University of São Paulo (1998), a professional development certificate in Neuropsychology from the Neurology unit at USP Medical School, a Master's degree in Experimental Psychology from the University of São Paulo (2004) and a PhD in Psychology (Neuroscience and Behavior) from the University of São Paulo (2007). He leads the Laboratory of Cognitive and Social Neuroscience at the Center for Health and Biological Sciences, Mackenzie Presbyterian University. He is a professor in Developmental Disorders and Psychology at Mackenzie Presbyterian University. In 2011, he was elected an affiliate member of the Brazilian Academy of Sciences. He is a research productivity Fellow at CNPq – Level 1C. He has wide experience in the field of psychology, with an emphasis on neuropsychology and cognitive, social and affective neuroscience. He also has experience with the use of non-invasive techniques of brain stimulation, high-density EEG, eye-tracking and other techniques.

Tanja S. H. Wingenbach holds a B.Sc. in Psychology (University of Luxembourg), a M.Sc. in Clinical Psychology (University of Basel), and a PhD in Psychology (University of Bath). From 2017–2019, she held a post-doctoral fellowship at the Social and Cognitive Neuroscience Laboratory, Mackenzie Presbyterian University in Sao Paulo. She was a post-doctoral senior research fellow at the University of Zurich / University Hospital Zurich from 2019–2022, after which she joined the University of Greenwich as a Lecturer in Psychology. Her research falls within emotion sciences and encompasses typical as well as atypical populations.

Marília Lira da Silveira Coêlho obtained her Bachelor's degree in Physiotherapy from University Center of Bahia in 2007, holds a Master's in Health and Medicine from Federal University of Bahia (2010) and a PhD in Developmental Disorders from Mackenzie Presbyterian University (2017). She is a professor in the Department of Physiotherapy and researcher at the Social and Cognitive Neuroscience Laboratory at the Center for Health and Biological Sciences, Mackenzie Presbyterian University. Her main research interests are multisensory integration, embodiment and ownership experiences and non-invasive techniques of brain stimulation.

William Edgar Comfort holds a BSc in Psychology from Bangor University (2007), an MSc in Neuropsychology from the University of Bristol (2012), with a clinical internship at the Head Injury Therapy Unit (HITU) of Frenchay Hospital, Bristol, and a PhD in Neuroscience and Cognition from the Universidade Federal do ABC (2015). He was previously an associate professor of Educational Psychology at Osasco University (UNIFIEO). He is currently a FAPESP-funded post-doctoral research fellow at the Social and Cognitive Neuroscience Laboratory of Mackenzie Presbyterian University and teaches Neuroscience and Applied Psychology at Mackenzie Presbyterian University. His research interests are in psychophysics, computational modelling, eyetracking, EMG, facial recognition, neural processing of facial expressions and visual perception impairments in schizophrenia.

Lucas Murrins Marques obtained his Bachelor in Psychology from Mackenzie Presbyterian University in 2014, holds a Master's degree in Basic Psychology from Minho University (2013) a Master's in Developmental Disorders from Mackenzie Presbyterian University (2016) and a PhD in Developmental Disorders from Mackenzie Presbyterian University (2020). He is currently a post-doctoral research fellow at IMREA from the Faculty of Medicine of University of São Paulo (Grant 2021/05897-5, São Paulo Research Foundation; FAPESP). His main research interests are in brain stimulation, moral judgement, psychophysiology, machine learning, emotion regulation and well-being.

Marcus Vinicius C. Alves is a Professor at Universidade Federal do Rio Grande do Norte (UFRN) and has a PhD in Psychobiology at Universidade Federal de São Paulo (Brazil) and experience as a visiting researcher at Memorial University of Newfoundland (Canada) and as a research and development specialist at Université du Luxembourg (Luxembourg). He obtained his Bachelor's degree in Psychology from the Federal University of Bahia in 2011 and holds a Master's in Psychobiology from the Federal University of São Paulo (2013). Alves' research interests include learning, memory, incidental and motivated forgetting, mental effort (processing systems, attention load, cognitive load theory and executive control) and social cognition, using mostly eye tracking and pupillometry.

Abbreviations

Aus	Action Units
ADT	Acute Tryptophan Depletion
ACEs	Adverse Childhood Experiences
ANOVA	Analysis of Variance
ACC	Anterior Cingulate Cortex
aIPS	Anterior Intraparietal Sulcus
pre-SMA	Anterior Supplementary Motor Area
AUC	Area Under the Curve
ANN	Artificial Neural Networks
ADHD	Attention Deficit Hyperactivity Disorder
ASD	Autistic Spectrum Disorder
BOLD	Blood Oxygenation Level Dependent
BCI	Brain-Computer Interfaces
CNS	Central Nervous System
CT	Computed Tomography
DMN	Default-Mode Network
DSM	Diagnostic and Statistical Manual of Mental Disorders
DA	Dopamine
DRN	Dorsal Raphe Nucleus
DLPFC	Dorsolateral Prefrontal Cortex
DEBQ	Dutch Eating Behavior Questionnaire
EPI	Echo Planar Imaging
EEG	Electroencephalography
EMG	Electromyography
ERD	Event-Related beta and mu Desynchronization
ERF	Event-Related Fields
ERPs	Event-Related Potentials
EBA	Extrastriate Body Area
EMDR	Eye Movement Desensitization and Reprocessing
FACS	Facial Action Coding System
fEMG	Facial Electromyography

fMRI	Functional Magnetic Resonance Imaging
fNIRS	Functional Near-Infrared Spectroscopy
FFA	Fusiform Face Area
FG	Fusiform Gyrus
IATs	Implicit Association Tests
racial IAT	Implicit Racial Attitude task
IFG	Inferior Frontal Gyrus
IPC	Inferior Parietal Cortex
IAPS	International Affective Picture System
ICD	International Classification of Diseases
LPC	Late Positive Component
LVF	Left Visual Field
LPN	Lexical Processing Negativity
LDA	Linear Discriminant Analysis
LORETA	LOw-REsolution brain electromagnetic TomogrAphy
ML	Machine Learning
MR	Magnetic Resonance
MRI	Magnetic Resonance Imaging
MEG	Magnetoencephalography
mdFC	Medial Frontal Cortex
mOFC	Medial Orbitofrontal Cortex
MPFC	Medial Prefrontal Cortex
μV	Microvolt
mdFG	Middle Frontal Gyri
MOG	Middle Occipital Gyrus
MTG	Middle Temporal Gyri
mV	Millivolt
MNS	Mirror Neuron System
MNs	Mirror Neurons
MFT	Moral Foundations Theory
MFVs	Moral Foundations Vignettes
MEPs	Motor Evoked Potentials
VIN−	Negative electrode
NAcc	Nucleus Accumbens
OFCA	Occipital Face Area
OFC	Orbitofrontal cortex
ORE	Other-Race Effect
ORB	Own-Race Bias
OXTR	Oxytocin Receptor
PPA	Parahippocampal Area
APTD	Phenylalanine/Tyrosine Depletion
VIN+	Positive electrode
PET	Positron Emission Tomography
pCC	Posterior Cingulate Cortex
PPC	Posterior Parietal Cortex

PMd	Premotor Dorsal Areas
M1	Primary Motor Cortex
P1	Purkinje Image
REM	Rapid Eye Movement
RSVP	Rapid Serial Visual Presentation
RT	Reaction Time
ROC	Receiver Operating Characteristic
RDoC	Research Domain Criteria
SSRIs	Serotonin Reuptake Inhibitors
SERT	Serotonin Transporter
SWS	Slow Wave Sleep
SEL	Social and Emotional Learning
SC	Somatosensory Cortex
swLORETA	Standardized Weighted LOw-Resolution Electromagnetic Tomography
SFG	Superior Frontal Gyrus
STC	Superior Temporal Cortex
STG	Superior Temporal Gyri
STS	Superior Temporal Sulcus
SMA	Supplementary Motor Area
SVM	Support Vector Machines
TPJ	Temporal/Parietal Junction
TDM	Theory of Dyadic Morality
TOM	Theory of Mind
TBS	Transcranial Brain Stimulation
tDCS	Transcranial Direct Current Stimulation
TMS	Transcranial Magnetic Stimulation
D2R	Type 2 Dopamine Receptors
VIP	Ventral Intraparietal Areas
PMv cortex	Ventral Premotor Cortex
VTA	Ventral Tegmental Area
VMPFC	Ventromedial Prefrontal Cortex
VPP	Vertex Positive Potential
VR	Virtual Reality
VVS	Visual Ventral Stream
MOR	μ-Opioid Receptor

List of Figures

Fig. 1.1   Statistical summary of brain regions involved in emotional
            processing based on the NeuroSynth database
            (Yarkoni et al., 2011)... 4
Fig. 1.2   Distribution of type-2 dopamine receptors, μ-opioid receptors,
            and 5-HT 1A transporters measured using PET radioligands........ 5
Fig. 1.3   Main dopamine pathways in the brain.. 8
Fig. 1.4   Organization of the human opioid system in the brain.
            Note that as specific opioid neuron projections
            cannot be established, this figure instead characterizes
            the relative expression of different receptor subtypes in
            some of the key nodes of the emotion circuit.............................. 11
Fig. 1.5   Main serotonin pathways in the brain 13

Fig. 2.1   Sato et al.'s (2014b) study. (Upper) An illustration
            of the trial sequence. The prime stimuli of dynamic
            and static facial expressions of fear and happiness were
            presented subliminally. (Lower) Mean (± *SE*) preference
            ratings. The asterisks indicate a significant difference
            between the fear and happiness conditions 25
Fig. 2.2   Sato et al.'s (2016) study. (Upper) The illustrations of
            food and mosaic stimuli. (Lower left) Mean (± *SE*) preference
            ratings in the subliminal condition. The asterisk indicates a
            significant difference between food and mosaic prime conditions.
            (Lower right) A scatter plot with a regression line showing
            a relationship between food preference scores under the
            subliminal condition and external eating tendency.
            The asterisk indicates a significant association 28
Fig. 2.3   Sato et al.'s (2019) study. (Upper left) Statistical
            parametric maps showing significant neural activation in
            response to food versus mosaic images under both the sublimi-
            nal and supraliminal conditions and mean (± *SE*) effect size
            differences between the food and mosaic conditions. The blue
            cross indicates the activation focus at the right amygdala.

(Upper right) Statistical parametric maps showing significantly stronger neural responses to food versus mosaic images under the supraliminal than subliminal condition and mean (± *SE*) effect size differences between the food and mosaic conditions. The blue cross indicates the activation focus at the right fusiform gyrus. (Lower) Models (left) and model comparison results (right) of dynamic causal modeling. The solid and dashed arrows indicate modulatory connections in the subcortical and cortical pathway models, respectively. The dual pathways model contains both pathways. The model comparison results in the right hemisphere are shown. *AMY* amygdala, *FG* fusiform gyrus, *PUL* pulvinar, *V1* primary visual cortex......... 30

Fig. 2.4   Sato et al.'s (2011) study. (Upper) Representative anatomical magnetic resonance image. The red cross indicates the location of the amygdala electrode. (Lower) Statistical parametric maps for amygdala gamma-band activation for fearful compared with neutral facial expressions (left) and mean (± *SE*) effect size at the peak activation focus (right) 33

Fig. 4.1   Salvador Dalí's lobster telephone (1938) 54
Fig. 4.2   The triad of aesthetic experience (Chatterjee & Vartanian, 2016).... 55

Fig. 5.1   Adjusted mean regional cerebral blood flow recorded by Rizzolatti et al. (1996) during grasping observation. The data are displayed as statistical maps overimposed on three planar projections (sagittal, coronal, and transverse) frames and as cortical rendering of the lateral cortical surfaces of the left hemisphere. The pixel values significantly higher than $p < 0.001$ are shown in red... 67

Fig. 5.2   Examples of pictures depicting bimanual and unimanual tools used as stimuli in Proverbio's et al. (2013) ERP study.......... 68

Fig. 5.3   Types of stimuli used in Iacoboni's et al. (2005) study. The same action (e.g. to take a mug) reveals an agent's different intention according to the context in which he/she is. Such an intention is encoded and inferred by means of fronto-parietal MNS activation. (Courtesy of Marco Iacoboni) 69

Fig. 5.4   Examples of congruent (left column) and incongruent (right column) stimuli used in Proverbio's et al. studies (2010, 2014a, 2015a), associated with the N400 component electrophysiological effect (third column), reflecting the violation of an expectation related to the aim of the action or of the gesture expressed by the actors, as referred to a shared grammar of gestures or to the context and to the pre-established use of a tool. The N400 effect is drawn as a red continuous line in the upper waveforms, as a blue continuous line in the middle waves, and as a red dotted line in the lower waves (where ERPs are shown in red for women and in blue for men). (Reproduced and modified with the permission of the authors)... 71

Fig. 5.5   Grand-average ERPs recorded in professional basketball
            players (**a**) and naïve viewers (**b**) in response to correct and
            incorrect basketball actions at frontal, parietal, and occipital
            scalp sites. (Taken and redrawn from Proverbio et al., 2012)   73
Fig. 5.6   (Above) Examples of stimuli used for the study on neurons
            sensory preference (i.e. the face of a conspecific emitting a
            vocalization vs. the opening and closing of a disc without any
            facial stimulus). (Below) Bioelectrical responses displayed by a
            multisensory cell of the associative auditory cortex of the
            macaque monkey. Note that the response to the combined voice
            and face conditions (red line) is far superior than the uni-sen-
            sory stimulation (in this case, the response to the incongruous
            coupling between disk and voice that did not stimulate the cell
            enough is also drawn as a yellow line). (Adapted from
            Ghazanfar and Schroeder (2006). Courtesy of the authors)...........   75
Fig. 5.7   Visual stimulation consisted in the silent presentation of
            pictures of animals and tools while the auditory stimulation
            consisted of the blind presentation of their verse or typical
            sound. The audiovisual stimulation involved the integration
            between the two modes. Brain images show the BOLD signals
            of neurometabolic activation obtained by fMRI in the various
            stimulation conditions. Note that the audiovisual condition
            activated the multimodal prefrontal regions, as well as the
            motor and premotor cortices, the posterior region of the STS,
            and the MTG. (Drawn and modified by Beauchamp et al.
            (2004a, b). Courtesy of the authors)..   77
Fig. 5.8   Some examples of 'sound' (top) and "silent' (centre) visual
            stimuli presented together with other hundreds of stimuli to
            unaware observers, instructed to detect and respond to infrequent
            images of cycling races. The analysis of ERP peaks, together
            with the reconstruction of their intracerebral generators by
            means of the swLORETA technique, demonstrated the activation
            of the left medial temporal cortex after only 110 ms from the
            presentation of the image. The extraction of sound information
            associated with the use of familiar tools after ~200 ms activated
            the primary (BA38) and secondary (BA41) auditory cortices.
            This information is responsible, for example, for auditory
            hallucinations, which, in this case, refer, in a dim way, to the call
            of the specific sound produced by the tool (in the figure, the
            sounds produced by the sax or by the infernal chainsaw). (Taken
            from Proverbio et al. (2011b). Courtesy of the authors)...................   78
Fig. 5.9   (**a**) Examples of visual stimuli used in the study by Hasegawa
            et al. (2004) (**b**) Activation of the left temporal region as a
            function of musical performance in the three groups of partici-
            pants. (**c**) fMRI activations in response to an exclusively visual
            stimulation in the brain of professional pianists.
            (Courtesy of the authors)...   79

Fig. 5.10   Coronal, sagittal, and axial views of the standardized and
             weighted LOw REsolution electromagnetic TomogrAphy
             (swLORETA) applied to the N400 bioelectric response
             generated only for one's own musical instrument. (Taken from
             Proverbio et al. (2014) and redrawn)... 81

Fig. 6.1   Isocolor voltage topographical maps (left- and right-side views)
             showing N170 scalp distribution in female and male
             observers. N170 response is relative to adult face processing.
             The time window corresponds to its peak (150–170 ms)
             of maximum activation. (Taken from Proverbio et al. (2012),
             with permission from the authors and the editor).......................... 88

Fig. 6.2   N170 latency values (along with SD) recorded in women
             and men in response to lateralized faces, as a function
             of cerebral hemisphere and stimulus contra-laterality
             (collapsed across occipito/temporal electrode sites). In this
             study, ERPs were recorded in strictly right-handed people (16
             men and 17 women) engaged in a face-sex categorization task.
             Occipital P1 and occipito/temporal N170 were left lateralized
             in women and bilateral in men. N170 to contralateral stimuli
             was larger over the RH in men and the LH in women. Inter-
             hemispheric transfer time (IHTT) was approximately 4 ms at
             the P1 level and approximately 8 ms at the N170 level. It was
             asymmetric in men, with faster latencies in the left visual field
             (LVF)/RH → LH (170 ms) direction than in the right-visual
             field (RVF)/LH → RH (185 ms) direction and symmetric in
             women. These findings suggest that the asymmetry in callosal
             transfer times might be due to faster transmission times of
             face-related information via fibers departing from the more
             efficient to the less efficient hemisphere
             (Proverbio et al., 2012)... 89

Fig. 6.3   Examples of photographs used as stimuli, as a function of facial
             expressions (Proverbio et al., 2007). The upper row shows
             positive emotional states with strongly positive emotions,
             such as joy on the left, and mildly positive ones, such as
             comfort or peacefulness, on the right. The lower row shows
             negative emotional states with the mildly negative emotions,
             such as discomfort or disappointment, on the left and strongly
             negative ones, such as displeasure or pain, on the right 90

Fig. 6.4   Mean latency (in ms) of the P1 component (along with SD)
             recorded at the lateral occipital area (independent of hemi-
             spheric site) and analyzed according to subjects' sex and type
             of facial expression. (Taken and modified from Proverbio et al.,
             2006b study, with permission of the authors and the editor).......... 91

Fig. 6.5   ERPs signals recorded over left and right lateral occipital sites
            following presentations of infant facial expressions exhibiting
            strongly negative emotions, according to viewer group. Smaller
            P300 amplitudes were recorded in fathers vs. mothers, espe-
            cially with infant expressions of suffering. (Taken from
            Proverbio et al., 2006a, with authors' and editors' permission).....   94
Fig. 6.6   Examples of social and nonsocial stimuli used to evaluate
            the interest in social information, regardless of stimulus
            color richness and perceptual complexity. (Taken from
            Proverbio et al. (2008)'s study)   95
Fig. 6.7   ERP difference waves obtained by subtracting ERPs to congru-
            ent from ERPs to incongruent actions separately for men and
            women, over anterior scalp sites. A much larger N400 response
            occurred to incongruent actions in women than men. (Taken
            and modified from Proverbio et al., 2010c)....................................   97
Fig. 6.8   (Top) ERP waveforms recorded in women and men as a
            function of stimulus type. VPP was much larger to faces and
            faces-in-things than objects in women. (Bottom) Mean ampli-
            tude of the N170 response recorded as a function of stimulus
            type and relative scalp distribution...   99
Fig. 6.9   Data obtained from the emotional impact scale (self-reporting
            questionnaire) administered to the 24 persons participating in
            the ERP experiment, separately for each image type, and
            according to their sex. Key: 0 = not at all; 1 = a little; 2 = fairly;
            3 = very much; 4 = extremely...   100
Fig. 6.10   ERPs recorded at right parietal sites as a function of stimulus
            content and valence and viewer's sex. A large effect of both
            emotional content of the stimulus is visible (evidenced by
            comparing ERPs to negative vs. positive unanimated scenes)
            and an effect of empathy for pain, especially in women
            (evidenced by comparing ERPs to negative scenes vs. ERPs to
            pictures portraying humans)...   101

Fig. 8.1   Trust minimizes apparent prediction error and facilitates
            the interpersonal sharing of priors...   125
Fig. 8.2   At a collective level, trust facilitates efficient group
            behavior, reducing group complexity and overall energy
            consumption   129

Fig. 9.1   Nature and functions of mind wandering.......................................   147
Fig. 9.2   Mind wandering processes   149

Fig. 12.1   Time course of cerebral activation during the processing of
            linguistic material as reflected by the latency of occurrence of
            various ERP components. Prelinguistic stimulus sensory
            processing occurs (P1 component) at about 100 ms poststimu-

lus; orthographic analysis of written words (posterior N1
component) at 150–200 ms; phonologic/phonetic analysis at
200–300 ms, as revealed by phonological mismatch negativity
(temporal and anterior pMMN) seen in response to phonologic
incongruities (both visual and auditory); and a large centroparietal negativity at about 400 ms (N400), recorded in response to
semantic incongruities and indexing lexical access mechanisms.
The comprehension of meaningful sentences reaches consciousness between 300 and 500 ms (P300 component); finally,
a second-order syntactic analysis is indexed by the appearance
of a late positive deflection (P600) at about 600 ms poststimu-
lus latency 197

Fig. 12.2   Visual perception of words activates the left occipitotemporal
cortex at about 170 ms poststimulus. This response is much
larger to words than non-orthographic strings. Lexical
processing reaches its peak at about 400 ms 198

Fig. 12.3   ERP waveforms recorded in response to words, pseudo-words,
and letter strings during a phonetic decision task (e.g., "Is
phone/k/present in oranges?"). The first lexical effect was
found at P2 level. (Adapted from Proverbio et al. (2004), with
the permission of MIT Press) 201

Fig. 12.4   ERPs recorded in response to terminal words completing
a previous context (see the text for specific sentences) determining a violation of semantic constraint (case 1), contextual
meaning (case 2), or pragmatic knowledge (case 3). Solid line,
congruent word; dotted line, incongruent word. (Taken and
adapted from studies of Hagoort and coauthors (Hagoorth et al.,
2004; Van Berkum et al., 1999), with permission of authors)........ 202

Fig. 12.5   N400 and LAN components elicited by incongruent (with
respect to stereotypes) sentences over anterior scalp sites in
Proverbio et al. (2017) study. (Courtesy of the authors) 205

Fig. 12.6   Examples of how Proverbio et al. (2016) induced a positive or
negative prejudice about previously unknown persons. In the
encoding task, faces were presented in association with a short
story that provided fictional information about the character,
such as an anecdote or personal information. The biographic
information could be positive, thereby inducing a positive
prejudice toward the depicted character, or vice versa, a
negative prejudice could induce a negative bias. (Courtesy of
Proverbio and coauthors).. 206

Fig. 12.7   Sagittal views of active sources during processing of negatively
biased, positively biased, and new faces according to
swLORETA analysis during the 450–550 ms time window. The
images highlight the strong activation of the left middle frontal
gyrus during memory recall of faces associated with a negative

prejudice. (Taken from Proverbio et al. (2016) with permission
from the authors. Creative Commons Public Domain picture) 206

Fig. 12.8   Isocolor topographical maps (front view) of surface voltage
measured in the 250–400 ms temporal window (N400 latency
range) to incongruent stimuli as a function of participants' sex.
It can be appreciated how N400 response to stereotypes
violation was not found in female participants. This suggests
that female participants were not surprised by final words that
violated sex stereotypes. (Adapted from Proverbio et al., 2018)..... 209

Fig. 12.9   Coronal and axial brain sections showing the location and
strength of electromagnetic dipoles explaining the surface
difference voltage obtained by subtracting ERPs to congruent
from ERPs to incongruent stimuli in the 250–400 ms latency
range, corresponding to the peak of N400. *L* left, *R*, right, *A*
anterior, *P* posterior, *MTG* middle temporal gyrus, *MFG* middle
frontal gyrus. (Taken for Proverbio et al., 2018) 209

Fig. 13.1   Electromagnetic-based imaging approaches (left) use electric or
magnetic sensors to capture the electromagnetic resultants
from the neuronal and synaptic activity. Hemodynamic-based
procedures (right) use light or magnetic sensors to measure the
cerebral blood flow and oxygen consumption levels...................... 214

Fig. 13.2   Supervised learning methods use labeled examples to learn
from data, while unsupervised learning methods extract
patterns from data using unlabeled inputs. The recently
proposed semi-supervised approach, otherwise, combines both
labeled and unlabeled inputs during the learning process.............. 216

Fig. 13.3   Different steps and approaches for data splitting. (**a**) The first
step of the validation process is to select a sample subset for
testing purposes. Then, cross-validation approaches are used to
split the remaining data into training and validation subsets. (**b**)
During the k-fold cross-validation, data is split into k-folds of
similar lengths. Then, the algorithm is validated k times, until
all folds were used as the validation subset. (**c**) The leave-one-
out cross-validation is a particular case of k-fold cross-validation, where each fold corresponds to a single example. (**d**) The
Monte Carlo cross-validation performs a predetermined number
of combinations, where the validation subset is composed of a
fixed quantity of randomly selected samples................................. 218

Fig. 13.4   Level of interaction between the feature selection algorithm
and the classifier. (**a**) During the filter approach, the feature
selection is performed before and apart from the classifier.
(**b**) During the wrapper approach, every single feature subset is
submitted to the classifier, and the classification performance
is used to evaluate the sample. (**c**) During the embedded
procedure, both the feature selection and the classifier
algorithms are merged and happen simultaneously........................ 220

Fig. 13.5   Examples of classifiers commonly applied to
             neuroimaging studies. (a) A decision tree, (b) artificial neural
             networks, (c) linear discrimination analysis, (d) support vector
             machines 221

Fig. 13.6   Illustrative example of (a) a confusion matrix and (b)
             three different examples of ROC curves representing
             classifiers with excellent (dotted line), good (dashed line),
             and bad (continuous line) performances 222

Fig. 14.1   Hyperscanning options: (a) using one data acquisition device
             for each subject and using synchrony devices to assure synchro-
             nized data acquisition; (b) sharing the same data acquisition
             device for all the subjects and providing synchrony between
             data from all subjects but limiting distance between subjects
             and reducing the number of channels (fNIRS). For schematic
             representation, fNIRS was presented in the image, but similar
             considerations are valid for fMRI hyperscanning. *LSL* lab
             streaming layer 237

Fig. 14.2   Types of social brain experiments and analysis possibilities. (a)
             Single subject brain signal acquisition; single subject task. (b)
             Single subject brain signal acquisition; multi-subject task. (c)
             Multi-subject brain signal acquisition; multi-subject task. 241

Fig. 16.1   Regions of interest in the face for emotion recognition.
             (This image and the regions of interest were based on the
             manuscript of Schurgin et al. (2014)) 273

Fig. 17.1   Facial muscle responses to facial expression of anger *Note*.
             This figure is a composite of Figures 2 and 4 in Wingenbach
             et al. (2020). The first column shows the five measured facial
             muscle sites and the second column the expected facial muscle
             responses when participants viewed angry facial expressions.
             Blue bars indicate an (expected) increase compared to a
             prestimulus baseline, and gold bars indicate an (expected)
             decrease. The third column shows the measured facial muscle
             responses to angry faces; the EMG data were range-corrected,
             and no increase in activity occurred in the corrugator. The
             fourth column shows the z-standardised means with positive
             z-values for the corrugator, which are in fact based on a
             decrease in corrugator activity in response to angry faces 295

List of Tables

Table 6.1   Main gender differences related to facial expression and
            decoding abilities... 91

Table 8.1   Glossary of terms... 122

Table 12.1 Example of sentence stimuli, relative to men or women,
            and violating or not current occupational gender stereotypes
            in which women engage more in care-related professions
            and men in strength–/power-related professions............................ 208

Table 17.1 Basic emotions with associated AUs and facial muscles............... 286

Part I
Affective Neuroscience

Chapter 1
Molecular Imaging of the Human Emotion Circuit

Lauri Nummenmaa, Kerttu Seppälä, and Vesa Putkinen

Abstract Emotions modulate behavioral priorities via central and peripheral nervous systems. Understanding emotions from the perspective of specific neurotransmitter systems is critical, because of the central role of affect in multiple psychopathologies and the role of specific neuroreceptor systems as corresponding drug targets. Here, we provide an integrative overview of molecular imaging studies that have targeted the human emotion circuit at the level of specific neuroreceptors and transmitters. We focus specifically on opioid, dopamine, and serotonin systems, given their key role in modulating motivation and emotions, and discuss how they contribute to both healthy and pathological emotions.

Keywords Molecular imaging · Human emotions · Dopamine system · Serotonin system · Opioid system

Introduction

Emotions prepare us for action. They coordinate systemic activation patterns at multiple physiological and behavioral scales to promote survival. Most modern emotion theories consider emotions as modulatory systems interacting with both lower-order systems, such as those involved in homeostasis, as well as higher-order cognitive circuits supporting decision-making. Categorical models of emotions propose that evolution has specified a set of basic emotions (usually including anger, fear, disgust, happiness, sadness, and surprise but possibly also others) that support specialized survival functions (Cordaro et al., 2018; Cowen & Keltner, 2017; Ekman, 1992; Nummenmaa & Saarimäki, 2017; Panksepp, 1982). These basic

L. Nummenmaa (✉)
Turku PET Centre and Turku University Hospital, Turku, Finland

Department of Psychology, University of Turku, Turku, Finland
e-mail: latanu@utu.fi

K. Seppälä · V. Putkinen
Turku PET Centre and Turku University Hospital, Turku, Finland

© The Author(s) 2023
P. S. Boggio et al. (eds.), *Social and Affective Neuroscience of Everyday Human Interaction*, https://doi.org/10.1007/978-3-031-08651-9_1

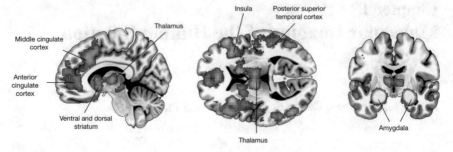

Fig. 1.1 Statistical summary of brain regions involved in emotional processing based on the NeuroSynth database (Yarkoni et al., 2011)

emotions are characterized by discrete neural and physiological substrates, distinctive subjective feelings (such as "I feel happy"), expressions, and a selective functionally dependent neural basis (Kreibig, 2010; Nummenmaa et al., 2014, 2018; Saarimäki et al., 2016; Tracy & Randles, 2011). Much of recent neuroimaging work has aimed at mapping the functional organization of the emotion circuits in the brain using functional magnetic resonance imaging (Hudson et al., 2020; Nummenmaa & Saarimäki, 2017; Wager et al., 2015), and these studies have been successful in delineating the neurobiological architecture of emotions (Fig. 1.1).

Meta-analyses of the BOLD-fMRI data have however yielded inconsistent support for the discrete neural basis of emotions. One proposed explanation for this is the low spatial resolution of BOLD-fMRI coupled with univariate analysis: if specific neural populations coding different emotions are intermixed within one voxel, their activation differences cannot be revealed by univariate techniques. In line with this view, multivariate pattern recognition studies have consistently provided support for a discrete neural basis of different basic and complex emotions (Kragel et al., 2016; Kragel & Labar, 2015; Putkinen et al., 2021; Saarimäki et al., 2016, 2018). Even though multivariate analysis techniques improve the discriminability and specificity of data patterns across different classes or conditions (Norman et al., 2006), they cannot resolve one of the main limitations of the BOLD-EPI data—that the signal is unspecific with respect to the underlying neurotransmitter circuits.

A single voxel in an echo-planar image may contain neurons operating with a multitude of different neurotransmitters, whose net activation is reflected in the BOLD signal. Understanding emotions from the perspective of specific neurotransmitter systems is however critical, because of the central role of affect in multiple psychopathologies and the role of specific neuroreceptor systems as drug targets. For example, the most commonly assumed working mechanism of antidepressants involves either increased neurotransmission by increasing synaptic neurotransmitter levels (such as norepinephrine or dopamine [DA]) or specific agonist effects of the targeted receptors. Thus, it is imperative to delineate not just the anatomical but also neuromolecular organization of the emotion circuits in the brain. Here, we provide an overview of the molecular mechanisms of emotions, with specific focus on *in vivo* imaging of specific neurotransmitter and neuroreceptor studies in humans. We

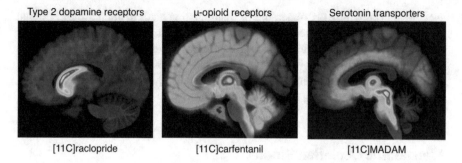

Type 2 dopamine receptors   μ-opioid receptors   Serotonin transporters

[11C]raclopride   [11C]carfentanil   [11C]MADAM

Fig. 1.2 Distribution of type-2 dopamine receptors, μ-opioid receptors, and 5-HT 1A transporters measured using PET radioligands

focus specifically on opioidergic, dopaminergic, and serotonergic mechanisms, as they can be readily studied *in vivo* in the human brain (Fig. 1.2).

Studying Human Neuroreceptor Systems *In Vivo*

Most commonly used functional imaging (fMRI) and electromagnetic (MEG / EEG) techniques for recording brain activation do not yield any information regarding the underlying mechanisms of neurotransmission. Because pharmacological microstimulation studies are not feasible in humans, main approaches for studying emotion-related neurotransmission involve different activation, blockade, and depletion studies, as well as nuclear medicine imaging techniques for direct *in vivo* measurements.

Pharmacological Activation and Blockage Studies

The classical behavioral pharmacological approach involves delivering specific receptor agonists or antagonists or other pharmacologically active agents into the circulatory system or directly into the target tissue in the case of animal studies. In humans, these studies are difficult to conduct, because oral or intravenous administration leads to systemic rather than regionally specific effects, and it has been well established through animal studies that the effects of receptor agonists/antagonists can be regionally highly selective (Berridge & Kringelbach, 2015). One way for overcoming this limitation is to use a pharmacological imaging approach, where functional imaging or electromagnetic recordings are performed during pharmacological treatment versus a placebo condition, which allows identifying the brain regions where the drug action influences neural responses. However, these regional responses may still be influenced by system-level effects, and pinpointing the specific regions whose pharmacological manipulation leads to altered BOLD signal is difficult. Furthermore, studies employing potent pharmacological agents such as morphine or dexamphetamine require strict clinical supervision. Finally, pharmacological manipulations may lead to physiological effects that directly confound the BOLD signal, such as respiratory depression caused by opioid agonists (Pattinson, 2008), further complicating their interpretation.

Monoamine Depletion Studies

A complementary approach to pharmacological activation and blockage studies involves techniques that temporarily lower the functioning of monoamines such as 5-HT, DA, and catecholamine, typically by blocking the synthesis or restricting the intake of amino acid precursors. The three most widely used techniques involve acute tryptophan depletion (ADT) to block 5-HT transporter synthesis by dietary restriction of the 5-HT precursor l-tryptophan. The effect is amplified by the consumption of a large quantity of other amino acids that compete with tryptophan at the blood–brain barrier (Booij et al., 2003). Phenylalanine/tyrosine depletion (APTD), in turn, targets the dopaminergic/catecholamic systems by restricting the dietary intake of its precursors, phenylalanine and tyrosine. Such techniques result in specific short-term effects in distinct neurotransmitter systems rather than on general protein metabolism in the brain (Booij et al., 2003); however, the interpretation of these results is complicated due to distinct system-level effects on transmitter synthesis. Nevertheless, these techniques are valuable when investigating the involvement of monoamine system function in specific mood disorders.

Molecular Imaging with Positron Emission Tomography

Functional molecular imaging using positron emission tomography (PET) is the current gold standard for *in vivo* molecular imaging in humans. It is based on injecting radiolabeled, biologically active molecules into the circulation. These molecules bind to specific target sites, and the unstable isotopes subsequently undergo positron emission decay. The radioisotope emits a positron—an antiparticle of an electron—which loses kinetic energy as it travels through brain tissue. After a certain degree of deceleration, the positron can interact with an electron, leading to an annihilation event producing two gamma photons (rays) moving in opposite directions. The gamma rays are recorded by the detector units of the PET camera, and on the basis of simultaneously detected gamma rays on the opposite sides of the detector ring, the location of the annihilation event can be computed. This subsequently allows reconstruction of the tracer uptake in the tissue. When combined with measurements of tracer input and output, these raw radioactivity counts can be transformed into biologically meaningful information such as radioligand binding at neuroreceptors.

This technique provides excellent biological resolution due to the potential for developing highly selective radioligands binding to different protein targets and spatial resolution up to a few millimeters. Despite its high sensitivity for *in vivo* biomarker tracing, PET lacks the capability for capturing the underlying tissue morphology at high spatial resolution; as such, this information usually needs to be acquired through separate MR or CT scans. Functional imaging of slow-acting neurotransmission is however possible (Backman et al., 2011; Zubieta et al., 2001), although temporal resolution is limited to tens of minutes for most neurotransmission studies. Modern integrated PET—MRI systems (Judenhofer et al., 2008) also allow for the simultaneous measurement of perfusion with both PET and arterial spin labeled MRI (Heijtel et al., 2014; Zhang et al., 2014), or perfusion with MRI and neuroreceptor occupancy (PET) significantly broadening the utility of PET (Sander et al., 2019). Furthermore, joint analysis of PET and structural MR images provide complementary information about the mesoscopic organization of the brain (Manninen et al., 2021). All in all, the PET technique is currently the most accurate and specific tool available for investigating *in vivo* neurotransmission in humans.

The Dopamine System

Rewards exert a powerful influence on our behavior. Both humans and animals are motivated to obtain various rewards ranging from food and sex to social contact, and the pleasurable sensations we experience on receiving the reward further reinforce our motivation to seek and consume the same reward in the future. The monoamine neurotransmitter dopamine (DA) and its receptors D1-D5 have been well-established as playing a key role in motor control and reward-related behavior and pleasure. There are multiple DA pathways in the brain that consist of neuronal projections which synthesize and release DA (Fig. 1.3). The mesolimbic pathway projects from the ventral tegmental area (VTA) to the ventral striatum. This pathway is particularly involved in processing incentive salience, generating pleasure responses and reinforcement learning. The mesocortical pathway projecting from the VTA to the prefrontal cortex is, in turn, more involved in executive functions although it also contributes to reward processing. The nigrostriatal pathway connects substantia nigra to the striatum (putamen and caudate) and contributes critically to motion control. Finally, the tuberoinfundibular pathway connects the hypothalamus and the pituitary gland. Importantly, all the main functions of the dopamine system are also central to reward processing, and it comes as no surprise that dopamine system has been implicated as one of the primary molecular pathways for reward (Wise & Rompre, 1989), and microinjection studies in animals have established that dopamine stimulation of the nucleus modulates incentive motivation (DiFeliceantonio & Berridge, 2016; Peciña & Berridge, 2013).

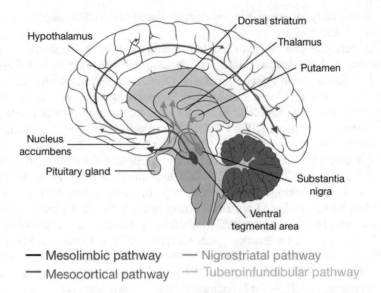

Fig. 1.3 Main dopamine pathways in the brain

PET studies using the radioligand [11C]raclopride in humans have consistently demonstrated DA release in central pathways during reward processing. Due to the poor temporal accuracy of PET, it is difficult to dissect the contribution of reward expectation and consumption phases to the release of DA: It is difficult to design sufficiently long (~45 min) tasks where rewards would be only anticipated but not delivered. As a result, studies conducted in this area mix both anticipation- and consumption-related effects. The PET analysis of DA transmission in reward has shown that feeding—one of the most salient biological rewards—triggers DA release primarily in the striatum. Because the magnitude of DA release is associated with the evaluation of the subjective pleasantness of the meal, this finding has been interpreted as evidence for hedonic (rather than homeostatic) responses to feeding (Small et al., 2003). This is further supported by another series of studies, which measured DA release during intravenous glucose/placebo delivery, thus precluding the subjective evaluation of the reward value of the glucose, yet systemically altering the blood glucose levels simulating a postprandial state (Haltia et al., 2007, 2008). These studies found no differences between the glucose and placebo conditions, suggesting that alterations in circulating glucose levels are not sufficient for central DA release. Instead, the hedonic responses driven by the orosensory and chemical taste pathways appear to be crucial for the DA response triggered by feeding.

There is less evidence for DA processing of other primary reward signals, but some studies suggest that romantic (Takahashi et al., 2015) and maternal attachment-related rewards (Atzil et al., 2017) are processed via the dopamine system in humans. However, these studies are difficult to interpret as the latter (Atzil et al., 2017) reported dopamine activations in regions where [11C]raclopride has either low or no specific binding and no sensitivity to even D2/D3R antagonist challenge (Svensson et al., 2019), and the former was based on an individual-differences approach (Takahashi et al., 2015) and failed to show significant main effects of DA release across the whole group of subjects. In addition, murine models typically show a decrease in DA release in response to social contact seeking (Manduca et al., 2014), rather than an increase as suggested by human PET data; this might however be due to cross-species differences in attachment circuits. Striatal DA reward signaling has however been shown to extend beyond biologically significant rewards. For example, more "cognitive" rewards such as listening to one's favorite music (Salimpoor et al., 2011), gambling (Joutsa et al., 2012), and playing video games (Koepp et al., 1998) lead to striatal dopamine release. In all of these tasks, the reward value is learned rather than intrinsic, suggesting that acquired reward signals are processed in comparable fashion via DA signaling as those with innate reward value. This is most clearly highlighted by data that shows that simple cognitive tasks such as task switching may trigger striatal DA release as soon as they are coupled with rewards (Jonasson et al., 2014).

Negative emotions also induce DA release. One study using [18F]fallypride revealed increased dopamine release in the amygdala and mediolateral frontal cortex during processing of negative emotional words (Badgaiyan et al., 2009), while a subsequent study using [11C] raclopride found similar effects in the caudate nucleus and putamen (Badgaiyan, 2010). There are multiple possibilities for the apparently contradicting findings showing that both pleasure and displeasure can lead to DA activation. For example, it is possible that the DA response to negative stimuli reflects preparatory avoidance behavior triggered by the aversive stimulus, consistent with the role of DA release in motor responses geared toward specific behavioral patterns. This might be reflected in similar activation as the preparatory approach for rewards during pleasurable events. Finally, type-2 DA receptors (D2R) have also been linked with executive control and working memory (Backman et al., 2011), and the emotion-dependent DA activations might reflect the prediction and planning of both escape (negative emotions) and seeking and exploration responses (positive emotions).

Recent PET–fMRI fusion imaging has also tried to dissect the specific role of DA in processing different aspects of emotions, specifically the pleasure-displeasure (valence) and arousal axes. This approach is based on separate PET measurement of neuroreceptor distribution, which can then be used to predict emotion-dependent BOLD responses in subsequent fMRI experiments (Karjalainen et al., 2017). The logic of these experiments is to examine whether interindividual variation in the regional BOLD responses is dependent on corresponding variability in neurotransmitter availability, which would be indicative of DA involvement in the emotional processes targeted in the fMRI experiment. However, this work has failed to establish associations between D2R availability and emotion-specific BOLD responses (Karjalainen et al., 2018) and instead suggests a key role of opioid system in modulating basic affective responses (see below).

Given the central role of dopamine in modulating motivation and reward, it is not surprising that dysregulated dopaminergic neurotransmission is the hallmark of numerous addictive disorders (Volkow et al., 2009). Human imaging studies have demonstrated that alcohol and drug dependence are associated with lowered D2R availability (Martinez et al., 2012; Volkow et al., 1996, 2001). Additionally, drug-induced striatal dopamine responses are blunted in methamphetamine abusers (Volkow et al., 2014). With behavioral addictions and addiction-like behaviors, the results are less clear. Animal studies on obesity suggest that striatal D2R is down-regulated in the obese brain (Johnson & Kenny, 2010), while human studies have yielded mixed results with some finding lower (de Weijer et al., 2011; Volkow et al., 2008; Wang et al., 2001) and others unaltered (Haltia et al., 2007, 2008; Steele et al., 2010) D2R availability in the striatum. Finally, pathological gambling is not associated with altered D2R availability (Joutsa et al., 2012). However, gambling-dependent dopamine signaling is amplified in pathological gamblers versus controls (Joutsa et al., 2012), in contrast to the blunting effect observed in amphetamine abusers upon drug administration (Volkow et al., 2014). In sum, substance abuse appears to markedly downregulate the D2R system possibly via direct pharmacological effects, whereas behavioral addictions and addiction-like states are modulated by at least partially independent pathways.

Opioid System

Endogenous opioids are expressed widely throughout the human central nervous system (Fig. 1.4) and numerous high-density receptor sites constitute central nodes in the human emotion circuit (Kantonen et al., 2020). Among the three classes of opioid receptors (μ, δ, and κ), the μ receptors mediate the effects of endogenous β-endorphins, endomorphins, enkephalins, and various exogenous opioid agonists (Henriksen & Willoch, 2008). The predominant action of μ-opioids in the central nervous system is inhibitory, but they can also exert excitatory effects. The neurons synthesizing β-endorphin are found in the arcuate nucleus in the hypothalamus and the nucleus tractus solitarii of the medulla, which projects extensively to regions throughout the CNS. Dopamine is oftentimes considered the primary neurotransmitter for reward processing (Wise & Rompre, 1989). Opioid and dopamine systems are however closely interlinked on cellular level (Tuominen et al., 2015), and opioids can produce reward independently of dopamine (Hnasko et al., 2005), likely via partially independent molecular pathways. Moreover, both opioidergic and dopaminergic microstimulation of the nucleus accumbens modulate incentive motivation (DiFeliceantonio & Berridge, 2016; Peciña & Berridge, 2013), suggesting complementary roles of these neurotransmitter systems in motivational and hedonic aspects of reward.

Opiates are commonly used illicit drugs, particularly in the United States, where the lifetime prevalence of opioid use disorder exceeds 2% (Grant et al., 2016). Such high misuse potential is attributed to the strong "liking" responses—the pleasurable subjective experiences produced by drug consumption (Comer et al., 2012). However, experiments with drug-naïve volunteers have not provided consistent

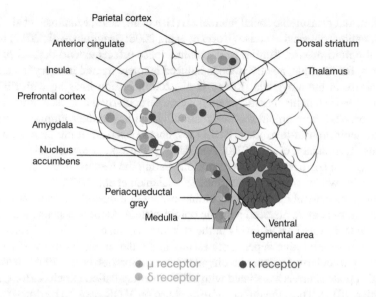

Fig. 1.4 Organization of the human opioid system in the brain. Note that as specific opioid neuron projections cannot be established, this figure instead characterizes the relative expression of different receptor subtypes in some of the key nodes of the emotion circuit

results on opioid agonists associated with liking or pleasure. Some studies report increased pleasure upon μ-receptor (MOR) agonist delivery (Riley et al., 2010; Zacny & Gutierrez, 2003, 2009), whereas others have not corroborated these findings (Ipser et al., 2013; Lasagna et al., 1955; Tedeschi et al., 1984). These discrepancies likely pertain to differences in the route of administration, receptor affinity, and genetically determined variation in receptor expression (Levran et al., 2012). Some recent experiments have found that opioid agonists shift the evaluation of external stimuli, making them seem more pleasant, without necessarily directly influencing tonic subjective emotional state per se (Heiskanen et al., 2019). Thus, it is possible that opioid agonists primarily influence the evaluative processing of emotions, rather than directly modulating the acute subjective feeling. Consequently, opioids might alleviate stress and dysphoria by shifting the evaluation of the internal and external world toward more positive directions.

By contrast, molecular imaging shows that reward consumption consistently triggers endogenous opioid release. Feeding leads to increased endogenous opioid release in the reward circuit and also elsewhere in the brain (Burghardt et al., 2015; Tuulari et al., 2017). However, this response is observed for both palatable and non-palatable meals and is actually stronger for fast-metabolizing, non-appetizing liquid meals than for palatable pizza. Thus, the response is likely a combination of the low-level homeostatic pleasure of feeding after fasting which is presumably more intense in response to a quickly metabolized liquid meal and possibly a partially independent effect of subjective hedonic responses. Corroborating evidence for the role of the opioid system in processing primary rewards comes from studies showing that pleasurable social interaction (Hsu et al., 2013; Manninen et al., 2017) and strenuous physical exercise (Boecker et al., 2008; Saanijoki et al., 2017) induce central opioid release. Similar to dopamine, these effects extend beyond primary rewards; for example, positive moods induced by mere mental imagery induce opioid release in the amygdala (Koepp et al., 2009). Fusion imaging with PET and fMRI suggests that the opioid system governs particularly the arousal dimension of emotions. The more opioid receptors an individual has in their limbic system, the weaker their arousal-dependent BOLD responses observed in the brain's emotion circuits (Karjalainen et al., 2018). Accordingly, the opioid system might act as a buffer against socioemotional stressors, alleviating the negative feelings associated with one's own or another's misfortune (Karjalainen et al., 2017).

While the general role of the dopamine system in drug addictions is fairly clear-cut, the story is more nuanced with the opioid system. Alcohol dependence is associated with elevated MOR levels in the striatum (Heinz et al., 2005; Weerts et al., 2011), whereas cocaine dependence results in similar effects in more widespread regions, particularly cortical and cingulate areas (Gorelick et al., 2005). However, chronic opiate abuse is associated with MOR downregulation (Koch & Hollt, 2008; Whistler, 2012). Thus, the effects of drug abuse on MOR seem to be drug-specific. More consistent data comes from studies on obesity that have implicated downregulated μ-receptor action as one of the key pathophysiological mechanisms (Burghardt et al., 2015; Karlsson et al., 2015, 2016; Tuominen et al., 2015). These effects are also specific to obesity rather than a general feature of behavioral addictions, as μ-receptor downregulation is not observed in pathological gambling for example (Majuri et al., 2016). Finally, despite the centrality of the opioid system in hedonia and affective functioning, there is no clear evidence of its involvement in the pathophysiology of mood disorders. PET imaging data are limited in scope, and the existing studies have yielded conflicting evidence on opioidergic alterations in major depression (Hsu et al., 2015; Kennedy et al., 2006). However, one recent large-scale study shows that subclinical depressive and anxious symptoms are consistently linked with MOR system downregulation (Nummenmaa et al., 2020). Finally, the opioid system may also contribute to affective pathophysiology due to its role in governing human attachment behavior whose disruptions are consistently linked with mood disorders (Mikulincer & Shaver, 2012). This is supported by PET studies that have consistently found that insecure attachment is linked with downregulated MOR in the limbic and paralimbic regions (Nummenmaa et al., 2015; Turtonen et al., 2021).

Serotonergic System

The monaomine neurotransmitter serotonin and its receptors $5HT_1$-$5HT_7$ are involved in the regulation of sleep, appetite, mood, and pleasure, but it is also involved in cognitive and physiological processes. In the central nervous system, serotonin is produced in the raphe nuclei in the brainstem, from where the

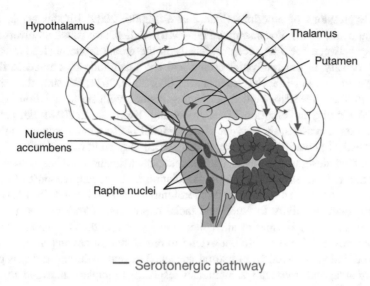

Fig. 1.5   Main serotonin pathways in the brain

serotonergic projections extend to the striatum and neocortex (Fig. 1.5). The brain's serotonergic systems also play a critical role in avoidance behaviors as well as fear and anxiety. Activation of the serotonergic system is critical for avoidance behavior in rodents (Deakin & Graeff, 1991), and genetic variations in serotonin transporter (SERT) expression influence the fear circuit's responsiveness to acute threat signals in humans (Hariri et al., 2002). Thus, major categories of anxiolytic drugs also inhibit SERT.

While dopamine and opioid systems are centrally involved in the pathophysiology of addictive disorders, the SERT system is consistently implicated in mood regulation and consequently in the pathogenesis of mood disorders (Mann, 1999). Although initial reports on 5-HTT in mood disorders have been variable, meta-analyses suggest that serotonin transporter availability is consistently lowered in depression (Ichimiya et al., 2002); but see Andrews et al. (2015), and altered serotonergic neurotransmission is also considered a hallmark of depression (Drevets et al., 1999). Accordingly, the most widely used and effective of antidepressants act by increasing extracellular serotonin levels. Importantly, individual differences in the expression of the serotonin transporter mediate the effects of stressful life events on the onset of depression (Risch et al., 2009). In a similar fashion, serotonin transporter availability varies seasonally, suggesting that altered serotonergic function may also underlie the pathophysiology of seasonal affective disorders (Praschak-Rieder et al., 2008).

Functional molecular imaging of the serotonergic system has been limited due to the lack of radioligands that show sensitivity to endogenous serotonin levels, essentially preventing serotonin activation studies with PET. However, fusion PET–fMRI imaging has elucidated the role of SERT in emotional processing. A number of studies indicate that the serotonergic system regulates amygdala responsiveness to facial expressions of emotions (Fisher et al., 2006, 2009; Rhodes et al., 2007; Selvaraj et al., 2015). For instance, PET–fMRI studies have found an inverse relationship between 5-HT1A receptor density in the dorsal raphe nucleus (DRN) or HT2A density in the prefrontal cortex and the magnitude of amygdala BOLD response to emotional faces (Fisher et al., 2006, 2009, 2011; Selvaraj et al., 2015). Some studies have also yielded conflicting results, with no association between 5-HT1A binding and emotional face processing (Kranz et al., 2018). For practical and economic reasons, these types of multimodal neuroimaging studies have limited statistical power (oftentimes n:s <30), which may yield inconsistent effects in correlational designs. However, pharmacological activation studies provide corroborating evidence for serotonergic modulation of amygdala responses to threat. Multiple studies have documented that serotonin reuptake inhibitors (SSRIs) modulate amygdala reactivity to emotional facial expressions (Anderson et al., 2007; Bigos et al., 2008; Harmer et al., 2006; Murphy et al., 2009). These effects are however not just face-specific but extend to emotional processing in general and also to emotions derived from natural speech. The serotonin and norepinephrine receptor antagonist mirtazapine attenuates responses to unpleasant events in sensorimotor and anterior areas while modulating responses to arousing events in cortical midline structures. These effects are paralleled by increased functional connectivity between cortical midline and limbic areas during pleasant events (Komulainen et al., 2017), suggesting large-scale modulation of affective processing by serotonergic drugs.

From a clinical viewpoint, subjective feelings linked with the neural and autonomic emotional response are also an important facet of mood disorders. In particular, negative self-concept and increased self-focus play an important role in the pathophysiology of depression. Some studies suggest that the serotonergic system can influence how subjects interpret and process self-relevant affective information. Mirtazapine attenuates self-referential emotional processing in healthy volunteers, as manifested in decreased cortical midline activation (Komulainen et al., 2016). This mechanism could underlie one form of serotonin-dependent antidepressant action. This is further evidenced in clinical trials, which show how short-term escitalopram treatment regulates self-referential processing in patients with major depressive disorder (Komulainen et al., 2018). Thus, serotonergic modulation seems to occur at multiple levels of the human emotion circuit, ranging from sensory to evaluative, cognitive and self-referential processes, and the serotonergic action of antidepressants likely impacts all these levels.

Conclusions

Recent advances in nuclear medicine imaging have helped to elucidate the role of opioid, dopamine, and serotonin systems in human emotions. There is clear evidence that dopamine and opioid systems modulate hedonic processes. However, both dopaminergic and opioidergic activation is observed during negative emotions too, suggesting that they may also support general motivational and arousal-modulation components of emotions. On pathophysiological level, the dopamine system is more clearly linked with substance abuse and addictive disorders, whereas opioidergic activations vary from substance to substance, with clear downregulation observed particularly in obesity. The serotonin system links more clearly with negative emotions including fear and sadness, yet outside pharmacological and clinical studies, the majority of these data come from pharmacological fMRI studies and those correlating transporter availability with BOLD–fMRI responses.

There is no clear one-to-one mapping between specific emotions or emotional behaviors and specific neurotransmitters. Obviously, numerous neurotransmitters have a wide variety of roles, and their specific actions are not limited to emotional behavior. Human imaging studies are challenging to conduct and are limited by radioligand pharmacokinetics and affinity. For the major neurotransmitter systems implicated in emotion, reliable radioligands exist for imaging serotonin, dopamine, opioid and endocannabinoid receptors and transmitters. For opioid and dopamine systems, there are also radioligands available that are sensitive to endogenous transmitter levels, whereas this has yet to be achieved for serotonin and endocannabinoid systems. In sum, targeting neurotransmitter mechanisms of emotions using PET is a powerful tool for dissecting the molecular mechanisms of emotions, further potentiated by next-generation PET–MRI devices which allow us to address the molecular specificity of emotion-related BOLD activation.

References

Anderson, I. M., Del-Ben, C. M., Mackie, S., Richardson, P., Williams, S. R., Elliott, R., & Deakin, J. F. W. (2007). Citalopram modulation of neuronal responses to aversive face emotions: A functional MRI study. *Neuroreport, 18,* 1351–1355.

Andrews, P. W., Bharwani, A., Lee, K. R., Fox, M., & Thomson, J. A. (2015). Is serotonin an upper or a downer? The evolution of the serotonergic system and its role in depression and the antidepressant response. *Neuroscience & Biobehavioral Reviews, 51,* 164–188.

Atzil, S., Touroutoglou, A., Rudy, T., Salcedo, S., Feldman, R., Hooker, J. M., Dickerson, B. C., Catana, C., & Barrett, L. F. (2017). Dopamine in the medial amygdala network mediates human bonding. *Proceedings of the National Academy of Sciences, 114,* 201612233.

Backman, L., Nyberg, L., Soveri, A., Johansson, J., Andersson, M., Dahlin, E., Neely, A. S., Virta, J., Laine, M., & Rinne, J. O. (2011). Effects of working-memory training on striatal Dopamine release. *Science, 333,* 718.

Badgaiyan, R. D. (2010). Dopamine is released in the striatum during human emotional processing. *Neuroreport, 21,* 1172–1176.

Badgaiyan, R. D., Fischman, A. J., & Alpert, N. M. (2009). Dopamine release during human emotional processing. *Neuroimage, 47,* 2041–2045.

Berridge, K. C., & Kringelbach, M. L. (2015). Pleasure systems in the Brain. *Neuron, 86,* 646–664.

Bigos, K. L., Pollock, B. G., Aizenstein, H. J., Fisher, P. M., Bies, R. R., & Hariri, A. R. (2008). Acute 5-HT Reuptake Blockade potentiates human Amygdala reactivity. *Neuropsychopharmacology, 33,* 3221–3225.

Boecker, H., Sprenger, T., Spilker, M. E., Henriksen, G., Koppenhoefer, M., Wagner, K. J., Valet, M., Berthele, A., & Tolle, T. R. (2008). The Runner's high: Opioidergic mechanisms in the Human Brain. *Cerebral Cortex, 18,* 2523–2531.

Booij, L., Van der Does, A. J. W., & Riedel, W. J. (2003). Monoamine depletion in psychiatric and healthy populations: Review. *Molecular Psychiatry, 8*, 951–973.

Burghardt, P. R., Rothberg, A. E., Dykhuis, K. E., Burant, C. F., & Zubieta, J. K. (2015). Endogenous Opioid mechanisms are implicated in obesity and weight loss in humans. *Journal of Clinical Endocrinology & Metabolism, 100*, 3193–3201.

Comer, S. D., Zacny, J. P., Dworkin, R. H., Turk, D. C., Bigelow, G. E., Foltin, R. W., Jasinski, D. R., Sellers, E. M., Adams, E. H., Balster, R., et al. (2012). Core outcome measures for opioid abuse liability laboratory assessment studies in humans: IMMPACT recommendations. *Pain, 153*, 2315–2324.

Cordaro, D. T., Sun, R., Keltner, D., Kamble, S., Huddar, N., & McNeil, G. (2018). Universals and cultural variations in 22 emotional expressions across five cultures. *Emotion, 18*, 75–93.

Cowen, A. S., & Keltner, D. (2017). Self-report captures 27 distinct categories of emotion bridged by continuous gradients. *Proceedings of the National Academy of Sciences.*

de Weijer, B. A., van de Giessen, E., van Amelsvoort, T. A., Boot, E., Braak, B., Janssen, I. M., van de Laar, A., Fliers, E., Serlie, M. J., & Booij, J. (2011). Lower striatal dopamine D2/3 receptor availability in obese compared with non-obese subjects. *EJNMMI Res, 1*, 37.

Deakin, J. F., & Graeff, F. G. (1991). 5-HT and mechanisms of defence. *Journal of psychopharmacology (Oxford, England), 5*, 305–315.

DiFeliceantonio, A. G., & Berridge, K. C. (2016). Dorsolateral neostriatum contribution to incentive salience: opioid or dopamine stimulation makes one reward cue more motivationally attractive than another. *European Journal of Neuroscience, 43*, 1203–1218.

Drevets, W. C., Frank, E., Price, J. C., Kupfer, D. J., Holt, D., Greer, P. J., Huang, Y., Gautier, C., & Mathis, C. (1999). Pet imaging of serotonin 1A receptor binding in depression. *Biological Psychiatry, 46*, 1375–1387.

Ekman, P. (1992). An argument for basic emotions. *Cognition & Emotion, 6*, 169–200.

Fisher, P. M., Meltzer, C. C., Ziolko, S. K., Price, J. C., Moses-Kolko, E. L., Berga, S. L., & Hariri, A. R. (2006). Capacity for 5-HT 1A –mediated autoregulation predicts amygdala reactivity. *Nature Neuroscience, 9*, 1362.

Fisher, P. M., Meltzer, C. C., Price, J. C., Coleman, R. L., Ziolko, S. K., Becker, C., Moses-Kolko, E. L., Berga, S. L., & Hariri, A. R. (2009). Medial prefrontal Cortex 5-HT2A density is correlated with Amygdala reactivity, response habituation, and functional coupling. *Cerebral Cortex, 19*, 2499–2507.

Fisher, P. M., Price, J. C., Meltzer, C. C., Moses-Kolko, E. L., Becker, C., Berga, S. L., & Hariri, A. R. (2011). Medial prefrontal cortex serotonin 1A and 2A receptor binding interacts to predict threat-related amygdala reactivity. *Biology of Mood & Anxiety Disorders, 1*, 2.

Gorelick, D. A., Kim, Y. K., Bencherif, B., Boyd, S. J., Nelson, R., Copersino, M., Endres, C. J., Dannals, R. F., & Frost, J. J. (2005). Imaging brain mu-opioid receptors in abstinent cocaine users: Time course and relation to cocaine craving. *Biological Psychiatry, 57*, 1573–1582.

Grant, B. F., Saha, T. D., Ruan, W. J., Goldstein, R. B., Chou, S. P., Jung, J., Zhang, H., Smith, S. M., Pickering, R. P., Huang, B., et al. (2016). Epidemiology of DSM-5 drug use disorder: Results from the national epidemiologic survey on alcohol and related conditions-III. *JAMA Psychiatry, 73*, 39–47.

Haltia, L. T., Rinne, J. O., Merisaari, H., Maguire, R. P., Savontaus, E., Helin, S., Nagren, K., & Kaasinen, V. (2007). Effects of intravenous glucose on dopaminergic function in the human brain in vivo. *Synapse, 61*, 748–756.

Haltia, L. T., Rinne, J. O., Helin, S., Parkkola, R., Nagren, K., & Kaasinen, V. (2008). Effects of intravenous placebo with glucose expectation on human basal ganglia dopaminergic function. *Synapse, 62*, 682–688.

Hariri, A. R., Mattay, V. S., Tessitore, A., Kolachana, B., Fera, F., Goldman, D., Egan, M. F., & Weinberger, D. R. (2002). Serotonin transporter genetic variation and the response of the human amygdala. *Science, 297*, 400–403.

Harmer, C. J., Mackay, C. E., Reid, C. B., Cowen, P. J., & Goodwin, G. M. (2006). Antidepressant drug treatment modifies the neural processing of nonconscious threat cues. *Biological Psychiatry, 59*, 816–820.

Heijtel, D. F. R., Mutsaerts, H., Bakker, E., Schober, P., Stevens, M. F., Petersen, E. T., van Berckel, B. N. M., Majoie, C., Booij, J., van Osch, M. J. P., et al. (2014). Accuracy and precision of pseudo-continuous arterial spin labeling perfusion during baseline and hypercapnia: A head-to-head comparison with O-15 H2O positron emission tomography. *Neuroimage, 92*, 182–192.

Heinz, A., Reimold, M., Wrase, J., Hermann, D., Croissant, B., Mundle, G., Dohmen, B. M., Braus, D. F., Schumann, G., Machulla, H. J., et al. (2005). Correlation of stable elevations in striatal mu-opioid receptor availability in detoxified alcoholic patients with alcohol craving: A positron emission tomography study using carbon 11-labeled carfentanil. *Archives of General Psychiatry, 62*, 57–64.

Heiskanen, T., Leppä, M., Suvilehto, J., Akural, E., Larinkoski, T., Jääskeläinen, I. P., Sams, M., Nummenmaa, L., & Kalso, E. (2019). The opioid agonist remifentanil increases subjective pleasure during emotional stimulation. *British Journal of Anaesthesiology, 19*, 435–458.

Henriksen, G., & Willoch, F. (2008). Imaging of opioid receptors in the central nervous system. *Brain, 131*, 1171–1196.

Hnasko, T. S., Sotak, B. N., & Palmiter, R. D. (2005). Morphine reward in dopamine-deficient mice. *Nature, 438*, 854–857.

Hsu, D. T., Sanford, B. J., Meyers, K. K., Love, T. M., Hazlett, K. E., Wang, H., Ni, L., Walker, S. J., Mickey, B. J., Korycinski, S. T., et al. (2013). Response of the mu-opioid system to social rejection and acceptance. *Molecular Psychiatry, 18*, 1211–1217.

Hsu, D. T., Sanford, B. J., Meyers, K. K., Love, T. M., Hazlett, K. E., Walker, S. J., Mickey, B. J., Koeppe, R. A., Langenecker, S. A., & Zubieta, J. K. (2015). It still hurts: Altered endogenous opioid activity in the brain during social rejection and acceptance in major depressive disorder. *Molecular Psychiatry, 20*, 193–200.

Hudson, M., Seppälä, K., Putkinen, V., Sun, L., Glerean, E., Karjalainen, T., Karlsson, H. K., Hirvonen, J., & Nummenmaa, L. (2020). Dissociable neural systems for unconditioned acute and sustained fear. *NeuroImage, 216*, 116522.

Ichimiya, T., Suhara, T., Sudo, Y., Okubo, Y., Nakayama, K., Nankai, M., Inoue, M., Yasuno, F., Takano, A., Maeda, J., et al. (2002). Serotonin transporter binding in patients with mood disorders: A PET study with [11C](+)McN5652. *Biological Psychiatry, 51*, 715–722.

Ipser, J. C., Terburg, D., Syal, S., Phillips, N., Solms, M., Panksepp, J., Malcolm-Smith, S., Thomas, K., Stein, D. J., & van Honk, J. (2013). Reduced fear-recognition sensitivity following acute buprenorphine administration in healthy volunteers. *Psychoneuroendocrinology, 38*, 166–170.

Johnson, P. M., & Kenny, P. J. (2010). Dopamine D2 receptors in addiction-like reward dysfunction and compulsive eating in obese rats. *Nature Neuroscience, 13*, 635–641.

Jonasson, L. S., Axelsson, J., Riklund, K., Braver, T. S., Ogren, M., Backman, L., & Nyberg, L. (2014). Dopamine release in nucleus accumbens during rewarded task switching measured by C-11 raclopride. *Neuroimage, 99*, 357–364.

Joutsa, J., Johansson, J., Niemela, S., Ollikainen, A., Hirvonen, M. M., Piepponen, P., Arponen, E., Alho, H., Voon, V., Rinne, J. O., et al. (2012). Mesolimbic dopamine release is linked to symptom severity in pathological gambling. *NeuroImage, 60*, 1992–1999.

Judenhofer, M. S., Wehrl, H. F., Newport, D. F., Catana, C., Siegel, S. B., Becker, M., Thielscher, A., Kneilling, M., Lichy, M. P., Eichner, M., et al. (2008). Simultaneous PET-MRI: A new approach for functional and morphological imaging. *Nature Medicine, 14*, 459–465.

Kantonen, T., Karjalainen, T., Isojärvi, J., Nuutila, P., Tuisku, J., Rinne, J., Hietala, J., Kaasinen, V., Kalliokoski, K., Scheinin, H., et al. (2020). Interindividual variability and lateralization of μ-opioid receptors in the human brain. *NeuroImage, 217*, 116922.

Karjalainen, T., Karlsson, H. K., Lahnakoski, J. M., Glerean, E., Nuutila, P., Jaaskelainen, I. P., Hari, R., Sams, M., & Nummenmaa, L. (2017). Dissociable roles of cerebral mu-Opioid and Type 2 Dopamine receptors in vicarious pain: A combined PET-fMRI study. *Cereb Cortex*, 1–10.

Karjalainen, T., Seppala, K., Glerean, E., Karlsson, H. K., Lahnakoski, J. M., Nuutila, P., Jaaskelainen, I. P., Hari, R., Sams, M., & Nummenmaa, L. (2018). Opioidergic regulation of emotional arousal: A combined PET-fMRI study. *Cerebral cortex (New York, N. Y.: 1991), 29*, 4006–4016.

Karlsson, H. K., Tuominen, L., Tuulari, J. J., Hirvonen, J., Parkkola, R., Helin, S., Salminen, P., Nuutila, P., & Nummenmaa, L. (2015). Obesity is associated with decreased mu-Opioid but unaltered Dopamine D-2 receptor availability in the brain. *Journal of Neuroscience, 35*, 3959–3965.

Karlsson, H. K., Tuominen, L., Tuulari, J. J., Hirvonen, J., Honka, H., Parkkola, R., Helin, S., Salminen, P., Nuutila, P., & Nummenmaa, L. (2016). Weight loss after bariatric surgery normalizes brain opioid receptors in morbid obesity. *Molecular Psychiatry, 21*, 1057–1062.

Kennedy, S. E., Koeppe, R. A., Young, E. A., & Zubieta, J. K. (2006). Dysregulation of endogenous opioid emotion regulation circuitry in major depression in women. *Archives of General Psychiatry, 63*, 1199–1208.

Koch, T., & Hollt, V. (2008). Role of receptor internalization in opioid tolerance and dependence. *Pharmacology & Therapeutics, 117*, 199–206.

Koepp, M. J., Gunn, R. N., Lawrence, A. D., Cunningham, V. J., Dagher, A., Jones, T., Brooks, D. J., Bench, C. J., & Grasby, P. M. (1998). Evidence for striatal dopamine release during a video game. *Nature, 393*, 266–268.

Koepp, M. J., Hammers, A., Lawrence, A. D., Asselin, M. C., Grasby, P. M., & Bench, C. J. (2009). Evidence for endogenous opioid release in the amygdala during positive emotion. *Neuroimage, 44*, 252–256.

Komulainen, E., Heikkila, R., Meskanen, K., Raij, T. T., Nummenmaa, L., Lahti, J., Jylha, P., Melartin, T., Harmer, C. J., Isometsa, E., et al. (2016). A single dose of mirtazapine attenuates neural responses to self-referential processing. *Journal of Psychopharmacology, 30*, 23–32.

Komulainen, E., Glerean, E., Meskanen, K., Heikkila, R., Nummenmaa, L., Raij, T. T., Lahti, J., Jylha, P., Melartin, T., Isometsa, E., et al. (2017). Single dose of mirtazapine modulates whole-brain functional connectivity during emotional narrative processing. *Psychiatry Research-Neuroimaging, 263*, 61–69.

Komulainen, E., Heikkila, R., Nummenmaa, L., Raij, T. T., Harmer, C. J., Isometsa, E., & Ekelund, J. (2018). Short-term escitalopram treatment normalizes aberrant self-referential processing in major depressive disorder. *Journal of Affective Disorders, 236*, 222–229.

Kragel, P. A., & Labar, K. S. (2015). Multivariate neural biomarkers of emotional states are categorically distinct. *Social Cognitive and Affective Neuroscience, 10*, 1437–1448.

Kragel, P. A., Knodt, A. R., Hariri, A. R., & LaBar, K. S. (2016). Decoding Spontaneous Emotional States in the Human Brain. *PLOS Biology, 14*, e2000106.

Kranz, G. S., Hahn, A., Kraus, C., Spies, M., Pichler, V., Jungwirth, J., Mitterhauser, M., Wadsak, W., Windischberger, C., Kasper, S., et al. (2018). Probing the association between serotonin-1A autoreceptor binding and amygdala reactivity in healthy volunteers. *NeuroImage, 171*, 1–5.

Kreibig, S. D. (2010). Autonomic nervous system activity in emotion: A review. *Biological Psychology, 84*, 394–421.

Lasagna, L., Vonfelsinger, J. M., & Beecher, H. K. (1955). Drug-induced mood changes in Man. 1. Observations on healthy subjects, chronically ill patients, and postaddicts. *JAMA-Journal of the American Medical Association, 157*, 1006–1020.

Levran, O., Yuferov, V., & Kreek, M. J. (2012). The genetics of the opioid system and specific drug addictions. *Hum Genet, 131*, 823–842.

Majuri, J., Joutsa, J., Johansson, J., Voon, V., Alakurtti, K., Parkkola, R., Lahti, T., Alho, H., Hirvonen, J., Arponen, E., et al. (2016). Dopamine and opioid neurotransmission in behavioral addictions: A comparative PET study in pathological gambling and binge eating. *Neuropsychopharmacology, 42*, 1169–1177.

Manduca, A., Campolongo, P., Palmery, M., Vanderschuren, L., Cuomo, V., & Trezza, V. (2014). Social play behavior, ultrasonic vocalizations and their modulation by morphine and amphetamine in Wistar and Sprague-Dawley rats. *Psychopharmacology, 231*, 1661–1673.

Mann, J. J. (1999). Role of the serotonergic system in the pathogenesis of major depression and suicidal behavior. *Neuropsychopharmacology, 21*, 99S–105S.

Manninen, S., Tuominen, L., Dunbar, R. I. M., Karjalainen, T., Hirvonen, J., Arponen, E., Jääskeläinen, I. P., Hari, R., Sams, M., & Nummenmaa, L. (2017). Social laughter triggers endogenous opioid release in humans. *The Journal of Neuroscience, 37*, 6125–6131.

Manninen, S., Karjalainen, T., Tuominen, L. J., Hietala, J., Kaasinen, V., Joutsa, J., Rinne, J., & Nummenmaa, L. (2021). Cerebral grey matter density is associated with neuroreceptor and neurotransporter availability: A combined PET and MRI study. *Neuroimage, 235*, 117968.

Martinez, D., Saccone, P. A., Liu, F., Slifstein, M., Orlowska, D., Grassetti, A., Cook, S., Broft, A., Van Heertum, R., & Comer, S. D. (2012). Deficits in dopamine D(2) receptors and presynaptic dopamine in heroin dependence: commonalities and differences with other types of addiction. *Biological Psychiatry, 71*, 192–198.

Mikulincer, M., & Shaver, P. R. (2012). An attachment perspective on psychopathology. *World Psychiatry, 11*, 11–15.

Murphy, S. E., Norbury, R., O'Sullivan, U., Cowen, P. J., & Harmer, C. J. (2009). Effect of a single dose of citalopram on amygdala response to emotional faces. *The British Journal of Psychiatry, 194*, 535–540.

Norman, K. A., Polyn, S. M., Detre, G. J., & Haxby, J. V. (2006). Beyond mind-reading: Multi-voxel pattern analysis of fMRI data. *Trends in Cognitive Sciences, 10*, 424–430.

Nummenmaa, L., & Saarimäki, H. (2017). Emotions as discrete patterns of systemic activity. *Neuroscience Letters, 693*, 3–8.

Nummenmaa, L., Glerean, E., Hari, R., & Hietanen, J. K. (2014). Bodily maps of emotions. *Proceedings of the National Academy of Sciences of the United States of America, 111*, 646–651.

Nummenmaa, L., Manninen, S., Tuominen, L., Hirvonen, J., Kalliokoski, K. K., Nuutila, P., Jääskeläinen, I. P., Hari, R., Dunbar, R. I. M., & Sams, M. (2015). Adult attachment style Is associated with cerebral μ-opioid receptor availability in humans. *Human Brain Mapping, 36*, 3621–3628.

Nummenmaa, L., Hari, R., Hietanen, J. K., & Glerean, E. (2018). Maps of subjective feelings. *Proceedings of the National Academy of Sciences of the United States of America, 115*, 9198–9203.

Nummenmaa, L., Karjalainen, T., Isojärvi, J., Kantonen, T., Tuisku, J., Kaasinen, V., Joutsa, J., Nuutila, P., Kalliokoski, K., Hirvonen, J., et al. (2020). Lowered endogenous mu-opioid receptor availability in subclinical depression and anxiety. *Neuropsychopharmacology, 45*, 1953–1959.

Panksepp, J. (1982). Toward a general psychobiological theory of emotions. *Behavioral and Brain Sciences, 5*, 407–422.

Pattinson, K. T. S. (2008). Opioids and the control of respiration. *BJA: British Journal of Anaesthesia, 100*, 747–758.

Peciña, S., & Berridge, K. C. (2013). Dopamine or opioid stimulation of nucleus accumbens similarly amplify cue-triggered 'wanting' for reward: entire core and medial shell mapped as substrates for PIT enhancement. *European Journal of Neuroscience, 37*, 1529–1540.

Praschak-Rieder, N., Willeit, M., Wilson, A. A., Houle, S., & Meyer, J. H. (2008). SEasonal variation in human brain serotonin transporter binding. *Archives of General Psychiatry, 65*, 1072–1078.

Putkinen, V., Nazari-Farsani, S., Seppälä, K., Karjalainen, T., Sun, L., Karlsson, H. K., Hudson, M., Heikkilä, T. T., Hirvonen, J., & Nummenmaa, L. (2021). Decoding music-evoked emotions in the auditory and motor cortex. *Cerebral Cortex, 31*, 2549–2560.

Rhodes, R. A., Murthy, N. V., Dresner, M. A., Selvaraj, S., Stavrakakis, N., Babar, S., Cowen, P. J., & Grasby, P. M. (2007). Human 5-HT transporter availability predicts Amygdala reactivity in vivo. *Journal of Neuroscience, 27*, 9233–9237.

Riley, J. L., Hastie, B. A., Glover, T. L., Fillingim, R. B., Staud, R., & Campbell, C. M. (2010). Cognitive-affective and somatic side effects of morphine and pentazocine: Side-effect profiles in healthy adults. *Pain Medicine, 11*, 195–206.

Risch, N., Herrell, R., Lehner, T., Liang, K. Y., Eaves, L., Hoh, J., Griem, A., Kovacs, M., Ott, J., & Merikangas, K. R. (2009). Interaction between the Serotonin Transporter Gene (5-HTTLPR), stressful life events, and risk of depression a meta-analysis. *JAMA-Journal of the American Medical Association, 301*, 2462–2471.

Saanijoki, T., Tuominen, L., Tuulari, J. J., Nummenmaa, L., Arponen, E., Kalliokoski, K., & Hirvonen, J. (2017). Opioid release after high-intensity interval training in healthy human subjects. *Neuropsychopharmacology, 43*, 246–254.

Saarimäki, H., Gotsopoulos, A., Jääskeläinen, I. P., Lampinen, J., Vuilleumier, P., Hari, R., Sams, M., & Nummenmaa, L. (2016). Discrete neural signatures of basic emotions. *Cerebral Cortex, 6*, 2563–2573.

Saarimäki, H., Ejtehadian, L. F., Glerean, E., Jääskeläinen, I. P., Vuilleumier, P., Sams, M., & Nummenmaa, L. (2018). Distributed affective space represents multiple emotion categories across the human brain. *Social Cognitive and Affective Neuroscience* 2018:nsy018-nsy018.

Salimpoor, V. N., Benovoy, M., Larcher, K., Dagher, A., & Zatorre, R. J. (2011). Anatomically distinct dopamine release during anticipation and experience of peak emotion to music. *Nature Neuroscience, 14*, 257–U355.

Sander, C. Y., Mandeville, J. B., Wey, H.-Y., Catana, C., Hooker, J. M., & Rosen, B. R. (2019). Effects of flow changes on radiotracer binding: Simultaneous measurement of neuroreceptor binding and cerebral blood flow modulation. *Journal of Cerebral Blood Flow and Metabolism, 39*, 131–146.

Selvaraj, S., Mouchlianitis, E., Faulkner, P., Turkheimer, F., Cowen, P. J., Roiser, J. P., & Howes, O. (2015). Presynaptic serotoninergic regulation of emotional processing: A multimodal brain imaging study. *Biological Psychiatry, 78*, 563–571.

Small, D. M., Jones-Gotman, M., & Dagher, A. (2003). Feeding-induced dopamine release in dorsal striatum correlates with meal pleasantness ratings in healthy human volunteers. *Neuroimage, 19*, 1709–1715.

Steele, K. E., Prokopowicz, G. P., Schweitzer, M. A., Magunsuon, T. H., Lidor, A. O., Kuwabawa, H., Kumar, A., Brasic, J., & Wong, D. F. (2010). Alterations of central dopamine receptors before and after gastric bypass surgery. *Obesity Surgery, 20*, 369–374.

Svensson, J. E., Schain, M., Plavén-Sigray, P., Cervenka, S., Tiger, M., Nord, M., Halldin, C., Farde, L., & Lundberg, J. (2019). Validity and reliability of extrastriatal [11-C]raclopride binding quantification in the living human brain. bioRxiv 2019, 600080.

Takahashi, K., Mizuno, K., Sasaki, A. T., Wada, Y., Tanaka, M., Ishii, A., Tajima, K., Tsuyuguchi, N., Watanabe, K., Zeki, S., et al. (2015). Imaging the passionate stage of romantic love by dopamine dynamics. *Frontiers in Human Neuroscience, 9*, 191.

Tedeschi, G., Smith, A. T., & Richens, A. (1984). Effect of meptazinol and ethanol on human psychomotor performance and mood ratings. *Human Toxicology, 3*, 37–43.

Tracy, J. L., & Randles, D. (2011). Four models of basic emotions: A review of Ekman and Cordaro, Izard, Levenson, and Panksepp and Watt. *Emotion Review, 3*, 397–405.

Tuominen, L., Tuulari, J., Karlsson, H., Hirvonen, J., Helina, S., Salminen, P., Parkkola, R., Hietala, J., Nuutila, P., & Nummenmaa, L. (2015). Aberrant mesolimbic dopamine-opiate interaction in obesity. *Neuroimage, 122*, 80–86.

Turtonen, O., Saarinen, A., Nummenmaa, L., Tuominen, L., Tikka, M., Armio, R.-L., Hautamäki, A., Laurikainen, H., Raitakari, O., Keltikangas-Jarvinen, L., et al. (2021). Adult attachment system links with brain μ-opioid receptor availability in vivo. *Biological Psychiatry, 6*, 360–369.

Tuulari, J. J., Tuominen, L., de Boer, F. E., Hirvonen, J., Helin, S., Nuutila, P., & Nummenmaa, L. (2017). Feeding releases endogenous opioids in humans. *Journal of Neuroscience, 37*, 8284–8291.

Volkow, N. D., Wang, G. J., Fowler, J. S., Logan, J., Hitzemann, R., Ding, Y. S., Pappas, N., Shea, C., & Piscani, K. (1996). Decreases in dopamine receptors but not in dopamine transporters in alcoholics. *Alcoholism: Clinical and Experimental Research, 20*, 1594–1598.

Volkow, N. D., Chang, L., Wang, G. J., Fowler, J. S., Ding, Y. S., Sedler, M., Logan, J., Franceschi, D., Gatley, J., Hitzemann, R., et al. (2001). Low level of brain dopamine D2 receptors in methamphetamine abusers: Association with metabolism in the orbitofrontal cortex. *The American Journal of Psychiatry, 158*, 2015–2021.

Volkow, N. D., Wang, G. J., Telang, F., Fowler, J. S., Thanos, P. K., Logan, J., Alexoff, D., Ding, Y. S., Wong, C., Ma, Y., et al. (2008). Low dopamine striatal D2 receptors are associated with prefrontal metabolism in obese subjects: Possible contributing factors. *Neuroimage, 42*, 1537–1543.

Volkow, N. D., Fowler, J. S., Wang, G. J., Baler, R., & Telang, F. (2009). Imaging dopamine's role in drug abuse and addiction. *Neuropharmacology, 56*, 3–8.

Volkow, N. D., Tomasi, D., Wang, G. J., Logan, J., Alexoff, D. L., Jayne, M., Fowler, J. S., Wong, C., Yin, P., & Du, C. (2014). Stimulant-induced dopamine increases are markedly blunted in active cocaine abusers. *Molecular Psychiatry, 19*, 1037.

Wager, T. D., Kang, J., Johnson, T. D., Nichols, T. E., Satpute, A. B., & Barrett, L. F. (2015). A Bayesian model of category-specific emotional brain responses. *PLOS Computational Biology, 11*, e1004066.

Wang, G. J., Volkow, N. D., Logan, J., Pappas, N. R., Wong, C. T., Zhu, W., Netusil, N., & Fowler, J. S. (2001). Brain dopamine and obesity. *Lancet, 357*, 354–357.

Weerts, E. M., Wand, G. S., Kuwabara, H., Munro, C. A., Dannals, R. F., Hilton, J., Frost, J. J., & McCaul, M. E. (2011). Positron emission tomography imaging of mu- and delta-opioid receptor binding in alcohol-dependent and healthy control subjects. *Alcoholism: Clinical and Experimental Research, 35*, 2162–2173.

Whistler, J. L. (2012). Examining the role of mu opioid receptor endocytosis in the beneficial and side-effects of prolonged opioid use: From a symposium on new concepts in mu-opioid pharmacology. *Drug and Alcohol Dependence, 121*, 189–204.

Wise, R. A., & Rompre, P. P. (1989). Brain dopamine and reward. *Annual Review of Psychology, 40*, 191–225.

Yarkoni, T., Poldrack, R. A., Nichols, T. E., Van Essen, D. C., & Wager, T. D. (2011). NeuroSynth: A new platform for large-scale automated synthesis of human functional neuroimaging data. *Frontiers in Neuroinformatics, 5*.

Zacny, J. P., & Gutierrez, S. (2003). Characterizing the subjective, psychomotor, and physiological effects of oral oxycodone in non-drug-abusing volunteers. *Psychopharmacology (Berl), 170*, 242–254.

Zacny, J. P., & Gutierrez, S. (2009). Within-subject comparison of the psychopharmacological profiles of oral hydrocodone and oxycodone combination products in non-drug-abusing volunteers. *Drug and Alcohol Dependence, 101*, 107–114.

Zhang, K., Herzog, H., Mauler, J., Filss, C., Okell, T. W., Kops, E. R., Tellmann, L., Fischer, T., Brocke, B., Sturm, W., et al. (2014). Comparison of cerebral blood flow acquired by simultaneous O-15 water positron emission tomography and arterial spin labeling magnetic resonance imaging. *Journal of Cerebral Blood Flow and Metabolism, 34*, 1373–1380.

Zubieta, J. K., Smith, Y. R., Bueller, J. A., Xu, Y. J., Kilbourn, M. R., Jewett, D. M., Meyer, C. R., Koeppe, R. A., & Stohler, C. S. (2001). Regional mu opioid receptor regulation of sensory and affective dimensions of pain. *Science, 293*, 311–315.

Chapter 2
The Neurocognitive Mechanisms of Unconscious Emotional Responses

Wataru Sato

Abstract The neurocognitive mechanism of emotion without conscious awareness has long been a subject of great interest (Pribram KH, Gill MM, Freud's "project" re-assessed: preface to contemporary cognitive theory and neuropsychology. Basic Books, 1976). Several pervious psychological studies have used subliminal presentations of emotional facial expressions in the context of the affective priming paradigm to investigate unconscious emotional processing (e.g., Murphy ST, Zajonc RB, J Person Soc Psychol 64:723–739, 1993; for a review, see Eastwood JD, Smilek D, Conscious Cognit 14:565–584, 2005). In a typical application of this paradigm, a facial expression depicting a negative or positive emotion is flashed briefly as a prime, then an emotionally neutral target (e.g., an ideograph) is presented. Participants are asked to make emotion-related judgments about the target. The studies reported that evaluations of the target were negatively biased by unconscious negative primes, compared to positive primes. This effect has been interpreted as evidence that unconscious emotion can be elicited and that it affects the evaluation of unrelated targets.

Keywords Unconscious emotional responses · Amygdala · Subcortical visual pathway · Emotional states

Introduction

The neurocognitive mechanisms for emotion without conscious awareness have been a long-standing topic of research (Pribram & Gill, 1976). Several pervious psychological studies have investigated unconscious emotional processing by

W. Sato (✉)
Psychological Process Research Team, Guardian Robot Project, RIKEN, Kyoto, Japan
e-mail: wataru.sato.ya@riken.jp

means of the paradigm of subliminal affective priming (e.g., Murphy & Zajonc, 1993; for a review, see Eastwood & Smilek, 2005). In these studies, a facial expression displaying negative or positive emotion is presented subliminally as a prime, then an emotionally neutral target, such as an ideogram, is presented supraliminally. Participants are instructed to evaluate the target. The studies showed that unconscious negative primes bias evaluations of the target more negatively than positive primes. This effect has been discussed as evidence that emotion can be unconsciously evoked, and that it modulates the evaluation of subsequent targets.

The subliminal affective priming paradigm, however, does not always produce clear effects, and several previous studies have failed to find the effects (e.g., Kemps et al., 1996). While Murphy and Zajonc (1993) found that the priming effect is stronger with subliminal than supraliminal emotional primes, the use of a very short presentation duration for stimuli may prevent even unconscious processing of the stimuli.

Furthermore, the neural mechanisms for unconscious emotional processing remain unclear. Although several neuroimaging (e.g., Morris et al., 1998) and neuropsychological (e.g., Kubota et al., 2000) studies have suggested that the amygdala plays an indispensable role in this process, previous findings are not consistent and debate remains in the literature (Pessoa & Adolphs, 2010). Additionally, the neural pathways underlying unconscious emotional processing remain unexplored. While some studies provided correlational data suggesting that the subcortical visual pathway sends information to the amygdala to implement unconscious emotional processing (e.g., Morris et al., 1999), there was no causal evidence. Besides, the accurate timing data of amygdala emotional processing was scarce.

In this chapter, I present the findings of our psychological and neuroscientific studies that investigated these issues. Our psychological experiments revealed that emotion arises rapidly and unconsciously. We identified the neural mechanisms for this process using functional magnetic resonance imaging (fMRI) and intracranial electroencephalography (EEG).

Psychological Study of the Unconscious Processing of Emotional Expressions

First, we tried to clearly demonstrate that emotional responses arise prior to the conscious awareness of the stimuli evoking such responses using the subliminal presentation of dynamic facial expressions. Dynamic facial expressions may be relevant in this regard because these are more ecologically valid than static expressions. Some previous psychological research has shown that dynamic facial expressions induce more obvious behavioral responses, such as subjective emotion elicitation (Sato & Yoshikawa, 2007b) and facial mimicry (Sato & Yoshikawa, 2007a), than static expressions. These data imply that it is advantageous to use dynamic rather than static facial expressions when attempting to elicit unconscious emotions.

We tested 22 healthy participants. As prime stimuli, we presented dynamic and static facial expressions of fear and happiness during 30 ms. The raw materials of the primes were grayscale photographs of facial expressions depicting fearful, happy, and neutral emotions, and they were used to create dynamic facial expressions by a morphing method. First, facial expressions with 34% and 66% intensities were created, and then 34%, 66%, and 100% facial expressions were displayed in succession to create a dynamic clip. The presentation duration for each image was 10 ms; therefore, the duration of each clip was 30 ms. The photographs of 100% facial expressions were presented as static expressions during 30 ms. A randomized mosaic image was made using a neutral face photograph by splitting the photo into pieces and randomly reordering them. The target stimuli were emotionally neutral ideograms. In each trial (Fig. 2.1), after a fixation cross, a prime stimulus was presented to either the left or right hemi-visual field; this was immediately replaced by a mask in the same place during 300 ms. Directly afterward, the target ideogram

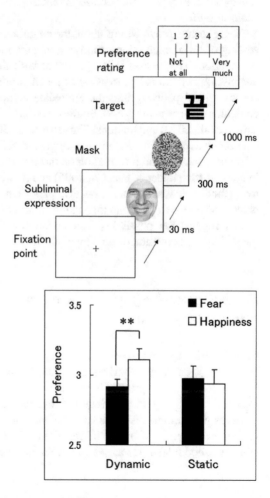

Fig. 2.1 Sato et al.'s (2014b) study. (Upper) An illustration of the trial sequence. The prime stimuli of dynamic and static facial expressions of fear and happiness were presented subliminally. (Lower) Mean (± *SE*) preference ratings. The asterisks indicate a significant difference between the fear and happiness conditions

was displayed at the same location during 1000 ms. Finally, the rating scale was displayed and participants rated their preference for the target ideogram. After the subliminal priming task, we conducted a forced-choice recognition session and confirmed that no participant had consciously perceived the prime stimuli.

For the preference ratings (Fig. 2.1), the results of our analysis of variance (ANOVA) with presentation condition (dynamic or static) and emotion (fear or happiness) as factors indicated that the interaction was significant. Follow-up simple effect analyses revealed that the effect of emotion was significant under the dynamic, but not static, presentation condition, indicating that the subliminal presentation of dynamic fearful versus happy facial expressions reduced preferences for targets.

The results demonstrated that dynamic facial expressions induce evident subliminal affective priming effects. These results extend our understanding of unconscious emotional processing and the boosting effect of dynamic facial expressions. No clear subliminal effects were detected under the static condition. The presentation duration may not have been sufficient to activate the emotional processing with static stimuli.

These results provide hints about the neural mechanism for the unconscious processing of facial expressions, implying that the mechanism is sensitive to dynamic information. This notion is in agreement with the neuroimaging finding that the unconscious emotional processing of facial expressions is performed via the subcortical visual pathway into the amygdala comprising the pulvinar and superior colliculus (Morris et al., 1999). Studies on anatomical connections in animals (Day-Brown et al., 2010) and humans (Tamietto et al., 2012) have revealed that the amygdala receives visual information through the subcortical pathway. Regarding the effect of visual motion information on these brain structures, a neuroimaging study in humans (Schneider & Kastner, 2005) and numerous physiological studies in animals (for a review, see Waleszczyk et al., 2004) indicated that the superior colliculus is more sensitive to dynamic than static information. Together with these data, our results suggest the possibility that unconscious emotional processing is implemented by the activation of the amygdala via the subcortical pathway.

Psychological Study of the Unconscious Emotional Processing of Food

Next, we tried to test the generalizability of unconscious emotional responses and their impact on daily behaviors using food stimuli. Emotional responses to food have important consequences for humans, both positively (e.g., facilitating wellbeing) and negatively (e.g., triggering overeating and lifestyle-related diseases). Previous psychological studies have shown that both the observation and ingestion of food evoke positive emotional reactions (Rodríguez et al., 2005), which in turn stimulate food intake (for a review, see Sørensen et al. 2003).

However, whether emotional responses to food could be unconsciously elicited remained unknown. As we discussed above, several psychological studies using the subliminal affective priming paradigm have shown that non-food emotional stimuli (e.g., facial expressions) induced unconscious emotional processing. On the basis of this evidence, we hypothesized that emotional responses to food would also be unconsciously activated.

In addition, we expected that unconscious food processing would have an influence on daily eating habits. Previous psychological studies have reported that eating habits can be assessed using self-reported questionnaires such as the Dutch Eating Behavior Questionnaire (DEBQ) (van Strien et al., 1986). The DEBQ assesses some eating habits related to overeating. Among the DEBQ sub-scales, several previous studies have shown that the external eating tendency, defined as eating behaviors in response to external (e.g., visual and olfactory) food stimuli, modulates automatic food processing (e.g., attentional shift to food; Brignell et al., 2009). Based on these data, we hypothesized that unconscious emotional reactions to food could be associated with external eating tendency.

To examine these hypotheses, we examined unconscious and conscious emotional responses to food and non-food stimuli and the relationships between these responses and eating habits (Sato et al., 2016). We tested 34 healthy participants. All participants fasted for more than 3 h prior to the experiment. Unconscious emotional responses were tested using the subliminal affective priming paradigm (Murphy & Zajonc, 1993). Food stimuli were color photographs of fast food (e.g., hamburgers) and Japanese diet (e.g., grilled teriyaki fish) (Fig. 2.2). Randomized mosaic stimuli were made from the food stimuli; all food stimuli were split into small squares and randomly sorted. A mask stimulus was also prepared by creating a randomized mosaic pattern. The photographs of neutral faces were used as targets under the subliminal condition. The target stimuli were randomly assigned to the experimental conditions. We used the Japanese version of the DEBQ (van Strien et al., 1986) to assess eating habits related to overeating. In each trial under the subliminal condition, a food or mosaic prime was displayed during 33 ms in the left or right hemi-visual field after a fixation cross; this was immediately replaced by a mask image during 167 ms. The target face was then displayed in the center during 1000 ms. Finally, the response panel was displayed and participants rated their preferences for the target faces. In each trial under the supraliminal condition, after the presentation of the fixation cross, a food or mosaic target was displayed during 200 ms in the left or right hemi-visual field. Participants rated their preferences for the target images. A following forced-choice recognition task was conducted to ensure that none of the participants had consciously recognized the primes.

Under the subliminal condition, the ANOVA with stimulus type (food or mosaic) as a factor revealed a significant main effect of stimulus type, demonstrating higher preference ratings for faces primed by food images than those for faces primed by mosaics (Fig. 2.2). Similarly, under the supraliminal condition, the main effect of stimulus type was significant, showing higher preference ratings for food images than for mosaics. Correlation analysis showed a significant positive correlation

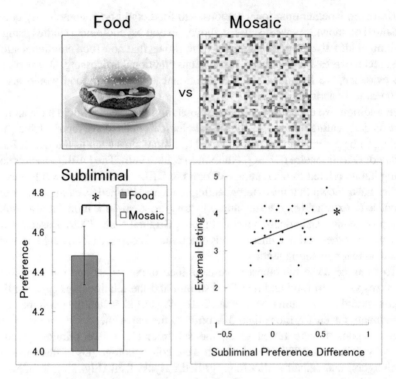

Fig. 2.2 Sato et al.'s (2016) study. (Upper) The illustrations of food and mosaic stimuli. (Lower left) Mean (± *SE*) preference ratings in the subliminal condition. The asterisk indicates a significant difference between food and mosaic prime conditions. (Lower right) A scatter plot with a regression line showing a relationship between food preference scores under the subliminal condition and external eating tendency. The asterisk indicates a significant association

between food preference scores under the subliminal condition and external eating tendency (Fig. 2.2).

These results revealed that unconscious emotional responses are elicited by food stimuli. The data, together with other evidence, suggest that unconscious emotional responses can be triggered by various types of stimuli, including emotional expressions and food. Moreover, the results demonstrated that the unconscious emotional responses to food are positively associated with the tendency for external eating. This suggests that unconscious emotional reactions play a key role in behaviors in daily life.

fMRI Study of the Neural Mechanisms for Unconscious Emotional Processing of Food

Next, we explored the neural mechanisms for unconscious emotional processing using visual food stimuli. Several prior fMRI studies have investigated neural activity in response to supraliminally presented food images. These studies consistently reported that some brain regions, including the neocortical visual areas (e.g., the fusiform gyrus) and limbic regions (e.g., the amygdala), are activated more strongly in response to food images than non-food images (e.g., Holsen et al., 2005; for a review, see van Meer et al., 2015). Accordingly, some scholars proposed that the neocortical visual areas are involved in the visual recognition of food images, which, in turn, activates the amygdala and other related regions for emotional processing (Chen et al., 2016).

However, the neural mechanisms underpinning the unconscious emotional responses to food remained unknown. In other literatures, several previous neuroimaging studies have reported that the unconscious processing of emotional expressions activates the amygdala (e.g., Morris et al., 1998). A few neuropsychological studies also found an indispensable role of the amygdala in the unconscious processing of emotional scenes (e.g., Kubota et al., 2000). Based on these findings, we hypothesized that the amygdala could be activated during both conscious and unconscious emotional processing of food.

Furthermore, prior neuroimaging studies investigating facial expression processing have found that neural pathways are different between conscious and unconscious emotional processing. Some studies provided evidence, though correlational and non-causal results, that emotional information in facial expressions is transmitted unconsciously through the subcortical pathway to the amygdala, such as the superior colliculus and pulvinar (e.g., Morris et al., 1999). It was also reported that the visual pathways involved in conscious and unconscious processing of emotional facial expressions differ (e.g., Vuilleumier et al., 2001). On the basis of such evidence, we hypothesized that the visual pathways to the amygdala for conscious and unconscious processing of food would differ and that subcortical structures would be involved in unconscious food processing.

In this study (Sato et al., 2019), we tested these hypotheses by measuring fMRI while participants viewed supraliminally or subliminally presented food images. We examined the commonalities and differences in neural responses to food versus mosaic images across presentation conditions. Furthermore, we conducted dynamic causal modeling and compared models with the subcortical, cortical, and dual visual pathways to the amygdala. We tested 22 healthy participants, all of whom had fasted for more than 3 h before the experiment. The stimuli presented were identical to those used in the above psychological experiment (Sato et al., 2016). Color photographs of fast food and Japanese diet and their corresponding randomized mosaic images were used. The participants completed two runs of 128 trials using a block design. Each run included one of the presentation conditions, and the order was fixed to the first subliminal and second supraliminal conditions. In each trial, a food or mosaic image was displayed in the center after a fixation cross. Under the subliminal conditions, the stimulus was displayed during 17 ms, immediately replaced by a mask for 1483 ms. Under the supraliminal condition, the stimulus was displayed during 1500 ms without mask presentation. In eight trials pseudo-randomly placed throughout the task blocks, a red cross was displayed during 1500 ms as the target, instead of the food or mosaic images. Participants were instructed to perform a dummy task to detect the red cross.

We performed a conjunction analysis to determine commonalities in neural responses to food versus mosaic images across presentation conditions. The results showed significantly stronger activation in the bilateral amygdala in response to food than mosaic images under both the subliminal and supraliminal conditions (Fig. 2.3). We conducted the interaction contrast between stimulus type and presentation condition to analyze differences in neural responses to food versus mosaic

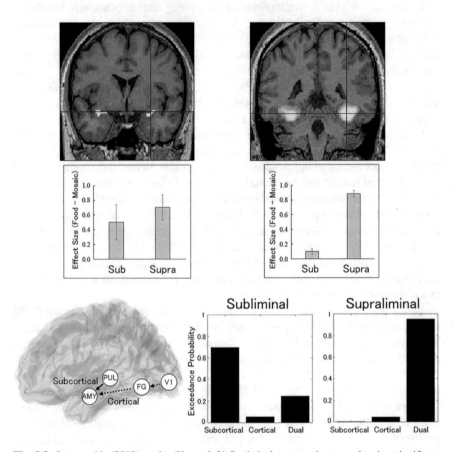

Fig. 2.3 Sato et al.'s (2019) study. (Upper left) Statistical parametric maps showing significant neural activation in response to food versus mosaic images under both the subliminal and supraliminal conditions and mean (± *SE*) effect size differences between the food and mosaic conditions. The blue cross indicates the activation focus at the right amygdala. (Upper right) Statistical parametric maps showing significantly stronger neural responses to food versus mosaic images under the supraliminal than subliminal condition and mean (± *SE*) effect size differences between the food and mosaic conditions. The blue cross indicates the activation focus at the right fusiform gyrus. (Lower) Models (left) and model comparison results (right) of dynamic causal modeling. The solid and dashed arrows indicate modulatory connections in the subcortical and cortical pathway models, respectively. The dual pathways model contains both pathways. The model comparison results in the right hemisphere are shown. *AMY* amygdala, *FG* fusiform gyrus, *PUL* pulvinar, *V1* primary visual cortex

images across presentation conditions. The results showed significantly stronger activation for food versus mosaic images under the supraliminal than subliminal condition in the broad bilateral posterior regions, including the fusiform gyrus (Fig. 2.3).

We performed dynamic causal modelling to determine the visual pathway to the amygdala in each hemisphere. We compared the models in which the subcortical (pulvinar–amygdala), cortical (primary visual cortex–fusiform gyrus–amygdala), and dual visual pathways were functionally coupled with the amygdala specifically during food processing (Fig. 2.3). In both hemispheres, the model comparison indicated that the subcortical pathway model was the most likely under the subliminal condition, while the dual pathways model was optimal under the supraliminal condition (Fig. 2.3).

Our results demonstrated that the amygdala is active in response to food images in both the subliminal and supraliminal conditions. These results imply that the amygdala is commonly involved in the unconscious and conscious emotional processing of food. These results are consistent with, and extend the substantial neuroimaging and neuropsychological evidence indicating, the involvement of the amygdala in the processing of stimuli with emotional significance (e.g., Sato et al., 2004). The visual areas were activated in response to supraliminally versus subliminally presented food. The neocortical visual areas may be related to the conscious perception of food.

Furthermore, our dynamic causal modeling analyses provide causal evidence that the amygdala is activated by visual food stimuli through the subcortical visual pathway before the conscious recognition of food occurs. Subsequently, the amygdala receives the processed visual information of food through the neocortical pathway. In addition to the aforementioned anatomical (Day-Brown et al., 2010; Tamietto et al., 2012) and neuroimaging (Morris et al., 1999) findings, our model of the subcortical visual input to the amygdala under the subliminal condition is consistent with data showing that a patient with damage in the neocortical visual areas showed amygdala activity, which was functionally coupled with pulvinar activity, in response to unseen emotional expressions (Morris et al., 2001).

Intracranial EEG Study of the Neural Processing of Emotional Expressions

Here, we tried to demonstrate rapid amygdala activation during emotional processing using facial expression stimuli. As described above, a number of neuroimaging studies have shown that the amygdala is active during the visual processing of emotional stimuli, such as emotional facial expressions and palatable food, even in the absence of conscious awareness of the stimuli (e.g., Morris et al., 1999). Some researchers proposed that the amygdala may be activated during an early stage of emotional processing because the amygdala receives sensory input from the subcortical pathway.

However, the temporal profile of the amygdala activation, specifically during the processing of emotional facial expressions, remained unclear. Some studies have examined this issue by recording magnetoencephalography while participants observed emotional facial expressions and found that stronger amygdala responses to fearful/threatening than neutral expressions occurred rapidly, approximately 100 ms after the stimulus onset (e.g., Luo et al., 2007). However, the results were inconsistent and there remains debate over whether the activity of such a deep and complex brain structure as the amygdala can be appropriately estimated from scalp-recorded electromagnetic signals (Papadelis et al., 2009).

Intracranial EEG recordings can offer direct evidence of electric neural activity with high temporal resolution. In this regard, a previous study examined amygdala activity while participants viewed negative, positive, and neutral scenes by employing intracranial EEG recordings and time–frequency analyses (Oya et al., 2002). The results showed stronger gamma-band (around 40 Hz) oscillations in the amygdala in response to negative scenes, as compared with both positive and neutral scenes, as early as 50–150 ms after stimulus onset. Based on these data, we hypothesized that the amygdala could reveal similar rapid gamma-band oscillations while viewing other emotional stimuli, i.e., fearful facial expressions. In this study (Sato et al., 2011), to test this hypothesis, we recorded the intracranial EEG from the amygdala while participants observed fearful, happy, and neutral facial expressions.

We tested six patients. All patients suffered from pharmacologically resistant epilepsy and their intracranial EEG was recorded in a presurgical evaluation. Surgical and electrophysiological assessments suggested that the main epileptic foci were outside the amygdala. Pre- and post-implantation anatomical assessments showed no structural abnormalities in any patient's bilateral amygdala. Implantation of intracranial electrodes was performed according to a stereotactic method (Mihara & Baba, 2001). Post-implantation anatomical MRI assessments ensured that the target electrodes were located in the amygdala (Fig. 2.4). The stimuli consisted of grayscale photographs of seven individuals' faces depicting fearful, happy, and neutral expressions. In each trial, after a fixation cross, the stimulus was displayed during 1000 ms in the center of the screen. The response panel was then displayed and the participants performed a dummy task to specify the gender of the displayed faces.

Time–frequency statistical parametric mapping analyses for the comparison between fearful and neutral expressions revealed significant gamma-band activity between 50 and 150 ms (starting before 100 ms; Fig. 2.4).

What are the implications of rapid amygdala activity triggered before 100 ms for our understanding of emotional processing? Numerous previous scalp- (e.g., Bentin et al., 1996) and subdurally-recorded (e.g., Sato et al., 2014a) EEG studies have reported that the first face-specific visual analysis in the neocortical visual areas occurs after 100 ms. Together with such findings, our data imply that the emotional processing of facial expressions in the amygdala is faster than the first visual analysis of faces in the neocortex. Furthermore, another line of scalp-recorded EEG research that investigated conscious awareness of visual stimuli has shown that the

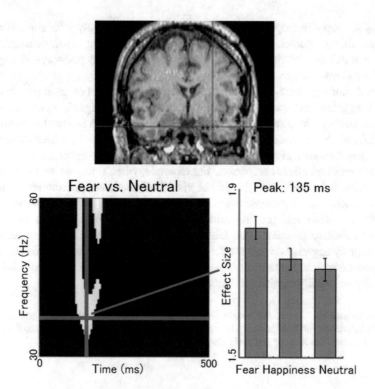

Fig. 2.4 Sato et al.'s (2011) study. (Upper) Representative anatomical magnetic resonance image. The red cross indicates the location of the amygdala electrode. (Lower) Statistical parametric maps for amygdala gamma-band activation for fearful compared with neutral facial expressions (left) and mean (± *SE*) effect size at the peak activation focus (right)

negative deflection at the posterior cortices from 200 to 400 ms was greater in response to seen than unseen stimuli (e.g., Genetti et al., 2009; for a review, see Koivisto & Revonsuo, 2010). Together with these findings, our results suggest that amygdala activity at about 100 ms reflects the emotional processing that takes place prior to the conscious perception of the stimuli.

Conclusion

In summary, our psychological data demonstrate that humans have psychological mechanisms for unconscious emotional processing. The findings presented in section "Psychological study of the unconscious emotional processing of food" suggest that such unconscious emotional responses are general and play important roles in daily life. The fMRI data presented reveal that the amygdala is involved in emotional processing via the subcortical pathway prior to conscious awareness of the stimuli. The intracranial EEG data described demonstrate that the amygdala is rapidly activated in response to emotional stimuli, specifically after approximately 100 ms. Taken together, these findings suggest that the amygdala implements rapid and unconscious emotional processing via the subcortical pathways at approximately 100 ms.

These findings have implications for human behavior. For example, first, the model suggests that rapidly evaluating stimulus emotional significance (via the subcortical visual pathway and unconscious and rapid amygdala activity) is mandatory, and difficult to consciously prevent. Therefore, people should acknowledge such psychological mechanisms and take precautions or slowly adjust their behaviors to mitigate rapid emotional responses. For example, when someone wants to control their eating behaviors, they should not visit food-abundant environments, such as supermarkets and convenience stores. Second, the model suggests that subjective emotional states could provide valuable information about the rapid and unconscious evaluative processes that take place in the amygdala. For example, if one feels slightly negative or positive feelings during social interaction, this subjective information may indicate that our amygdala has automatically detected subtle biologically or socially significant messages.

Acknowledgments The author would like to thank Dr. Paulo Sérgio Boggio and Dr. Tanja S. H. Wingenbach for their advice. This study was supported by funds from Research Complex Program from Japan Science and Technology Agency.

References

Bentin, S., Allison, T., Puce, A., Perez, E., & McCarthy, G. (1996). Electrophysiological studies of face perception in humans. *Journal of Cognitive Neuroscience, 8*, 551–565.

Brignell, C., Griffiths, T., Bradley, B. P., & Mogg, K. (2009). Attentional and approach biases for pictorial food cues. Influence of external eating. *Appetite, 52*, 299–306.

Chen, J., Papies, E. K., & Barsalou, L. W. (2016). A core eating network and its modulations underlie diverse eating phenomena. *Brain and Cognition, 110*, 20–42.

Day-Brown, J. D., Wei, H., Chomsung, R. D., Petry, H. M., & Bickford, M. E. (2010). Pulvinar projections to the striatum and amygdala in the tree shrew. *Frontiers in Neuroanatomy, 4*, 143.

Eastwood, J. D., & Smilek, D. (2005). Functional consequences of perceiving facial expressions of emotion without awareness. *Consciousness and Cognition, 14*, 565–584.

Genetti, M., Khateb, A., Heinzer, S., Michel, C. M., & Pegna, A. J. (2009). Temporal dynamics of awareness for facial identity revealed with ERP. *Brain and Cognition, 69*, 296–305.

Holsen, L. M., Zarcone, J. R., Thompson, T. I., Brooks, W. M., Anderson, M. F., Ahluwalia, J. S., Nollen, N. L., & Savage, C. R. (2005). Neural mechanisms underlying food motivation in children and adolescents. *Neuroimage, 27*, 669–676.

Kemps, E. B. F., Erauw, K., & Vandierendonck, A. (1996). The affective primacy hypothesis: Affective or cognitive processing of optimally and suboptimally presented primes? *Psychologica Belgica, 36*, 209–219.

Koivisto, M., & Revonsuo, A. (2010). Event-related brain potential correlates of visual awareness. *Neuroscience & Biobehavioral Reviews, 34*, 922–934.

Kubota, Y., Sato, W., Murai, T., Toichi, M., Ikeda, A., & Sengoku, A. (2000). Emotional cognition without awareness after unilateral temporal lobectomy in humans. *Journal of Neuroscience, 20*, RC97.

Luo, Q., Holroyd, T., Jones, M., Hendler, T., & Blair, J. (2007). Neural dynamics for facial threat processing as revealed by gamma band synchronization using MEG. *Neuroimage, 34*, 839–847.

Mihara, T., & Baba, K. (2001). Combined use of subdural and depth electrodes. In H. O. Luders & Y. G. Comair (Eds.), *Epilepsy surgery* (2nd ed., pp. 613–621). Lippincott Williams and Wilkins.

Morris, J. S., Ohman, A., & Dolan, R. J. (1998). Conscious and unconscious emotional learning in the human amygdala. *Nature, 393*, 467–470.

Morris, J. S., Ohman, A., & Dolan, R. J. (1999). A subcortical pathway to the right amygdala mediating "unseen" fear. *Proceedings of the National Academy of Sciences of the United States of America, 96*, 1680–1685.

Morris, J. S., de Gelder, B., Weiskrantz, L., & Dolan, R. J. (2001). Differential extrageniculostriate and amygdala responses to presentation of emotional faces in a cortically blind field. *Brain, 124*, 1241–1252.

Murphy, S. T., & Zajonc, R. B. (1993). Affect, cognition, and awareness: Affective priming with optimal and suboptimal stimulus exposures. *Journal of Personality and Social Psychology, 64*, 723–739.

Oya, H., Kawasaki, H., Howard, M. A., & Adolphs, R. (2002). Electrophysiological responses in the human amygdala discriminate emotion categories of complex visual stimuli. *Journal of Neuroscience, 22*, 9502–9512.

Papadelis, C., Poghosyan, V., Fenwick, P. B., & Ioannides, A. A. (2009). MEG's ability to localise accurately weak transient neural sources. *Clinical Neurophysiology, 120*, 1958–1970.

Pessoa, L., & Adolphs, R. (2010). Emotion processing and the amygdala: From a 'low road' to 'many roads' of evaluating biological significance. *Nature Review Neuroscience, 11*, 773–783.

Pribram, K. H., & Gill, M. M. (1976). *Freud's "Project" re-assessed: Preface to contemporary cognitive theory and neuropsychology*. Basic Books.

Rodríguez, S., Fernández, M. C., Cepeda-Benito, A., & Vila, J. (2005). Subjective and physiological reactivity to chocolate images in high and low chocolate cravers. *Biological Psychology, 70*, 9–18.

Sato, W., & Yoshikawa, S. (2007a). Spontaneous facial mimicry in response to dynamic facial expressions. *Cognition, 104*, 1–18.

Sato, W., & Yoshikawa, S. (2007b). Enhanced experience of emotional arousal in response to dynamic facial expressions. *Journal of Nonverbal Behavior, 31*, 119–135.

Sato, W., Yoshikawa, S., Kochiyama, T., & Matsumura, M. (2004). The amygdala processes the emotional significance of facial expressions: An fMRI investigation using the interaction between expression and face direction. *Neuroimage, 22*, 1006–1013.

Sato, W., Kochiyama, T., Uono, S., Matsuda, K., Usui, K., Inoue, Y., & Toichi, M. (2011). Rapid amygdala gamma oscillations in response to fearful facial expressions. *Neuropsychologia, 49*, 612–617.

Sato, W., Kochiyama, T., Uono, S., Matsuda, K., Usui, K., Inoue, Y., & Toichi, M. (2014a). Rapid, high-frequency, and theta-coupled gamma oscillations in the inferior occipital gyrus during face processing. *Cortex, 60*, 52–68.

Sato, W., Kubota, T., & Toichi, M. (2014b). Enhanced subliminal emotional responses to dynamic facial expressions. *Frontiers in Psychology, 5*, 994.

Sato, W., Sawada, R., Kubota, Y., Toichi, M., & Fushiki, T. (2016). Unconscious affective responses to food. *PLoS One, 11*, e0160956.

Sato, W., Kochiyama, T., Minemoto, K., Sawada, R., & Fushiki, T. (2019). Amygdala activation during unconscious visual processing of food. *Scientific Reports, 9*, 7277.

Schneider, K. A., & Kastner, S. (2005). Visual responses of the human superior colliculus: A high-resolution functional magnetic resonance imaging study. *Journal of Neurophysiology, 94*, 2491–2503.

Sørensen, L. B., Møller, P., Flint, A., Martens, M., & Raben, A. (2003). Effect of sensory perception of foods on appetite and food intake: A review of studies on humans. *International Journal of Obesity and Related Metabolic Disorders, 27*, 1152–1166.

Tamietto, M., Pullens, P., de Gelder, B., Weiskrantz, L., & Goebel, R. (2012). Subcortical con-
    nections to human amygdala and changes following destruction of the visual cortex. *Current
    Biology, 22*, 1449–1455.
van Meer, F., van der Laan, L. N., Adan, R. A., Viergever, M. A., & Smeets, P. A. (2015). What you
    see is what you eat: An ALE meta-analysis of the neural correlates of food viewing in children
    and adolescents. *Neuroimage, 104*, 35–43.
van Strien, T., Frijters, J. E. R., Bergers, G. P. A., & Defares, P. B. (1986). The Dutch Eating
    Behavior Questionnaire (DEBQ) for assessment of restrained, emotional, and external eating
    behavior. *International Journal of Eating Disorders, 5*, 295–315.
Vuilleumier, P., Sagiv, N., Hazeltine, E., Poldrack, R. A., Swick, D., Rafal, R. D., & Gabrieli,
    J. D. (2001). Neural fate of seen and unseen faces in visuospatial neglect: A combined event-
    related functional MRI and event-related potential study. *Proceedings of the National Academy
    of Sciences of the United States of America, 98*, 3495–3500.
Waleszczyk, W. J., Wang, C., Benedek, G., Burke, W., & Dreher, B. (2004). Motion sensitivity in
    cat's superior colliculus: Contribution of different visual processing channels to response prop-
    erties of collicular neurons. *Acta Neurobiologiae Experimentalis, 64*, 209–228.

Chapter 3
Social and Affective Neuroscience of Embodiment

Marília Lira da Silveira Coêlho ⓘ**, Tanja S. H. Wingenbach** ⓘ**, and Paulo Sérgio Boggio** ⓘ

Abstract Embodiment has been discussed in the context of social, affective, and cognitive psychology, and also in the investigations of neuroscience in order to understand the relationship between biological mechanisms, body and cognitive, and social and affective processes. New theoretical models have been presented by researchers considering not only the sensory–motor interaction and the environment but also biological mechanisms regulating homeostasis and neural processes (Tsakiris M, Q J Exp Psychol 70(4):597–609, 2017). Historically, the body and the mind were comprehended as separate entities. The body was considered to function as a machine, responsible for providing sensory information to the mind and executing its commands. The mind, however, would process information in an isolated way, similar to a computer (Pecher D, Zwaan RA, Grounding cognition: the role of perception and action in memory, language, and thinking. Cambridge University Press, 2005). This mind and body perspective (Marmeleira J, Duarte Santos G, Percept Motor Skills 126, 2019; Marshall PJ, Child Dev Perspect 10(4):245–250, 2016), for many years, was the basis for studies in social and cognitive areas, in neuroscience, and clinical psychology.

Keywords Embodiment · Empathy · Racial bias · Social embodiment · Emotion embodiment

M. L. da Silveira Coêlho (✉) · P. S. Boggio
Social and Cognitive Neuroscience Laboratory, Developmental Disorders Program,
Center for Health and Biological Sciences, Mackenzie Presbyterian University,
São Paulo, Brazil
e-mail: marilialira.coelho@mackenzie.br

T. S. H. Wingenbach
School of Human Sciences, Faculty of Education, Health, and Human Sciences,
University of Greenwich, Greenwich, London, UK

© The Author(s) 2023
P. S. Boggio et al. (eds.), *Social and Affective Neuroscience of Everyday Human Interaction*, https://doi.org/10.1007/978-3-031-08651-9_3

Introduction

Embodiment has been discussed in the context of social, affective, and cognitive psychology, and also in the investigations of neuroscience in order to understand the relationship between biological mechanisms, body and cognitive, and social and affective processes. New theoretical models have been presented by researchers considering not only the sensory–motor interaction and the environment but also biological mechanisms regulating homeostasis and neural processes (Tsakiris, 2017).

Historically, the body and the mind were comprehended as separate entities. The body was considered to function as a machine, responsible for providing sensory information to the mind and executing its commands. The mind, however, would process information in an isolated way, similar to a computer (Pecher & Zwaan, 2005). This mind and body perspective (Marmeleira & Duarte Santos, 2019; Marshall, 2016), for many years, was the basis for studies in social and cognitive areas, in neuroscience, and clinical psychology.

However, the dichotomous discussion of mind and body has been replaced by an approach that considers the individual's integrality. Embodiment, in turn, arises from the connection between body, emotions, brain, and environment (Marshall, 2016). Thus, the body is no longer seen as a simple sensory–motor interface, neither is the mind seen as a set of logical functions and isolated cognitive abilities. Together, the body and mind become an integral biological system modulated by experiences provided by homeostatic self-regulation interconnected with interactions with other individuals and with the environment (Marmeleira & Duarte Santos, 2019). In this perspective, embodiment is understood as a representation of the self and its interaction with the world. In this chapter, we are discussing embodiment in both social and affective processes.

Neuroscience of Embodiment

Embodiment is experienced through representations in the brain based on simulations of predictions and patterns constructed by our experiences both at the perceptual and motor level (Barrett, 2017; Longo & Tsakiris, 2013). Our perceptual experience occurs through sensory inputs, such as auditory, visual, or vestibular sensations, and also through somatic experiences, such as touch, pain, vibration, and the position of the body itself. To exemplify, let us consider the action of grasping a pen with the fingers. The tactile sensation when touching the pen is temporally and spatially congruent with seeing the fingers grasping the pen. Incoming visual information about the location of the body (i.e., fingers grasping the pen) is processed by the visual cortex and is related to a somatic representation of the perception of the visual space around the body parts (Holmes & Spence, 2004; Kilteni et al., 2015). The execution of the motor action (here: grasping a pen) includes efferent motor signals and the associated touch sensation includes afferent feedback. The synchronous integration of the visuo-tactile and proprioceptive signals promotes the experience of the moving body parts being perceived as one's own (Longo & Tsakiris, 2013; Tsakiris, 2010). These integration processes allow to differentiate between one's own perceptual experiences and those of others but also serve as the basis for experiences being grounded in one's body (hence, embodiment).

The brain areas of the posterior parietal cortex (PPC) and ventral premotor cortex (MPCv) play a fundamental role in the perception of the body and the surrounding space (Holmes & Spence, 2004). Visual–somatosensory coordination includes encoding the position of the body in space and comparing the felt with the seen position. Multisensory neurons respond to tactile and proprioceptive stimulation (e.g., touch sensation when grasping a pen and knowledge of the hand's location in space), but also to visual stimulation (seeing the hand moving and the fingers grasping the pen) (Graziano, 1999; Zopf et al., 2010). The representation of an action can be used in simulations to predict sensations and to track mismatches between sensory predictions and real perception of the sensory environment (Barrett, 2017). The continuous coupling of visuo-tactile and proprioceptive signals can explain the strong neural connections between the visual, motor, and somatosensory cortex.

These processes can be facilitated by specific neurons that fire both during action observation and action execution. Early monkey studies showed that some neurons (in the pre-motor brain area F5) fire during action observation as well as action execution (Gallese, 2007; Rizzolatti et al., 1996) which serves as a potential explanation for simulation processes and understanding others' actions. These neurons are now called mirror neurons. In the study conducted by Rizzolatti et al. (1996), it was discovered that some neurons fired when the monkey saw a grasping action and it were the same neurons that fired when the monkey was performing a grasping action. Another experiment included a second monkey and a human experimenter and a similar response of this group of neurons was found (Rizzolatti et al., 1996). These results demonstrate the activation of the mirror neuron system when observing movement-related action. Mirror neurons were found to be somatotopically arranged in the premotor cortex and reciprocally connected in the posterior parietal cortex; these areas are considered analog to the areas containing mirror neurons in monkeys (Rizzolatti et al., 1996).

The experiments in monkeys revealed that in addition to the activation of the F5 area for observation of the action and execution of the action, this brain area is also active during partially hidden observation, when it is possible to predict the result of the action, even in the absence of complete visual information of the execution of the action and interaction with the target object. Umiltà et al. (2001) conducted a study with monkeys with two experimental conditions: "total" vision condition, when the monkey was shown a fully visible action directed at an object (hand–object interaction), and the "partial" vision condition, when the same action was shown, but the final part of the action was hidden. The results showed that there was activation of mirror neurons in the F5 area in both experimental conditions (Umiltà et al., 2001), which provides support to suggest that the understanding of the action can be based on predictions of the internal motor representation of the action, through the anticipation of the final objective of the action performed by others, and, therefore, this mechanism can be understood as a precursor of more sophisticated skills of understanding the intention of others (Gallese, 2007).

Gallese (2007) calls the mechanism of mirror neurons capable of helping us understand others "incorporated simulation." The incorporated simulation theory by Gallese (2007) proposes that the mirror neuron system may be involved in processes of social cognition, such as understanding others' actions and intentions, attributing mental states to others, and language. Other studies suggest that the mirror neuron system is involved in social cognition processes, such as facial expression recognition and ultimately empathy (Mier et al., 2010; Schulte-Rüther et al., 2007); the mirror neuron system is thought to include the fusiform gyrus, superior temporal sulcus, posterior parietal cortex, ventral premotor cortex, and tonsil (Schmidt et al., 2021).

Overall, such evidence suggests that there is an embodied nature to actions and cognitive processes. This embodiment makes it possible to run simulations to guide action and to use such internal models to give meaning and coherence to sensations. (Barrett, 2017). Thus, brain simulations function as filters for sensory stimulus inputs, driving action, and constructs perception of both cognition and emotions (see Barret review, 2017). Conversely, the manipulation of multisensory stimuli can modulate representations of the body and create perceptual illusions of body parts and embodiment illusions of the self and self-other (see next section). Having touched upon the neuroscience of embodiment, this chapter continues by delving into social and affective processes that can be explained by embodiment.

Embodiment and Social Embodiment

Embodiment is centered on our subjective experiences grounded in our physical body (Gillihan & Farah, 2005). It is through this bodily self-awareness that we understand that we have a body, that we feel it as our own, that it occupies a place in space, and that there is a space around it. The formation of this body self-awareness depends on the integration of bodily signs of different sensory modalities, which signal the location of body parts and of the entire body in space, as well as providing information that we are within this body. Therefore, this body assumes the perspective of the "self" in experimentation and interaction with the world (Blanke et al., 2015; Mul et al., 2019).

Embodiment from the internal body representation perspective can be expressed through the sensations of body ownership and of motor agency. The sensation of agency precedes a motor action, and it involves the efferent component because centrally generated motor commands precede a voluntary movement (Tsakiris et al., 2006). It is the intention and execution of actions that allow the sensation of movement control of the body in a given task (Gallagher, 2000; Tsakiris et al., 2006). Body ownership is related to the sensation of the presence of the body itself. According to Gallagher (2000), it is the feeling that "my body" belongs to me, and it is always present in one's mental life. This feeling of embodiment is present during motor actions in performing a task, as well as during passive bodily experiences such as being touched (Tsakiris et al., 2006). The body scheme's neural construction is formed throughout life: a dynamic update based on sensory cues experienced by the body and its interaction with the environment (Cardinali et al., 2009). Hence, we learn cognitive and motor skills and the perception of our own body based on these sensory experiences. Embodiment is modulated by bodily experiences, but also by affective experiences and internal body representation (Braun et al., 2018; Marmeleira & Duarte Santos, 2019).

Therefore, the sense of body ownership should be considered as a result of external sensory stimuli that integrate different sensory signals (somatosensory, vestibular, visual, somatosensory) to the formation of body perception (Botvinick & Cohen, 1998; Kilteni et al., 2015; Tsakiris, 2010), and internal, interoceptive stimuli, which form the internal body representation. This multisensory information interacts with motor systems in motor action, making it possible for the body scheme to locate and perceive a body part's position in space (Margolis & Longo, 2014; Medina & Coslett, 2010), contributing to the implementation of actions involved in the interaction with the environment (Assaiante et al., 2014).

The plasticity of the multisensory integration, through simultaneous sensory stimuli of spatial and temporal congruence, has been vastly studied, showing that bodily representations and peripersonal space can be modulated after seconds of sensory manipulation, incorporation of instruments, mirror images, and use of inanimate objects such as a rubber hand. Synchronous visuo-tactile or visuomotor interactions make it possible to change one's perception of peripersonal and body space, which can modify the body scheme and induce the sensation of body ownership, including someone else's body part, as in the rubber hand illusion (Botvinick & Cohen, 1998; Holmes & Spence, 2004; Kilteni et al., 2015). In this illusion, the participant's hand is occluded from their vision and replaced by a prosthesis with similar characteristics, positioned close to the body aligned with the shoulder. In order for the illusion to occur both the real and the fake, hands must be touched synchronously in time and precisely in the same location. This visuo-tactile–proprioceptive interaction generates a conflict of what is seen in the prosthesis and what is felt in the hand, and it promotes incorporation of the rubber hand by the body scheme and the sensation of body ownership (Botvinick & Cohen, 1998). Thus, the illusions that manipulate the sense of body ownership are potentially experimental tools for investigating body representation and peripersonal space (Costantini & Haggard, 2007).

In this context, it is possible to suggest that self-awareness is highly malleable and influenced by external sensory information as evidenced by several studies. However, in addition to external sensory information, we have internal representations formed by interoceptors that allow us to have consciousness of our body (Tsakiris, 2017). Craig (2009) presented in a review that interoceptive representations contained in the insular cortex provide a basis for the subjective feelings of body and consciousness. The insula is the interoceptive center in the brain, and it plays a fundamental role in the representation of self-awareness involving the integration of external stimuli arising from the environment and the feeling of agency and control of one's own body. The insula is also linked to the affective processing of the self and the other and of processes of social cognition, such as empathy, representation of oneself, and sense of identity (Craig, 2009; Tsakiris, 2017). Thus, interoception plays a fundamental role in the self-awareness and in the stability of the internal representation that, despite the influences of exteroceptive signs and social interaction with others, maintains the representation of the body's self-awareness as being "mine" (Tsakiris, 2017).

From this perspective, social neuroscience began studying embodiment in order to have a better understanding of social perception, attitudes, and emotion of the self and the others (Niedenthal & Barsalou, 2005). Studies have shown that embodiment can be influenced by social experiences and by the processes of perceived social information, which makes us susceptible to experiencing overlap of body representation of the other (Tsakiris, 2017). Sforza et al. (2010) demonstrated that by synchronic touching, the face of people who were seeing simultaneous touches on a partner's face, induced by the "enfacement" illusion, the partner's facial characteristics were incorporated in the representation of the participant's own face; the same did not happen during asynchronic touch. Similar results were found in the study carried out by Tajadura-Jiménez and Tsakiris (2014); in addition, the authors showed the role of individual interoceptive sensitivity in the modulation of exteroceptor signals by stimulation multisensory synchronic recognition. These findings suggest that the sense of body ownership is malleable through multisensory integration, and it is possible to induce the sense of ownership of the part of the body of the other as being my own body, yet the perception of the recognition of the body itself, as distinct from others, is weighted by individual interoceptive sensitivity.

From studies on the embodiment of the self and the other, it is possible to demonstrate how perceptual illusions can modulate multisensory integration, but also the social perception of the other. In the study conducted by Paladino et al. (2010), the sensation of being touched synchronously to the observed touch of another person provoked more positive affective reactions than in the asynchronous condition. In addition, participants felt closer to the other person and perception of face resemblance was increased. Other studies were conducted in order to understand whether the modulation of social perception in the embodiment of the other can influence racial bias. Peck et al. (2013), through virtual reality, investigated whether the embodiment of light-skinned people in virtual bodies of dark skin, light skin, purple skin, and without virtual body modulated the implicit racial bias. The results revealed that the implicit racial bias decreased when the dark-skinned virtual body was incorporated. Farmer et al. (2014) used the rubber hand illusion with black-and-white hands in Caucasian participants. The synchronous stimulation in the dark-skinned rubber hand was demonstrated to have a more positive implicit attitude toward black people and induced a sensation of body ownership. However, the authors observed that the most favorable results of the illusion of the rubber hand were influenced in the participants with low racial attitudes implied in relation to dark-skinned. Similarly, Lira et al. (2017) revealed that the increase in racial bias implied in relation to dark-skinned affected the temporal dynamics of multisensory integration during the rubber hand illusion and promoted delay in assigning the sense of body ownership of the hand of another racial group in Caucasian participants. These results together show that social embodiment and recognition of the self and the other are influenced by the way we are connected to the other, which involves cultural, emotional, and affective aspects.

Finally, perceptive illusions have been shown to be an important tool to manipulate the embodiment of the body itself and the body of the other. Interestingly, the embodiment of the other has helped to understand social processes such as empathy, racial bias, change of the negative valence for the judgment of the other, social perception, among other aspects. The studies have shown us the malleability and the rapid adaptability to the judgment of the implicit social attitude when we experience the body of the other despite the existing cultural differences. Perhaps the advancement of studies of social embodiment allows us to better understand categorization, prejudice, and discrimination from the embodiment of the other and its neural and physiological correlates.

Embodiment of Emotion

Embodied cognition accounts postulate that there are interrelations between the body (e.g., body posture, gestures) and cognition, and it is assumed that emotions are also embodied. Darwin (1872) observed that physical bodily actions are closely related to an emotional experience and that an experienced emotion seems to result in a particular behavioral pattern. The assumption of embodiment is that we acquire memory and thus knowledge on concrete objects or abstract concepts (e.g., emotions) through experience and store all information of the specific experience (i.e., context, affect, behavior, etc.) together in a representation (Barsalou, 2008). Sensory experiences from all modalities (motor, sensory, and affective) are stored in these representations. When knowledge is required of a concrete object or abstract concept, the memory stored in its representation can get activated and a simulation of the initial state when the knowledge was acquired takes place in sensory–motor brain areas and can initiate responses across the body, although this can be a partial re-enactment of lesser intensity (Barsalou, 2008; Niedenthal, 2007). Using functional magnetic resonance imaging, Wicker et al. (2003) showed that the same brain region (i.e., insula) is activated in participants when they are seeing a facial expression of disgust as when they are experiencing disgust themselves, demonstrating that the same neural network is involved in the representation as in the experience of this emotion. It is very likely that a triggered representation of an emotion presents itself beyond the neural activation and changes occur across the body.

Representations are indeed not solely localized in the brain but encompass the whole body. Nummenmaa et al. (2014) conducted multiple experiments on the representation of emotions across the body. In one experiment, participants were asked to localize specific emotions in the body (by coloring in body maps) where the emotion would be felt. In another experiment, emotions were elicited in participants, and they were asked to report the accompanying bodily sensations. In yet another experiment, participants were asked to link observed facial emotion to parts of the body where the emotion would be felt by the person displaying the emotion facially. The results showed that bodily sensations are linked to discrete emotions reflecting the representation of emotion concepts across the body. For example, the emotion of sadness was portrayed as a reduction in bodily sensations in the limbs in line with the lowered muscle tone and drive in activity experienced during sadness. In a further study, Nummenmaa et al. (2018) showed there are neural activation patterns associated with emotional states and demonstrated again that emotions are embodied.

The various aspects stored in a representation of an emotion, i.e., body postures, facial expressions, physiological responses (e.g., pulse), can each trigger the other parts of the representation. As such, verbally reporting about a joyful experience and thereby accessing the conceptual knowledge on joy is likely to activate the representation of joy, leading to the experience of positive affect (that was felt when the situation initially occurred), an associated facial expression of smiling and other physical components. This occurrence has been demonstrated experimentally. Providing participants with one-sentence descriptions of emotional situations and prompting them to imagine the scenario leads to respective subjective feelings and facial muscle activation associated with the emotion imagined (Brown & Schwartz, 1980). Likewise, research has shown that the mere production of a facial expression associated with a specific emotion can activate the representation of this emotion and lead to subjective experience of said emotion. Hess et al. (1992) asked participants to either feel (to generate the feeling but to keep it inside and not show it) the emotions anger, sadness, happiness, and peacefulness, or to merely express these emotions, or to express and feel the emotions. Self-report ratings of felt emotions were obtained and showed that even the experimental condition of mere production of facial emotional expression led to emotion experience, despite the instruction to not feel and only express the emotion. A recent study further demonstrated that emotions are represented across the body. Participants observed facial expressions of fear and anger while electromyography was recorded from muscles in the face and arm each associated with expressions of fear and anger and the results showed congruent muscle activity in face and arm for the emotions investigated (Moody et al., 2017). Such results demonstrate that individual aspects of conceptual knowledge can activate other parts of the emotion representation including changes across the body.

The literature presented in this section thus far has included an explicit emotional stimulus which activated emotion representations. However, activation of emotion representations also take effect across the body when people are unaware of the activated emotion representation, that is, without explicit emotional stimulus. In a study, participants believed brain lateralization was measured using electroencephalography while they listened to music and were told they had to relax/contract facial muscles as a conflicting task (Duclos et al., 1989). However, the facial muscle activation manipulations actually resulted in facial expressions associated with individual emotional expressions and no brain activity was measured. Self-ratings on emotional experience were obtained but covered up as a necessity to control for interference with the obtained electroencephalography recordings. Results showed that facial expression manipulations associated with anger, disgust, fear, and sadness resulted in higher emotional experience reports for each of these emotions. In a second experiment, the body posture of participants was manipulated to represent fear, anger, and sadness and resulted in the respective emotional experiences. Since emotion representations can be triggered without us perceiving an emotional stimulus, it might be the case that elicited emotion representations have further effects, in that bodily states might also affect our behavior and cognitions related to an emotion experience, and this without our awareness.

Embodied aspects of emotions indeed affect cognitions that are related to a current bodily state even in the case that the relationship between the bodily action and the emotion is unknown to a person. Probably the most famous study on the effects of bodily action on cognition was conducted by Strack et al. (1988) who manipulated participants' mouth position and examined the effects on evaluations of cartoons regarding their funniness. When participants were holding a pen with their teeth, sticking out of their mouth, and so unconsciously simulating a smile, participants rated cartoons as funnier than participants holding a pen with their lips in a way that smiling was prevented. The experimentally induced smile was not an expression of truly felt positive affect but elicited the respective representation and could so influence the evaluations of the cartoons. In both experimental conditions included in the study by Strack et al. (1988), muscular feedback from the face influenced the evaluations of the cartoons. One explanation is that the experimentally induced smile was perceived by participants as resulting from the cartoons and interpreted as being amused, which is a more cognitive explanation. An alternative explanation, rooted in the body, is that the experimentally induced smile created muscular feedback which elicited the respective emotion representation, and thereby altered evaluations, respectively. This study constitutes one example of how the body can influence cognitions without being aware of this influence. However, it should be noted that a multicentre replication study did not consistently reproduce the same results (Wagenmakers et al., 2016). Nonetheless, a preceding study also demonstrated that manipulations of facial expression toward frowning and smiling without participants' awareness affected participants' emotional experience as well as funniness evaluations of cartoons (Laird, 1974). Such findings further align with aforementioned literature in this chapter on bodily state manipulations related to emotions and respective emotional experiences. It is clear that physical changes occur within the body during emotional experience, but it has also been demonstrated that these changes serve a purpose, in that they prepare for subsequent action, e.g., increased blood flow to skeletal muscles during fear to prepare for flight (e.g., Balters & Steinert, 2017; de Gelder et al., 2004). Consequently, it is no long stretch to assume that bodily states would also affect cognitions, which is the fundamental proposition of grounded cognition or embodiment theories (Barsalou, 2008) and many research findings support this assumption (Winkielman et al., 2015).

A further example of how embodiment of emotions can affect cognitions provides a study on memory. Participants enacted body postures associated with specific emotions (but were unaware of this emotion-related manipulation), which facilitated recalling of personal experiences containing these emotions (Schnall & Laird, 2003). In this case, the facilitation of the performance resulted from the congruency between the triggered emotion representation and the emotion in the task. Hence, incongruence between bodily state and emotion stimulus should lead to hampered performance. That bodily states incongruent with an observed stimulus in a task can affect cognitions was demonstrated in a study where facial muscle activations were manipulated and its effect on facial emotion recognition investigated. When participants were holding a pen in their mouth in a way that antagonist muscles to the observed facial emotional expressions were activated, this induced facial muscle feedback that was incongruent with the muscle activation underlying the observed facial expression and lowered recognition rates compared to a control condition without mouth movement manipulation (Wingenbach et al., 2018). These results can be explained with embodiment of emotion. Given that the observation of a specific facial emotional expression should elicit its representation (including the respective facial muscle activity), motor information incongruent with the visual information, i.e., observed emotion, should cause interference with the elicited representation, and thus hamper recognition. Similarly, an electroencephalography study demonstrated that interfering with the simulation of observed facial emotion through facial muscle activation manipulation impairs processing of the observed facial emotion as evidenced by greater semantic retrieval demand, i.e., larger N400 amplitude (Davis et al., 2017). In a similar fashion, botulinum toxin-a injections in the corrugator muscle of the face (associated with frowning) impaired language comprehension of the emotional content of sad and angry nature (both emotional expressions include corrugator activation) as measured by reading time compared to pre-injection (Havas et al., 2010). These results exemplify the effect bodily states can have on cognitions and highlight the all-encompassing nature of emotion representations across the body.

The relationship between bodily states and cognitions within the framework of embodiment of emotion is bidirectional. That is, cognitive processes related to emotion can have an effect on our bodily states just as bodily states can affect cognitions. For example, participants' posture was measured in vertical height during generation of terms associated with pride and disappointment and a significant decrease in height was found during the disappointment condition compared to the pride condition (Oosterwijk et al., 2007). The experience of pride is generally accompanied with a straightened body posture, whereas disappointment usually results in a slumping position and the conceptual understanding of these terms reflected in participants' posture. Further evidence for the effect of cognition on bodily states based on embodiment of emotion comes from a study where participants had to pull or push a lever while seeing positive and negative stimuli and were found to push faster for negative valenced stimuli than they pulled and vice versa for positive valance stimuli (Chen & Bargh, 1999). The results can be explained by an evaluation of a stimulus as positive or negative that is embodied in the bodily behavior by facilitating approach for positive stimuli and avoidance for negative stimuli. Similar results were obtained from a study where participants were faster at pulling a slider when the content of a read sentence was positive compared to negative in content and pushed faster for negative compared to positive content (Filik et al., 2015). The literature demonstrates that conceptual knowledge on emotion-related stimuli reflects in bodily actions and facilitates corresponding actions.

Interestingly, embodiment of emotions goes beyond the own body and can even reflect in our language and the physical space surrounding us. For example, when describing emotional states, an individual that is currently in a sad mood might describe themselves as "feeling down" and an individual in good spirits might "feel elevated." The arousal level and valence associated with both of these affective states (low/negative and high/positive, respectively) reflect in the language used to communicate about these affective states. A study asking participants to place negative, neutral, and positive valenced terms within a three-dimensional space found their valence to reflect the placement (Marmolejo-Ramos et al., 2018). Words associated with positive valence were placed high up and close to the participants, words of negative valence were placed low and farther away from participants, and neutral words in between. The evaluation of a term as positive vs negative thus affected the vicinity of proximity. It seems that embodiment of emotion does not only entail our own bodies but also the physical space surrounding our bodies.

Neuroscientific methods can also be used to demonstrate the effect embodiment of emotion has on our cognitions, behavior, and body itself. Price et al. (2012) conducted a study displaying positive and neutral images to participants while electroencephalography was recorded, and the position of participants was manipulated to reclining or leaning. Results showed that the late positive potentials were larger when participants were leaning toward positive images, but no effect of body posture was found during the viewing of neutral images. This study demonstrates that even in the absence of a cognitive task, embodiment of emotion takes effect as specific body postures modulated brain activity. Such findings suggest that embodiment of emotion might be the result of primal reactions like approach and avoidance of emotional stimuli taking effect also in higher order processing of emotional stimuli. To conclude this section, embodiment of emotion can be investigated on a behavioral, peripheral–physiological, and neural level, individually or in combination as has been shown in the various parts of this chapter.

Conclusion

Embodiment is a subject that has broadened the scientific discussion about the biological system, self-regulation, and neural processes. As shown in this chapter, perceptive illusions have been demonstrated as an important tool to manipulate the corporation of the body itself and the overlap with the body of the other. Interestingly, the embodiment of the other has helped to understand social processes such as empathy, racial bias, change of the negative valence for the judgment of the other, social perception, among other aspects. The studies have shown us the malleability and the rapid adaptability to the judgment of the implicit social attitude when we experience the body of the other despite the existing cultural differences. The advancement of research of social embodiment has allowed us to better understand categorization, prejudice, and discrimination from the embodiment of the other and its neural and physiological correlates. Moreover, neuroscientific methods help us demonstrate the similarity in neural patterns during emotional experience and during the simulation of an emotional experience, and can thus provide evidence for the embodiment of emotion. This neuroscientific evidence is in addition to the vast evidence from behavioral studies on embodiment of emotion, such as embodied emotion expressed through body posture, facial expression, language, and cognitive processes (e.g., stimulus evaluations).

References

Assaiante, C., Barlaam, F., Cignetti, F., & Vaugoyeau, M. (2014). Body schema building during childhood and adolescence: A neurosensory approach. *Neurophysiologie Clinique*. https://doi.org/10.1016/j.neucli.2013.10.125

Balters, S., & Steinert, M. (2017). Capturing emotion reactivity through physiology measurement as a foundation for affective engineering in engineering design science and engineering practices. *Journal of Intelligent Manufacturing, 28*(7), 1585–1607. https://doi.org/10.1007/s10845-015-1145-2

Barrett, L. F. (2017). The theory of constructed emotion: An active inference account of interoception and categorization. *Social Cognitive and Affective Neuroscience, 12*(1), 1–23. https://doi.org/10.1093/scan/nsw154

Barsalou, L. W. (2008). Grounded cognition. *Annual Review of Psychology, 59*(1), 617–645. https://doi.org/10.1146/annurev.psych.59.103006.093639

Blanke, O., Slater, M., & Serino, A. (2015). Behavioral, neural, and computational principles of bodily self-consciousness. *Neuron*. https://doi.org/10.1016/j.neuron.2015.09.029

Botvinick, M., & Cohen, J. (1998). Rubber hands "feel" touch that eyes see [8]. *Nature*. https://doi.org/10.1038/35784

Braun, N., Debener, S., Spychala, N., Bongartz, E., Sörös, P., Müller, H. H. O., & Philipsen, A. (2018). The senses of agency and ownership: A review. *Frontiers in Psychology, 9*(APR), 1–17. https://doi.org/10.3389/fpsyg.2018.00535

Brown, S.-L., & Schwartz, G. E. (1980). Relationships between facial electromyography and subjective experience during affective imagery. *Biological Psychology, 11*(1), 49–62. https://doi.org/10.1016/0301-0511(80)90026-5

Cardinali, L., Brozzoli, C., & Farnè, A. (2009). Peripersonal space and body schema: Two labels for the same concept? *Brain Topography*. https://doi.org/10.1007/s10548-009-0092-7

Chen, M., & Bargh, J. A. (1999). Consequences of automatic evaluation: Immediate behavioral predispositions to approach or avoid the stimulus. *Personality and Social Psychology Bulletin, 25*(2), 215–224. https://doi.org/10.1177/0146167299025002007

Costantini, M., & Haggard, P. (2007). The rubber hand illusion: Sensitivity and reference frame for body ownership. *Consciousness and Cognition, 16*, 229–240. https://doi.org/10.1016/j.concog.2007.01.001

Craig, A. D. (2009). How do you feel – Now? The anterior insula and human awareness. *Nature Reviews Neuroscience, 10*(1), 59–70. https://doi.org/10.1038/nrn2555

Darwin, C. (1872). *The expression of the emotions in man and animals*. John Murray. https://doi.org/10.1037/10001-000

Davis, J. D., Winkielman, P., & Coulson, S. (2017). Sensorimotor simulation and emotion processing: Impairing facial action increases semantic retrieval demands. *Cognitive, Affective, & Behavioral Neuroscience, 17*(3), 652–664. https://doi.org/10.3758/s13415-017-0503-2

de Gelder, B., Snyder, J., Greve, D., Gerard, G., & Hadjikhani, N. (2004). Fear fosters flight: a mechanism for fear contagion when perceiving emotion expressed by a whole body. *Proceedings of the National Academy of Sciences of the United States of America, 101*(47), 16701–16706. https://doi.org/10.1073/pnas.0407042101

Duclos, S. E., Laird, J. D., Schneider, E., Sexter, M., Stern, L., Van Lighten, O., & Hiatt, F. (1989). Emotion-specific effects of facial expressions and postures on emotional experience. *Journal of Personality and Social Psychology, 57*.

Farmer, H., Maister, L., Tsakiris, M., & Paulus, F. M. (2014). Change my body, change my mind: The effects of illusory ownership of an outgroup hand on implicit attitudes toward that outgroup. *Frontiers in Psychology, 4*(January), 1–10. https://doi.org/10.3389/fpsyg.2013.01016

Filik, R., Hunter, C. M., & Leuthold, H. (2015). When language gets emotional: Irony and the embodiment of affect in discourse. *Acta Psychologica, 156*, 114–125. https://doi.org/10.1016/J.ACTPSY.2014.08.007

Gallagher, S. (2000). Philosophical conceptions of the self: Implications for cognitive science. *Trends in Cognitive Sciences*.

Gallese, V. (2007). Before and below "theory of mind": Embodied simulation and the neural correlates of social cognition. *Philosophical Transactions of the Royal Society, B: Biological Sciences, 362*(1480), 659–669. https://doi.org/10.1098/rstb.2006.2002

Gillihan, S. J., & Farah, M. J. (2005). *Is self special? Psychological Bulletin*. https://doi.org/10.1037/0033-2909.131.1.76

Graziano, M. S. A. (1999). Where is my arm? The relative role of vision and proprioception in the neuronal representation of limb position. *Proceedings of the National Academy of Sciences of the United States of America, 96*(18), 10418–10421. https://doi.org/10.1073/pnas.96.18.10418

Havas, D. A., Glenberg, A. M., Gutowski, K. A., Lucarelli, M. J., & Davidson, R. J. (2010). Cosmetic use of botulinum toxin-a affects processing of emotional language. *Psychological Science, 21*(7), 895–900. https://doi.org/10.1177/0956797610374742

Hess, U., Kappas, A., McHugo, G. J., Lanzetta, J. T., & Kleck, R. E. (1992). The facilitative effect of facial expression on the self-generation of emotion. *International Journal of Psychophysiology, 12*(3), 251–265. https://doi.org/10.1016/0167-8760(92)90064-I

Holmes, N. P., & Spence, C. (2004). The body schema and multisensory representation(s) of peripersonal space. *Cognitive Processing, 5*(2), 94–105. https://doi.org/10.1007/s10339-004-0013-3

Kilteni, K., Maselli, A., Kording, K. P., & Slater, M. (2015). Over my fake body: Body ownership illusions for studying the multisensory basis of own-body perception. *Frontiers in Human Neuroscience*. https://doi.org/10.3389/fnhum.2015.00141

Laird, J. D. (1974). Self-attribution of emotion: The effects of expressive behavior on the quality of emotional experience. *Journal of Personality and Social Psychology, 29*(4), 475–486. https://doi.org/10.1037/h0036125

Lira, M., Egito, J. H., Dall'Agnol, P. A., Amodio, D. M., Gonçalves, Ó. F., & Boggio, P. S. (2017). The influence of skin colour on the experience of ownership in the rubber hand illusion. *Scientific Reports, 7*(1), 1–13. https://doi.org/10.1038/s41598-017-16137-3

Longo, M. R., & Tsakiris, M. (2013). Merging second-person and first-person neuroscience. *Behavioral and Brain Sciences, 36*(4), 429–430). Cambridge University Press. https://doi.org/10.1017/S0140525X12001975

Margolis, A. N., & Longo, M. R. (2014). Visual detail about the body modulates tactile localisation biases. *Experimental Brain Research*. https://doi.org/10.1007/s00221-014-4118-3

Marmeleira, J., & Duarte Santos, G. (2019). Do not neglect the body and action: The emergence of embodiment approaches to understanding human development. *Perceptual and Motor Skills, 126*. https://doi.org/10.1177/0031512519834389

Marmolejo-Ramos, F., Tirado, C., Arshamian, E., Vélez, J. I., & Arshamian, A. (2018). The allocation of valenced concepts onto 3D space. *Cognition and Emotion, 32*(4), 709–718. https://doi.org/10.1080/02699931.2017.1344121

Marshall, P. J. (2016). Embodiment and human development. *Child Development Perspectives, 10*(4), 245–250. https://doi.org/10.1111/cdep.12190

Medina, J., & Coslett, H. B. (2010). From maps to form to space: Touch and the body schema. *Neuropsychologia, 48*, 645–654. https://doi.org/10.1016/j.neuropsychologia.2009.08.017

Mier, D., Lis, S., Neuthe, K., Sauer, C., Esslinger, C., Gallhofer, B., & Kirsch, P. (2010). The involvement of emotion recognition in affective theory of mind. *Psychophysiology, 47*(6), 1028–1039. https://doi.org/10.1111/j.1469-8986.2010.01031.x

Moody, E. J., Reed, C. L., Van Bommel, T., App, B., & McIntosh, D. N. (2017). Emotional mimicry beyond the face? *Social Psychological and Personality Science*, 194855061772683. https://doi.org/10.1177/1948550617726832

Mul, C. L., Cardini, F., Stagg, S. D., Sadeghi Esfahlani, S., Kiourtsoglou, D., Cardellicchio, P., & Aspell, J. E. (2019). Altered bodily self-consciousness and peripersonal space in autism. *Autism*. https://doi.org/10.1177/1362361319838950

Niedenthal, P. M. (2007). Embodying emotion. *Science, 316*(5827), 1002–1005. https://doi.org/10.1126/science.1136930

Niedenthal, P. M., & Barsalou, L. W. (2005). Embodiment in attitudes, social perception, and emotion Silvia Krauth-Gruber and François Ric. *Personality and Social Psychology Review, 9*(3), 184–211.

Nummenmaa, L., Glerean, E., Hari, R., & Hietanen, J. K. (2014). Bodily maps of emotions. *Proceedings of the National Academy of Sciences, 111*(2), 646–651. https://doi.org/10.1073/pnas.1321664111

Nummenmaa, L., Hari, R., Hietanen, J. K., & Glerean, E. (2018). Maps of subjective feelings. *Proceedings of the National Academy of Sciences, 115*(37), 9198–9203. https://doi.org/10.1073/pnas.1807390115

Oosterwijk, S., Rotteveel, M., Fischer, A. H., & Hess, U. (2007). Embodied emotion concepts: How generating words about pride and disappointment influences posture. *European Journal of Social Psychology, 39*, 457–466. https://doi.org/10.1002/ejsp.584

Paladino, M. P., Mazzurega, M., Pavani, F., & Schubert, T. W. (2010). Synchronous multisensory stimulation blurs self-other boundaries. *Psychological Science, 21*(9), 1202–1207. https://doi.org/10.1177/0956797610379234

Pecher, D., & Zwaan, R. A. (Eds.). (2005). *Grounding cognition: The role of perception and action in memory, language, and thinking.* Cambridge University Press.

Peck, T. C., Seinfeld, S., Aglioti, S. M., & Slater, M. (2013). Putting yourself in the skin of a black avatar reduces implicit racial bias. *Consciousness and Cognition, 22*, 779–787.

Price, T. F., Dieckman, L. W., & Harmon-Jones, E. (2012). Embodying approach motivation: Body posture influences startle eyeblink and event-related potential responses to appetitive stimuli. *Biological Psychology, 90*(3), 211–217. https://doi.org/10.1016/J.BIOPSYCHO.2012.04.001

Rizzolatti, G., Fadiga, L., Gallese, V., & Fogassi, L. (1996). Premotor cortex and the recognition of motor actions. *Cognitive Brain Research, 3*(2), 131–141. https://doi.org/10.1016/0926-6410(95)00038-0

Schmidt, S. N. L., Hass, J., Kirsch, P., & Mier, D. (2021). The human mirror neuron system – A common neural basis for social cognition? *Psychophysiology, 58*(5), e13781. https://doi.org/10.1111/psyp.13781

Schnall, S., & Laird, J. (2003). Keep smiling: Enduring effects of facial expressions and postures on emotional experience and memory. *Cognition & Emotion, 17*(5), 787–797. https://doi.org/10.1080/02699930302286

Schulte-Rüther, M., Markowitsch, H. J., Fink, G. R., & Piefke, M. (2007). Mirror neuron and theory of mind mechanisms involved in face-to-face interactions: A functional magnetic resonance imaging approach to empathy. *Journal of Cognitive Neuroscience, 19*(8), 1354–1372. https://doi.org/10.1162/jocn.2007.19.8.1354

Sforza, A., Bufalari, I., Haggard, P., & Aglioti, S. M. (2010). My face in yours: Visuo-tactile facial stimulation influences sense of identity. *Social Neuroscience, 5*(2), 148–162. https://doi.org/10.1080/17470910903205503

Strack, F., Martin, L. L., & Stepper, S. (1988). Inhibiting and facilitating conditions of the human smile: A nonobtrusive test of the facial feedback hypothesis. *Journal of Personality and Social Psychology, 54*(5), 768–777. https://doi.org/10.1037/0022-3514.54.5.768

Tajadura-Jiménez, A., & Tsakiris, M. (2014). Balancing the "inner" and the "outer" self: Interoceptive sensitivity modulates self–other boundaries. *Journal of Experimental Psychology: General, 143*(2), 736–744. https://doi.org/10.1037/a0033171

Tsakiris, M. (2010). My body in the brain: A neurocognitive model of body-ownership. *Neuropsychologia, 48*, 703–712. https://doi.org/10.1016/j.neuropsychologia.2009.09.034

Tsakiris, M. (2017). The multisensory basis of the self: From body to identity to others. *Quarterly Journal of Experimental Psychology, 70*(4), 597–609. https://doi.org/10.1080/1747021 8.2016.1181768

Tsakiris, M., Prabhu, G., & Haggard, P. (2006). Having a body versus moving your body: How agency structures body-ownership. *Consciousness and Cognition, 15*, 423–432. https://doi. org/10.1016/j.concog.2005.09.004

Umiltà, M. A., Kohler, E., Gallese, V., Fogassi, L., Fadiga, L., Keysers, C., & Rizzolatti, G. (2001). I know what you are doing: A neurophysiological study. *Neuron, 31*(1), 155–165. https://doi. org/10.1016/S0896-6273(01)00337-3

Wagenmakers, E.-J., Beek, T., Dijkhoff, L., Gronau, Q. F., Acosta, A., Adams, R. B., … Zwaan, R. A. (2016). Registered replication report. *Perspectives on Psychological Science, 11*(6), 917–928. https://doi.org/10.1177/1745691616674458

Wicker, B., Keysers, C., Plailly, J., Royet, J.-P., Gallese, V., & Rizzolatti, G. (2003). Both of us disgusted in my insula: The common neural basis of seeing and feeling disgust. *Neuron, 40*(3), 655–664. https://doi.org/10.1016/S0896-6273(03)00679-2

Wingenbach, T. S. H., Brosnan, M., Pfaltz, M. C., Plichta, M., & Ashwin, C. (2018). Incongruence between observers' and observed facial muscle activation reduces recognition of emotional facial expressions from video stimuli. *Frontiers in Psychology, 9*, 864. https://doi.org/10.3389/ FPSYG.2018.00864

Winkielman, P., Niedenthal, P., Wielgosz, J., Eelen, J., & Kavanagh, L. C. (2015). Embodiment of cognition and emotion. In *APA handbook of personality and social psychology, Volume 1: Attitudes and social cognition* (pp. 151–175). https://doi.org/10.1037/14341-004

Zopf, R., Savage, G., & Williams, M. A. (2010). Crossmodal congruency measures of lateral distance effects on the rubber hand illusion. *Neuropsychologia, 48*(3), 713–725. https://doi. org/10.1016/j.neuropsychologia.2009.10.028

Chapter 4
The Neuroscience of Beauty

William Edgar Comfort ⓘ and Ana Luísa Freitas

Abstract Appreciating beauty is part of everyday life, when we contemplate fine arts, architecture, music, and natural scenes. Aesthetic appreciation, like any ordinary phenomenon of human life, triggers affective and cognitive processes that can provide the subject with sensations of hedonic pleasure and cognitive self-reward (Leder H, Belke B, Oeberst A, Augustin D. Br J Psychol 95(4):489–508, 2004). Although humans share several neuropsychological processes, the experience of aesthetic appreciation is undeniably idiosyncratic, and sometimes it is not that simple to find beauty where we were supposed to find it, and more often the same object can elicit different reactions amongst observers.

Keywords Aesthetic stimuli · Aesthetic appreciation · Halo effect · Default-mode network

Introduction

Appreciating beauty is part of everyday life, when we contemplate fine arts, architecture, music, and natural scenes. Aesthetic appreciation, like any ordinary phenomenon of human life, triggers affective and cognitive processes that can provide the subject with sensations of hedonic pleasure and cognitive self-reward (Leder et al., 2004). Although humans share several neuropsychological processes, the experience of aesthetic appreciation is undeniably idiosyncratic, and sometimes it is not that simple to find beauty where we were supposed to find it, and more often the same object can elicit different reactions amongst observers. Let us take the following situation as an example.

I once had a conversation with an artist friend (WC) about Salvador Dalí's Surrealist object, Lobster Telephone (1938). I told her that, though I had a great

W. E. Comfort (✉) · A. L. Freitas
Social and Cognitive Neuroscience Laboratory, Developmental Disorders Program, Center for Health and Biological Sciences, Mackenzie Presbyterian University, São Paulo, Brazil
e-mail: 9032936@mackenzie.br

appreciation for Dalí's other work, that particular piece irritated me no end: I saw it as a lazy juxtaposition of two randomly selected items. She replied that the role of art was just that – to provoke; to elicit a reaction, any reaction, even an emotionally-negative one, in the viewer. Although I still hold that initial aversive reaction to Dalí's five versions of the telephone, my friend was right to emphasise the 'experience' of the artwork rather than its objective attributes, and consequently the subjectivity implicit in both the intensity and valence of my emotional reaction. Beyond the oft-quoted expression that 'beauty is in the eye of the beholder', a Brazilian variant contends that 'he who loves the ugly holds the perception of beauty' (*quem ama o feio, bonito lhe parece*), that is, we are able to distinguish and even empathise with a positive aesthetic experience in another even when we objectively view the object of their desire as ugly. As such, aesthetic experience goes beyond merely a shared and plastic conception of beauty and can elicit conflicting and often contradictory mental and emotional states in the same observer (Fig. 4.1).

In light of this perceptual variability in response to aesthetic stimuli, there has been increasing interest in neuroaesthetics, a relatively recent field which studies the biological bases underlying aesthetic experiences. Such experiences may include the evaluation of facial attractiveness, the appraisal of paintings, sculptures and other works of art, and complex emotional reactions to beauty, either in the natural environment or to man-made structures. Contributions to the study of neuroaesthetics are wide-ranging, drawing from such dispersed disciplines as visual perception, art theory and emotion, and hold important insights for more established areas of study in attention, face recognition, and cognitive ergonomics.

Beauty can be defined as a property or value of an object, natural scene, or person which engenders a physiological and psychological experience of pleasure and satisfaction. Cognitive neuroscience as a whole is well-placed to provide greater understanding of how humans form and process the experience of beauty, from a predominantly dopaminergic network for encoding hedonic value to the effect of context and long-term memory on the modulation of our individualised experience of beautiful stimuli. Beauty is often predominantly measured in terms of positive affective response to aesthetic stimuli, such as paintings, physically attractive faces and natural scenes, or even the activation of a subset of distinct brain regions, such

Fig. 4.1 Salvador Dalí's lobster telephone (1938)

Fig. 4.2   The triad of
aesthetic experience
(Chatterjee & Vartanian,
2016)

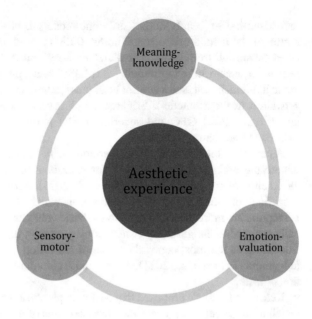

as the orbitofrontal cortex, a frequent criticism of scientific reductionism levied by colleagues in the humanities (Brown & Dissanayake, 2009).

Within the field of neuroaesthetics, however, the aesthetic experience is not merely reduced to the perception and appreciation of beauty however the global affective and cognitive valuation of external stimuli, either for their artistic or other intrinsic qualities. Pearce et al. (2016) effectively distinguish between the cognitive neuroscience of art and aesthetics, arguing that an appropriate conceptualisation of neuroaesthetics must consider artistic stimuli not merely in aesthetic terms but through a broader context of modulating factors, such as expertise, perceived value, and complex emotional states.

In a series of groundbreaking syntheses, Anjan Chatterjee, Oshin Vartanian, and colleagues (Chatterjee, 2011, 2014; Chatterjee & Vartanian, 2014, 2016; Pearce et al., 2016) have set forth the key aspects within the neuroscience of aesthetics, in what they denominate the triad of aesthetic experience: distinct brain networks for sensory-motor, emotion-valuation, and meaning-knowledge functions in the appraisal of aesthetic stimuli (see Fig. 4.2).

Defining Neuroaesthetics

Any workable definition of neuroaesthetics must proceed from work in perception, more especially visual perception, given the primacy of this sense in human perception (Kupers et al., 2011). To this end, Ishizu and Zeki (2011) investigated whether similar patterns of brain activity were correlated across different sensory modalities. Their hypothesis, in accordance with Burke's (2014) assertion of a single representation of beauty across different sensory modalities, was that a similar region of the medial orbitofrontal cortex (OFC) would be activated in response to aesthetic stimuli from both a visual and auditory source. Stimuli classified through ratings as 'beautiful', 'indifferent', and 'ugly' were presented as pairs in an aesthetic judgement task and a control brightness judgement task. The contrast between activation in the aesthetic > brightness task revealed significant activation in the lateral and medial OFC and superior frontal gyrus, indicating a selective OFC response to aesthetic judgement.

A neural system for aesthetic perception has been posited as relying on the same underlying brain structures as those for emotional processing as well as a generalised object-appraisal system (Brown et al., 2011). According to this view, a 'naturalization' of neuroaesthetics is required, in which the neural bases of aesthetic perception are more directly linked to the basic valuation of sensory stimuli, specifically mapped to the gustatory cortex, consistent with the identification of the anterior insula as the most concordant area of activation between studies of aesthetic judgement (Brown et al., 2011), an area more commonly associated with the valuation of taste.

Reber et al. (2004) proposed that aesthetic pleasure is related to the fluency with which an observer can process the characteristics of a given object. The easier the integration of the plastic elements, the greater the congruence construction, so the more fluent this observation can be, the greater the chances of positive attribution to what one observes. The fluidity of this perceptual dynamics is associated with the elicitation of pleasure, and negative reactions are common when observing asymmetric or disharmonic combinations (Ikeda et al., 2015). An absence of or interruption to this fluidity may have been one of the main contributions to our sometimes negative affective experiences with aesthetic stimuli, as in the author's reaction to Dalí's Lobster Telephone described at the beginning of this chapter.

In a study with functional magnetic resonance imaging (fMRI), to understand the neural mechanisms underlying the aesthetic and emotional aspects of colour perception, Ikeda et al. (2015) verified activation of the left medial orbitofrontal cortex (mOFC) during the observation of visually congruent stimuli and the left amygdala during the observation of incongruent stimuli (Ikeda et al., 2015). Their results led them to conclude that stimulus valuation is conditioned by automatic visual processes of stimuli features mediated by the amygdala, and the aesthetic values measured by the mOFC. These results suggest that differences in appraisal aspects of object valence may be due to separate brain regions.

Additionally, while studying different behavioural and electrophysiological responses to aesthetic experiences with modern art, Pihko et al. (2011), Leder et al. (2014), and Else et al. (2015) suggested that the observers' backgrounds should be considered when interpreting differences in response, as both the expertise level of the observer and semantic content, such as labels and titles (Gerger & Leder, 2015), can interfere in the implicit evaluation of art. For example, Gerger and Leder (2015) found varying activation in the corrugator muscle of the eyebrow and the zygomatic major facial muscle of viewers according to whether the artwork is titled or untitled, whether semantically congruent or not. These different, often discrete, muscle activations can be recorded by facial electromyography (fEMG), and the data collected by Gerger and Leder (2015) suggests that when the title was absent or incongruent, observers reported less interest and subjective aesthetic appreciation.

Although the results of Gerger and Leder (2015) appear to corroborate the relationship between the valuation of experience and perceptual fluency (Reber et al., 2004), this pattern of response may be more directly associated with the fact that study participants may have been influenced mainly by the automatic emotional processes underlying perceptual fluidity, as previously proposed by Brown et al. (2011). Corrugator and zygomatic contraction is associated with greater cognitive effort and, therefore, it is also assumed to be reflective of a reduction in perceptual fluency and consequent increase in negative experience reports. What the authors point out, however, is that fluidity of processing in the conjunction of characteristics and contextual information alone is not sufficient for positive aesthetic evaluations, as often moderate levels of cognitive effort can contribute to the positivity of aesthetic experiences, a fact observed in cases of expert evaluations that focus on art in a more elaborate way (Gerger & Leder, 2015).

Motivation and Facial Attractiveness

Another fundamental aspect of beauty research concerns its importance for the maintenance of the species. Evolutionarily, attractive faces constitute an important factor both for the establishment of sexual relations (reproductive ends) and for parental behaviour (Hahn & Perrett, 2014), because they usually symbolise fertility, gene quality, and health (Chatterjee & Vartanian, 2016).

Thinking in reproductive aspects, in heterosexuals, there is the activation of a neural network involved in motivation and reward systems, involving structures such as the nucleus accumbens, the medial prefrontal cortex, the anterior dorsal cingulate and the orbitofrontal cortices, which shows a high level of response to attractive rather than unattractive faces of people of the opposite sex. Amongst both heterosexuals and homosexuals, Hahn and Perrett (2014) point out that there are comparative studies in the literature that indicate greater activation of the orbitofrontal cortex and dorsomedial thalamus when observers are presented with faces of people of the desired sex, regardless of sexual orientation or even the gender of the observer. These results are consistent with previous research by Ishai (2007), who found greater OFC activation in response to attractive male faces in heterosexual women and homosexual men and greater OFC activation in response to attractive female faces in heterosexual men and homosexual women.

Similar to mate bonding behaviours amongst adults, the attractiveness of an individual infant's face appears to influence both caregiver behaviour and the quality of care (Langlois et al., 1995). Cute or attractive children are more likely to receive care and positive affects, such as tenderness, and less likely to suffer aggression (Hahn & Perrett, 2014), which is possibly related to the fact that cute children are usually seen as healthy and worthy of parental investment. In an fMRI study,

Glocker et al. (2009b) found that the baby schema, i.e., 'a set of infantile physical features' (Glocker et al., 2009a), activates the nucleus accumbens (NAcc), a key structure of the reward system, in nulliparous women.

Sexual and gender differences in neural responses modulated by attractive faces and infant cuteness must be better investigated, but what we know so far leads us to the hypothesis that baby facial attractiveness has its influence in human caregiving, regardless of kinship, and attractiveness in adult faces continues to play a key role in mating bonds.

Beauty and the Beast? Aesthetics and Complex Emotions

The evaluation of aesthetic stimuli is not only related to aspects of positive or negative valuation or reward processing. It is often related to the development of complex emotions such as awe, envy, and anxiety (Armstrong & Detweiler-Bedell, 2008), closely linked to the perception of the sublime in art, a philosophical concept underpinning our understanding of subjective emotional responses to aesthetic stimuli (Chatterjee & Vartanian, 2016). In a fMRI study, Cupchik et al. (2009) found significant bilateral activation in the insula when participants viewed artworks, consistent with an emotional aspect in aesthetic evaluation, suggesting that the aesthetic experience emerges from a top-down guiding of attention and bottom-up perceptual cues for fluency and visual organisation.

One key network in the subjective evaluation of aesthetic stimuli is the default-mode network (DMN), implicated in mind-wandering, autobiographical memory, and other processes. Vessel and colleagues tested fMRI response in participants rating artworks on how 'moving' they perceived them to be, on a scale of 1–4 (Vessel et al., 2012). They found an increase in activation in several areas within the DMN, including the anterior medial PFC, lateral orbitofrontal cortex, posterior cingulate cortex, and hippocampus, in response to more emotionally 'moving' artworks but, importantly, only the most highly rated images for emotional resonance led to an activation in the aMPFC, in contrast to other studies (Kawabata & Zeki, 2004; Ishizu & Zeki, 2011) which found activation varied linearly in response to the emotional response to aesthetic stimuli. This pattern of results suggests that a 'sublime' experience to aesthetic stimuli depends on an intense emotional reaction and corresponding activation in the anterior medial PFC in conjunction with the more commonplace aesthetic valuation occurring primarily in the OFC.

Judging Books By Their Covers

Despite its prevalence and the automaticity with which it occurs, the so-called 'halo effect' is a frequently employed cognitive bias that guides and directs our decision-making and judgement, causing relevant aspects to be relegated and others, such as aesthetics, to stand out and interfere with one's general judgement, even when there is sufficient information available to form an independent evaluation of extraneous salient attributes (Nisbett & Wilson, 1977).

Individuals whose faces are more attractive are often judged more positively in various dimensions and also receive different treatment in different domains of social life (Liang et al., 2010). The evaluation and personality traits attributed to facial attractiveness seem to have emerged as an adaptive response, in the light of evolutionary hypotheses that misshapen faces were associated with parasites, disease, and a lack of biogenic immunity. This evolutionarily driven preference for attractive, unblemished faces affects even otherwise healthy individuals, who nevertheless have a facial asymmetry conveying inferior levels of health, intelligence, and sociability (Zebrowitz et al., 2002; Liang et al., 2010).

Conclusion

The perceptual, behavioural, and neural mechanisms involved in the perception of aesthetic stimuli are key to our understanding of the everyday interactions and motivations which characterise our relationships with the natural and man-made world. The aim of understanding the cognitive processes underlying aesthetic appreciation has been a key driver for applying the tools and methods of cognitive neuroscience to the field of art and aesthetic stimuli and establishing links with the parallel study of perceptual, emotional, attentional, semantic, memory, and decision-making processes. Investigating neural networks engaged in decoding and valuing aesthetic content is an important challenge that should involve other areas of knowledge. For the phenomena of aesthetic appreciation to be understood in all their complexity, it is necessary to integrate into the physiological aspects also the historical, social, and cultural aspects, which holistically make up the person.

References

Armstrong, T., & Detweiler-Bedell, B. (2008). Beauty as an emotion: The exhilarating prospect of mastering a challenging world. *Review of General Psychology, 12*(4), 305–329.

Brown, S., & Dissanayake, E. (2009). The arts are more than aesthetics: Neuroaesthetics as narrow aesthetics. In M. Skov & O. Vartanian (Eds.), *Neuroaesthetics* (pp. 43–57). Baywood.

Brown, S., Gao, X., Tisdelle, L., Eickhoff, S. B., & Liotti, M. (2011). Naturalizing aesthetics: Brain areas for aesthetic appraisal across sensory modalities. *Neuroimage, 58*(1), 250–258.

Burke, E. (2014). *A philosophical enquiry into the origin of our ideas of the sublime and beautiful.* Cambridge University Press. (Original work published 1757) https://doi.org/10.1017/CBO9781107360495

Chatterjee, A. (2011). Neuroaesthetics: a coming of age story. *Journal of Cognitive Neuroscience, 23*(1), 53–62.

Chatterjee, A. (2014). Scientific aesthetics: Three steps forward. *British Journal of Psychology, 105*(4), 465–467.

Chatterjee, A., & Vartanian, O. (2014). Neuroaesthetics. *Trends in Cognitive Sciences, 18*(7), 370–375.

Chatterjee, A., & Vartanian, O. (2016). Neuroscience of aesthetics. *Annals of the New York Academy of Sciences, 1369*(1), 172–194.

Cupchik, G. C., Vartanian, O., Crawley, A., & Mikulis, D. J. (2009). Viewing artworks: Contributions of cognitive control and perceptual facilitation to aesthetic experience. *Brain and Cognition, 70*(1), 84–91.

Dalí, S. (1938). *Lobster Telephone [Object]. Tate Modern*. London, United Kingdom. https:// www.tate.org.uk/art/artworks/dali-lobster-telephone-t03257.

Else, J. E., Ellis, J., & Orme, E. (2015). Art expertise modulates the emotional response to modern art, especially abstract: An ERP investigation. *Frontiers in Human Neuroscience, 9*, 525.

Gerger, G., & Leder, H. (2015). Titles change the esthetic appreciations of paintings. *Frontiers in Human Neuroscience, 9*, 464.

Glocker, M. L., Langleben, D. D., Ruparel, K., Loughead, J. W., Gur, R. C., & Sachser, N. (2009a). Baby schema in infant faces induces cuteness perception and motivation for caretaking in adults. *Ethology, 115*(3), 257–263.

Glocker, M. L., Langleben, D. D., Ruparel, K., Loughead, J. W., Valdez, J. N., Griffin, M. D., … Gur, R. C. (2009b). Baby schema modulates the brain reward system in nulliparous women. *Proceedings of the National Academy of Sciences, 106*(22), 9115–9119.

Hahn, A. C., & Perrett, D. I. (2014). Neural and behavioral responses to attractiveness in adult and infant faces. *Neuroscience & Biobehavioral Reviews, 46*, 591–603.

Ikeda, T., Matsuyoshi, D., Sawamoto, N., Fukuyama, H., & Osaka, N. (2015). Color harmony represented by activity in the medial orbitofrontal cortex and amygdala. *Frontiers in Human Neuroscience, 9*, 382.

Ishai, A. (2007). Sex, beauty and the orbitofrontal cortex. *International Journal of Psychophysiology, 63*(2), 181–185.

Ishizu, T., & Zeki, S. (2011). Toward a brain-based theory of beauty. *PLoS One, 6*(7), e21852.

Kawabata, H., & Zeki, S. (2004). Neural correlates of beauty. *Journal of Neurophysiology, 91*(4), 1699–1705.

Kupers, R., Pietrini, P., Ricciardi, E., & Ptito, M. (2011). The nature of consciousness in the visually deprived brain. *Frontiers in Psychology, 2*, 19.

Langlois, J. H., Ritter, J. M., Casey, R. J., & Sawin, D. B. (1995). Infant attractiveness predicts maternal behaviors and attitudes. *Developmental Psychology, 31*(3), 464.

Leder, H., Belke, B., Oeberst, A., & Augustin, D. (2004). A model of aesthetic appreciation and aesthetic judgments. *British Journal of Psychology, 95*(4), 489–508.

Leder, H., Gerger, G., Brieber, D., & Schwarz, N. (2014). What makes an art expert? Emotion and evaluation in art appreciation. *Cognition and Emotion, 28*(6), 1137–1147.

Liang, X., Zebrowitz, L. A., & Zhang, Y. (2010). Neural activation in the "reward circuit" shows a nonlinear response to facial attractiveness. *Social Neuroscience, 5*(3), 320–334.

Nisbett, R. E., & Wilson, T. D. (1977). The halo effect: Evidence for unconscious alteration of judgments. *Journal of Personality and Social Psychology, 35*(4), 250–256.

Pearce, M. T., Zaidel, D. W., Vartanian, O., Skov, M., Leder, H., Chatterjee, A., & Nadal, M. (2016).Neuroaesthetics: The cognitive neuroscience of aesthetic experience. *Perspectives on Psychological Science, 11*(2), 265–279.

Pihko, E., Virtanen, A., Saarinen, V. M., Pannasch, S., Hirvenkari, L., Tossavainen, T., … Hari, R. (2011). Experiencing art: The influence of expertise and painting abstraction level. *Frontiers in Human Neuroscience, 5*, 94.

Reber, R., Schwarz, N., & Winkielman, P. (2004). Processing fluency and aesthetic pleasure: Is beauty in the perceiver's processing experience? *Personality and Social Psychology Review, 8*(4), 364–382.

Vessel, E. A., Starr, G. G., & Rubin, N. (2012). The brain on art: Intense aesthetic experience activates the default mode network. *Frontiers in Human Neuroscience, 6*, 66.

Zebrowitz, L. A., Hall, J. A., Murphy, N. A., & Rhodes, G. (2002). Looking smart and looking good: Facial cues to intelligence and their origins. *Personality and Social Psychology Bulletin, 28*(2), 238–249.

Part II
Social Neuroscience and Moral Emotions

Chapter 5
Mirror Neurons in Action: ERPs and Neuroimaging Evidence

Alice Mado Proverbio and Alberto Zani

Abstract According to V.S. Ramachandran (inaugural 'Decade of the Brain' lecture at Society for Neuroscience meeting), 'mirror neurons are to neuroscience what DNA was to biology'. Their discovery (by Rizzolatti's group) led to the understanding of how hominids rapidly evolved through imitation and cultural transmission in the last 100,000 years. In this chapter, we will review the role of human mirror neuron system (MNS) in several mental and brain functions including: interacting with the environment, grasping objects, empathy and compassion for others, empathizing, emulation and emotional contagion, observing and imitating, learning sports, motor skills and dance, motor rule understanding, understanding the intentions of others, understanding gestures and body language, lip reading, recognizing actions by their sounds, learning to play a musical instrument. The chapter is enriched with a discussion of possible criticalities and caveats.

Keywords Mirror neuron system · Empathy · Audio-visuomotor neurons

Introduction

Many neuroimaging studies have searched for the human correlate of the monkey 'mirror neuron system' (MNS) and tried to isolate mirror neurons by using noninvasive imaging techniques such as fMRI (e.g. Dinstein, 2008; Iacoboni, 2005; Iacoboni et al., 2005; Schmidt et al., 2021). The results provided evidence of a strong activation of the anterior intraparietal area (AIP) and the ventral premotor areas (F5) when subjects passively observed others performing movements, actively executed movements themselves, or imitated movements made by others. In addition, AIP and F5 areas were frequently found engaged in tasks involving empathy,

A. M. Proverbio (✉)
Department of Psychology, University of Milano-Bicocca, Milan, Italy
e-mail: mado.proverbio@unimib.it

A. Zani
School of Psychology, Vita-Salute San Raffaele University, Milan, Italy

© The Author(s) 2023
P. S. Boggio et al. (eds.), *Social and Affective Neuroscience of Everyday Human Interaction*, https://doi.org/10.1007/978-3-031-08651-9_5

social cognition, and theory of mind, along with the inferior frontal gyrus, inferior parietal cortex, fusiform gyrus, posterior superior temporal sulcus, and amygdala. These data seem to suggest the existence of a shared neural mechanism for social cognition.

Despite the long line of research, studies on the human MNS still suffer from some severe methodological problems. Electrophysiological single unit recordings, which are required for a clear-cut demonstration of mirror neurons properties, are not feasible in humans. Therefore, the majority of studies approaching MNS in humans rely on methods with low temporal resolution (e.g., fMRI), which is an indirect method based on blood oxygenation signal (BOLD) and not directly measuring neuronal activity (see also the paragraph devoted to pitfalls at the end of the chapter). In this regard, ERPs can be excellent tools for providing the necessary temporal resolution for studying action and gesture recognition processes in healthy humans.

Visuomotor Neurons and Action Encoding

Mirror neurons (MNs) were first discovered in the ventral premotor cortex (PMv cortex) of the macaque monkey (F5 area) by Rizzolatti and Luppino (2001). These neurons were activated both when the animal performed a specific motor action and when observed another simian or human individual performing that same action. The MNs do not respond to the simple presentation of food or other objects that also affect the animal, nor they are activated by the observation of a mimed action without the presence of the objects. In order for the MNs to activate (or 'to fire', i.e., to show an intense discharge frequency), an actual interaction of the hand with a target object of the action is essential. Despite being motor neurons, MNs are not activated by single movements (e.g. of the fingers) comprising a whole motor act, but, like all the other neurons in the PM cortex, are instead activated in association with goal-directed and purposeful motor actions. The MNs are stimulated by the execution/observation of motor actions performed with the hand, but also with the mouth. They are very sensitive to the type of grip (i.e. precision grip, power grip, grip of small or large objects, grip of little seeds, etc.) and encode the *actions goal*. For instance, the neural micro-population that encodes the gesture of taking an apple will not be the same if its purpose is to eat it (i.e. the animal takes the apple and then brings it to the mouth), or to throw it away (i.e. the monkey takes the apple and throws it). After the discovery of MNs in the premotor cortex (PM), other studies have shown their presence in the inferior parietal lobe (IPL), in particular in the rostral portion of this brain lobe. These neurons would be more involved in the representation of the actions associated with an object or a tool of which they process the motor properties (i.e. *affordances*), such as, for instance, its graspability and/or usability, while dealing with information coming from the fronto-parietal-occipital visual ventral stream (VVS).

Many studies have shown that the mirror neuron system (MNS) is also present in humans (Rizzolatti & Craighero, 2004; Rizzolatti & Sinigaglia, 2016). Fine examples of these studies are the EEG investigations on the reactivity of brain rhythms during actions observation. Many studies have shown that the sight of actions performed by other individuals (with hands, legs, fingers, etc.) induces a block of observers' sensory-motor EEG rhythm (or so-called *mu* rhythm) recorded at scalp sites, which would reflect a state of relative inactivity in the Rolandic region (e.g. Lelord et al., 1998). An important PET study on human volunteers is the one carried out by Rizzolatti and colleagues (Rizzolatti et al., 1996), which allowed a first localization of the areas involved with the MNS during the observation of grasping movements. Volunteers were tested in three different conditions. In the first, they observed grasping gestures of common objects performed by the experimenter; in the second, they proceeded to reach and grasp the objects themselves, while in the third, they simply observed the objects. The results showed that only action observation activated significantly the inferior parietal lobule (IPL) and the ventral premotor area (PMv) together with the posterior portion of the inferior frontal gyrus (IFG) (Fig. 5.1).

Other studies have shown that the MNS is not only activated at the sight of gestures but also of manageable objects. By means of fMRI evidence, Creem-Regehr and Lee (2005) demonstrated that graspable tool shapes activated motor-related regions of the cortex, including the PMv area and the posterior parietal cortex

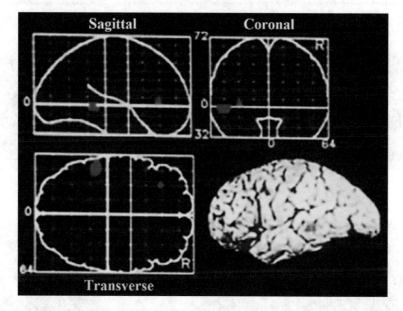

Fig. 5.1 Adjusted mean regional cerebral blood flow recorded by Rizzolatti et al. (1996) during grasping observation. The data are displayed as statistical maps overimposed on three planar projections (sagittal, coronal, and transverse) frames and as cortical rendering of the lateral cortical surfaces of the left hemisphere. The pixel values significantly higher than *p < 0.001* are shown in red

(PPC). The event-related potentials (ERPs) study by Proverbio et al. (2011a, b) provided the possible time course of this activation showing that the earliest neural tool/non-tool discrimination was indexed by an increased anterior negativity in the 210–270 ms post-stimulus latency range in response to tools rather than to objects. Source reconstructions for these findings highlighted the contribution of left-sided brain premotor and somatosensory cortices, possibly including the anterior intraparietal sulcus (aIPS). Further studies demonstrated that the cortical representation of actions (especially tools manipulation and use) is asymmetrically represented over the left hemisphere. Indeed, a lesion of the left inferior parietal cortex (IPC, BA40) is often associated with apraxic deficits, whilst a right-sided lesion rarely causes these deficits (Goldenberg & Spatt, 2009). The question of whether this hemispheric asymmetry depends on right-hand use or a hemispheric functional specialization for fine-grained, precision movements has been explored in another ERP study by Proverbio et al. (2013). The authors recorded ERPs to pictures depicting unimanual (e.g. a hammer) or bimanual (e.g. a bicycle handlebar) tools, while participants were instructed to respond motorically to infrequent images of green plants (Fig. 5.2). A prefrontal N400 component (elicited by non-targets) was much larger over the left scalp sites to bimanual than unimanual tools. swLORETA (acronym for *standardized weighted LOw-REsolution electromagnetic TomogrAphy*) sources reconstruction revealed that besides the left and right parietal cortices (BA39,

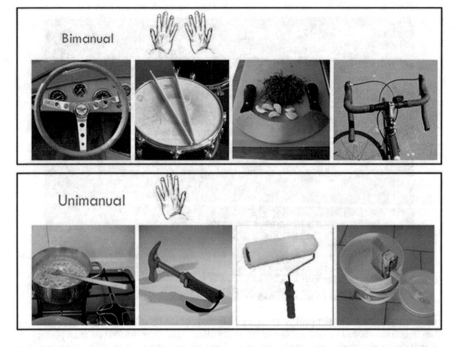

Fig. 5.2 Examples of pictures depicting bimanual and unimanual tools used as stimuli in Proverbio's et al. (2013) ERP study

BA40), tools observation always activated the left premotor cortex (BA6) regardless of the hand involved in their manipulation/use. Overall, these data suggest that looking at tools automatically activates mental representations associated with their manipulation, with a left-sided hemispheric asymmetry for this brain activation.

Mirror Neurons and Understanding the Intentions of Others: Empathy

An fMRI study by Iacoboni et al. (2005) has robustly demonstrated that the activation of visuomotor MNs makes it possible to share behavioral goals and to understand other people's intentions (a multifaceted capacity called *mentalizing* or *theory of mind*). In this famous experiment, participants observed three types of stimuli: grasping actions without context (the box in the middle in Fig. 5.3), the context without actions (the left box in Fig. 5.3), and manual actions performed in two different contexts (the right upper or lower boxes in Fig. 5.3). In this last condition, the context suggested the intention associated with the grasping action (i.e. drinking or clearing). Actions associated with the specific contexts produced a significant increase of the bold signals at the back of the IFG and in the PMv, being part of the MNS. Furthermore, the activation revealed to be greater for drinking (biologically more relevant) than for clearing. These data showed how these regions, active

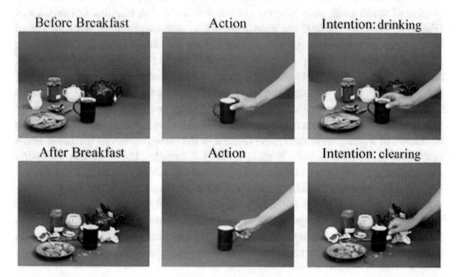

Fig. 5.3 Types of stimuli used in Iacoboni's et al. (2005) study. The same action (e.g. to take a mug) reveals an agent's different intention according to the context in which he/she is. Such an intention is encoded and inferred by means of fronto-parietal MNS activation. (Courtesy of Marco Iacoboni)

during the execution and observation of an action, were also involved in understanding the intentions of others.

For its ability of understanding visual gestures and their aims, the fronto-parietal MNS is involved in a multiplicity of mental functions including:

(i) Understanding motor events
(ii) Understanding actions and intentions of others
(iii) Understanding mental/emotional states of others: empathy (adopting another person's point of view)
(iv) Imitation, in yawning, scratching, crossing legs, posture
(v) Learning: visuomotor processes (e.g. music, sport)
(vi) Social cohesion, group behaviour, disgust (emulation, what the other feels, I also feel it: fear, embarrassment, shame or head emulation)
(vii) Empathy for pain (when looking at someone who is suffering, feeling vicarious pain, compassion)

It has been shown how recognition of body language, both symbolic and affective, as well as the congruence of people's gestures are strongly related to the fronto-parietal MNS receiving and processing information from brain regions specialized in recognizing faces (i.e. fusiform face area or FFA), facial expressions (i.e. FFA and superior temporal sulcus, STS), and bodies (i.e. extrastriate body area, EBA). In a series of electrophysiological studies by Proverbio et al. (2010, 2014a, 2015a), visual ERPs were recorded in different samples of volunteers viewing hundreds of images depicting actors and actresses mimicking a symbolic gesture (iconic, deictic, or emblematic, such as, for instance, those in Fig. 5.4, top) or an emotional display of mood using body language (as shown in Fig. 5.4, middle), or using a tool (Fig. 5.4, bottom). In half the cases, the scene was incongruent with its verbal description and/or with respect to pragmatics or standard knowledge about tools use. In all cases, the perception of incongruent images (from the points of view of the gesture or of the action meaning and/or aim) elicited a wide negative response (i.e. N400) tending to be larger at anterior scalp sites. Applying swLORETA inverse solution to the N400 potential (within its time window of occurrence), it emerged that the incongruity between actions and their presumed intentions stimulated the activation of slightly different neural circuits in the three conditions (certainly more emotional in the case of body language; Fig. 5.4, middle), but invariably including the inferior regions of both the frontal premotor and parietal areas (i.e. the fronto-parietal MNs system, in addition to the anterior cingulate cortex (ACC), the superior temporal cortex (STC), and the visual (FFA and EBA areas).

All in all, these data suggest how the MNS underpins the ability to recognize the intentions of an agent, through the observation of a gesture and the motor simulation of that same gesture by an observer.

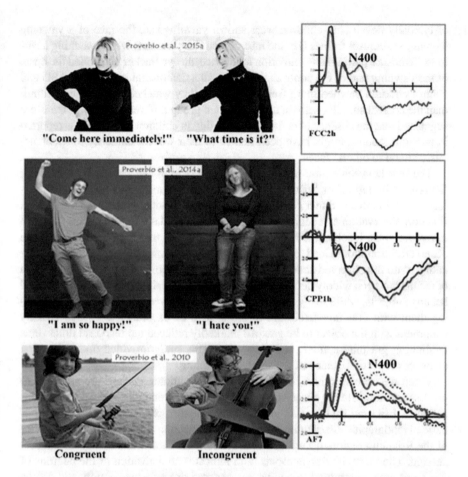

Fig. 5.4 Examples of congruent (left column) and incongruent (right column) stimuli used in Proverbio's et al. studies (2010, 2014a, 2015a), associated with the N400 component electrophysiological effect (third column), reflecting the violation of an expectation related to the aim of the action or of the gesture expressed by the actors, as referred to a shared grammar of gestures or to the context and to the pre-established use of a tool. The N400 effect is drawn as a red continuous line in the upper waveforms, as a blue continuous line in the middle waves, and as a red dotted line in the lower waves (where ERPs are shown in red for women and in blue for men). (Reproduced and modified with the permission of the authors)

Observation and Imitation

The ability to imitate the gestures of others, either unconsciously (e.g. as in yawning or in posture, such as crossing the legs) or consciously (e.g. when we imitate the master's gesture to successfully learn to play tennis) is strongly based on the MNS. The imitative ability of yawning, for example, has been investigated by Usui et al. (2013) in a study in which children with autistic spectrum disorder (ASD) and/

or typically developing children were shown yawning (i.e. the face of a yawning woman) vs. control frames (i.e. the face of a smiling woman) while watching a cartoon. To ensure participants' attention to the face, an eye tracker controlled the onset of the yawning and of the control stimuli. Results demonstrated that both ASD and control children yawned more frequently when they watched the yawning stimuli than the control stimuli (without any significant group differences). It was therefore suggested that the absence of contagious yawning in children with ASD, as reported in previous studies, might have been related to their weaker tendency to spontaneously attend to others' faces.

The link between action production and observation has also been explored in 'automatic imitation' or 'visuomotor priming' paradigms, where participants perform an action that is either congruent or incongruent with an observed movement. If action observation and action production employed shared mechanisms (namely, mirror neurons, Iacoboni, 2005), performing an action that is compatible with the observed action should lead to facilitation, while performing an action that is incompatible with the observed action should result in an interference effect. This pattern of results has been widely documented. For example, Craighero et al. (1996) primed healthy subjects, while ready to execute a grasping movement, by visually presenting them with drawings irrelevant to the task to be executed. Drawings visually congruent with the object to be grasped markedly reduced the response times, thus facilitating grasping actions, and vice versa. This study provided one of the first evidences for the existence of a visuomotor priming.

When we observe others, the motor and sensorimotor systems are activated to process and simulate the observed gesture. This activation induces the desynchronization of EEG *mu* rhythm (i.e. an oscillation rhythm of 8–12 Hz with a central-parietal topographic distribution over the scalp) reflecting a state of relative inactivity of the Rolandic region, a kind of *stand-by* from the motor or somatosensory processing. Therefore, its desynchronization indicates an activation of the neurons of this same area, committed to coding an observed or performed action, and can be used to measure MNs activity in both human adults (Pfurtscheller et al., 2006) and infants (Nyström et al., 2011). For example, Proverbio (2012) provided evidence that watching manipulable objects automatically activates their motor properties as indexed by the EEG desynchronization of *mu* rhythm over centro-parietal scalp sites during perception of tools vs. non-manipulable objects. Other studies have shown a lack of reduction of event-related beta and mu desynchronization (ERD) in ASD children during perception of actions, as opposed to comparable ERD responses during action execution (Oberman et al., 2008). Interestingly, Van Elk et al. (2008) showed that the longer is the motor experience of infants with crawling, the stronger is the *mu* rhythm desynchronization during observation of other children's crawling. This piece of findings indicates that experience strongly modulates MNS responsivity. As proof of this, it has been shown that the skills acquired in a certain athletic or sporting discipline, or, for instance, in dance, strongly modulates MNS responsivity. Proverbio et al. (2012) compared EEG/ERP signals relative to the visual processing of actions that violated basketball rules (e.g. in defense, blocking, and shooting actions) with that of correct basketball actions in professional basketball players and controls. They found that incorrect actions elicited anterior N400 responses reflecting the automatic detection of action incorrectness only in professional players (see ERP waveforms in Fig. 5.5). According to source reconstruction, N400 generators included the fronto-parietal MNS, the cerebellum, the EBA, and the STS. Similarly, the detection of incorrect dance gestures has been shown to elicit a response in the fronto/parietal MNS circuits in professional dancers vs. controls (Calvo-Merino et al., 2005; Orlandi et al., 2017).

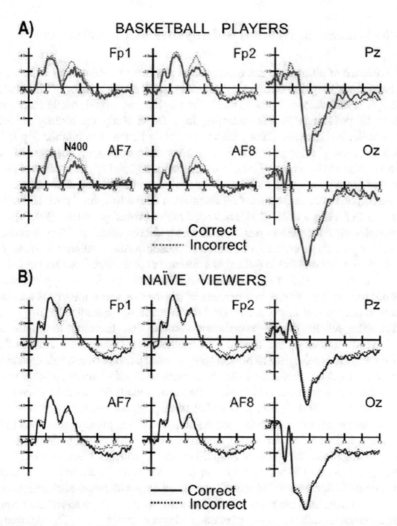

Fig. 5.5 Grand-average ERPs recorded in professional basketball players (**a**) and naïve viewers (**b**) in response to correct and incorrect basketball actions at frontal, parietal, and occipital scalp sites. (Taken and redrawn from Proverbio et al., 2012)

Audio-Visuomotor Neurons

The existence of multimodal audiovisual cortical regions has been demonstrated both for phonetic/articulatory language (i.e. verbal language) and for human and animal vocalizations (e.g. a chirp, a whinny, a cry, a laughter), as well as for encoding of noises typically produced using objects (e.g. the noise produced by crushing nuts, or by chewing). These multimodal neurons are a particular class of MNs that encode both visual and auditory information.

Audio-Visuomotor Neurons in Language and Vocalizations

The existence of a link between motor and perceptual representations of language has been since long demonstrated. According to Liberman's theory (Liberman & Mattingly, 1985), knowing how to understand a phoneme would strictly correspond to how to pronounce it. For example, in a fMRI study on healthy subjects, Pulvermüller and Shtyrov (2006) found that, while listening to bilabial (/ p /) and dental occlusive phonemes (/ t /), simultaneous activations were observed of both auditory areas of the temporal lobe (for understanding) and of the precentral motor areas (for production), with a difference in the locus of activation depending on the processed phoneme: at the motor representation of the lips, for / p / e of the language for / t /. Fadiga et al. (2002) recorded motor evoked potentials (MEPs) from the muscles of the tongue in participants who had been asked to listen to acoustic stimuli. These stimuli consisted of words or pseudowords containing a double / f / (e.g. *baffo* (i.e. *moustache*) in Italian) or a double / r / (e.g. *birra* (i.e. *beer*) in Italian) and bitonal sounds. The / f / is a labiodental consonant that, for being pronounced, does not require a particular involvement of the tongue, while the / r / is a linguo-palatal consonant that involves a marked involvement of the tongue for its pronunciation. The results of the experiment showed that listening to words and pseudowords containing the double / r / resulted in a significant increase of the MEPs, compared to the case of bitonal sounds, words, and pseudowords containing the double / f /. As a whole, these data demonstrated that, in humans, a MNS would exist dedicated to the comprehension of linguistic sounds (i.e. an *echo mirror system*): when an individual listens to verbal stimuli, an automatic activation would occur of motor centers responsible for the emission of the phonemes present in the words heard. These data are highly consistent with other findings deriving from fMRI investigations. Wright et al., (2003) evaluated whether speech accompanied by both auditory and visual information (as it normally does), induced a higher activation of STS, compared to speech associated only with mono-sensory information. In this study, the volunteers watched an actor speaking in three different conditions: audiovisual speech, auditory speech, and visual speech. The STS was strongly activated in all conditions, but, above all, and in a super addictive way, in the audiovisual condition; apparently, these results confirmed the multisensory nature of the STS.

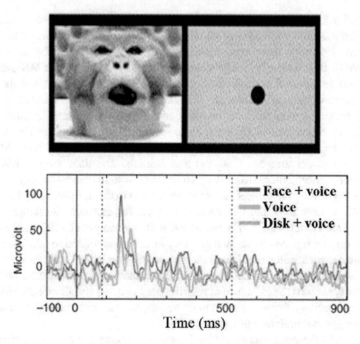

Fig. 5.6 (Above) Examples of stimuli used for the study on neurons sensory preference (i.e. the face of a conspecific emitting a vocalization vs. the opening and closing of a disc without any facial stimulus). (Below) Bioelectrical responses displayed by a multisensory cell of the associative auditory cortex of the macaque monkey. Note that the response to the combined voice and face conditions (red line) is far superior than the uni-sensory stimulation (in this case, the response to the incongruous coupling between disk and voice that did not stimulate the cell enough is also drawn as a yellow line). (Adapted from Ghazanfar and Schroeder (2006). Courtesy of the authors)

A very similar, but more direct demonstration of the existence of audio visuomotor neurons derives from a single-cell recording, neurophysiological study carried out by Ghazanfar and Schroeder (2006). The authors identified neurons in the STS that not only responded to faces or voices but also exhibited a far greater responsivity to the audiovisual association, thus demonstrating their multisensory specialization (Fig. 5.6).

Audio-Visuomotor Neurons and the Sound of Objects

In a famous study by Kohler et al. (2002), published in *Science*, it was demonstrated that the brain retains specific neural representations of the actions performed on objects (e.g. beating eggs, hammering) and of the sounds typically produced by their use. Congruently, the research group coordinated by Giacomo Rizzolatti discovered neurons in the PMv of the macaque monkey that 'fired' both when the animal performed a specific action and when it only heard its sound. Most neurons also fired when the monkey simply watched an action. These audiovisual MNs encoded the actions regardless of whether they were performed, listened to, or simply seen; altogether, these observations led to the discovery of the audio-visuo-motor MNs. Besides the PMv cortex, hosting the audio-visuo-motor MNs, there are interesting audiovisual neurons that conjointly encode the objects and the sound they produce (which, of course, reveal of fundamental importance for music learning and for the regulation of sensory feedback). Many neuroimaging studies have long shown the existence of multisensory audiomotor neurons in the posterior region of the STS and in the middle temporal gyrus (MTG) that respond to the sounds and visual images of objects and animals. The data showed how these regions are activated more strongly by audiovisual stimuli than by uni-sensory stimuli, thus suggesting the crucial role of these regions in the multi-sensory integration of inputs coming from the two modalities (see, for instance, Beauchamp et al., 2004a, b, and Tranel et al., 2003). For instance, Beauchamp et al. (2004a, b) explored how the brain integrated visual and auditory information related to familiar animals and objects, presenting them individually or in association with each other, by means of fMRI scanning of cerebral activity in a sample of participants. Their findings clearly showed the existence of multisensory systems simultaneously encoding visual and auditory features linked to an action, such as a phonatory gesture of an animal or the manipulation of tools (see Fig. 5.7).

Because of the repeated association between an object and its typical sound, and of the fact that the brain represents the so-called *object-sound knowledge*, we can activate the image of a sound based on the object's view. It is for this reason that a musician can visually recognize the sound associated with a gesture or knows how to predict the sound that will be emitted, before it is played, observing, for instance, the tension of the hair of a bow, the position of the fingers on a keyboard, or the key pressed down.

In an electrophysiological study by Proverbio et al. (2011b), it was shown that the only view of objects or actions associated with a sound can activate brain temporal cortex, a region overseeing auditory perception. In this study, high-density ERPs were recorded in 15 students who were required to look at hundreds of images associated with a given sound or to silence (see Fig. 5.8 for some examples of stimuli). ERP signals analysis showed that, despite stimulation being only visual, sound-related stimuli were distinguished from non-sound-related stimuli already after only 110 milliseconds post-stimulus processing. According to the authors, this happened because perception and recognition of objects, agents, and stimulus-contexts stimulated the access to conjoined auditory information. Indeed, as it was well known to silent movies filmmakers, there is no need for a real auditory stimulus to activate the sensation of hearing sounds typically associated with what we are seeing: This is how in a silent movie you will almost hear the whistle of the steam train or his rattling on the tracks.

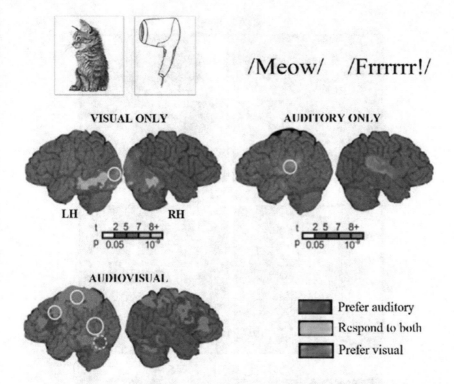

Fig. 5.7 Visual stimulation consisted in the silent presentation of pictures of animals and tools while the auditory stimulation consisted of the blind presentation of their verse or typical sound. The audiovisual stimulation involved the integration between the two modes. Brain images show the BOLD signals of neurometabolic activation obtained by fMRI in the various stimulation conditions. Note that the audiovisual condition activated the multimodal prefrontal regions, as well as the motor and premotor cortices, the posterior region of the STS, and the MTG. (Drawn and modified by Beauchamp et al. (2004a, b). Courtesy of the authors)

Audio-Visuomotor Neurons in the Coding of Musical Actions and Sounds

While investigating how professional pianists could identify the musical piece performed in silent scenes by looking at the movements of the musicians' hands on the keys (i.e. looking at actions performed on objects), Hasegawa et al. (2004) hypothesized that visuomotor representation of musical gestures was strictly associated with the auditory representation following a specific learning. In this study, seven participants without any musical experience (control group), ten participants with some experience of the piano (not very experienced), and nine professional pianists were tested. During fMRI scanning, the participants observed silent videos showing bimanual movements of a pianist pressing the keys of a piano keyboard (Fig. 5.9a: Right), or, in a basic condition, only random, sliding across keyboard, key touches

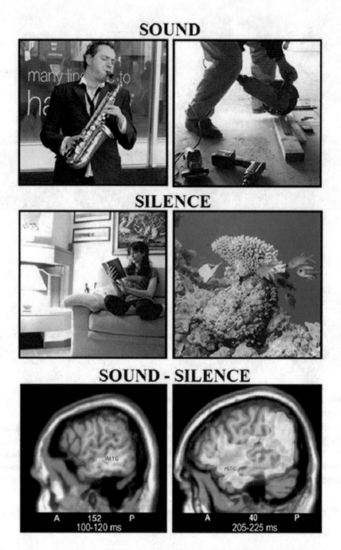

Fig. 5.8 Some examples of 'sound' (top) and ″silent' (centre) visual stimuli presented together with other hundreds of stimuli to unaware observers, instructed to detect and respond to infrequent images of cycling races. The analysis of ERP peaks, together with the reconstruction of their intra-cerebral generators by means of the swLORETA technique, demonstrated the activation of the left medial temporal cortex after only 110 ms from the presentation of the image. The extraction of sound information associated with the use of familiar tools after ~200 ms activated the primary (BA38) and secondary (BA41) auditory cortices. This information is responsible, for example, for auditory hallucinations, which, in this case, refer, in a dim way, to the call of the specific sound produced by the tool (in the figure, the sounds produced by the sax or by the infernal chainsaw). (Taken from Proverbio et al. (2011b). Courtesy of the authors)

(Fig. 5.9a: Left). Pressure movements could be completely random, that is, not at all combined with a musical piece or related to the execution of a more or less famous piece. Professional pianists were able to identify these pieces, but, above all, the

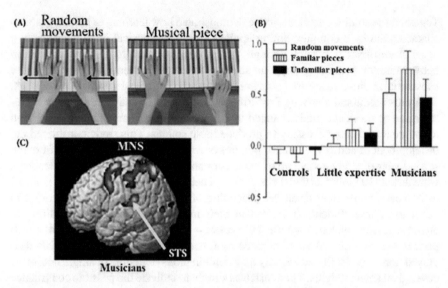

Fig. 5.9 (**a**) Examples of visual stimuli used in the study by Hasegawa et al. (2004) (**b**) Activation of the left temporal region as a function of musical performance in the three groups of participants. (**c**) fMRI activations in response to an exclusively visual stimulation in the brain of professional pianists. (Courtesy of the authors)

view of the musical performance – regardless of the piece – activated their fronto-parietal MNS (i.e. motor simulation) and STS, thus demonstrating that seeing familiar musical gestures activates the stored memory of the associated sounds, but only in those who actually know how to perform them. This study clearly demonstrated the role of audio-visuomotor neurons in musical learning (Paraskevopoulos et al., 2012; Schulz et al., 2003).

A similarly interesting study on audio-visuomotor coding is the one carried out by Lahav et al. (2007). In this study, naïve participants (i.e. non-musicians) were trained to play a short musical sequence by ear. Their cerebral activity was then tested by means of fMRI while they listened to the newly learned piece. The authors found that, despite the participants not making any kind of movement while listening, both motor and mirror regions were activated, including the bilateral frontoparietal motor circuit, along with the IFG and the PMv, the IPS and the IPG. Moreover, the presentation of the same musical notes organized in a different order, activated in a much less measure the same regions, whereas listening to a familiar musical sequence whose motor program was unknown, did not activate these regions at all. These data supported the hypothesis of the existence of a "*hearing-doing*" (or "*hearing-action*") system, strongly dependent on the individual's motor's repertoire. In this regard, with a study combining Transcranial Magnetic Stimulation (TMS) and MEP recordings, Candidi et al. (2014) showed that, in expert pianists, the observation of a piano fingering error – a visual gesture shown without any audio – induced a significant motor effect, and in particular a somatotopic corticospinal facilitation concerning the finger of the hand engaged in the fingering error.

Together, the studies described above demonstrated how learning of skilled gestures characterized by a complex timing applied to a given musical instrument (or to a vocal performance) occurs through the progressive and long-term association between motor, somatosensory, and auditory functional patterns, namely through a substantial audio-visuomotor coding of the musical gesture, which takes many years.

A cross-sectional study by Proverbio et al. (2015b) investigated how the representation of musical sounds changed as a function of the years of study in relation to the motor gesture necessary to produce these sounds. This study considered the development of audio-visuomotor mirror systems in young students going from the second year of study course up to the master and beyond. In all, 19 music students were tested: 10 violinists and 9 clarinetists. Their chronological age ranged from 14 to 24 years, while their academic practicing of their instrument ranged from 2 to 18 years. These students (recruited in their instrument classes while waiting to attend a lesson) watched – on a PC screen – and listened – by means of headphones – a total of 400 video clips of professional violinists and clarinetists who played non-melodically 200 totally new combinations of double or single notes that covered all sound heights. Their task was simply to indicate the possible congruence between the gesture and the sound reproduced in each video clip on the basis of their senses. Half of the time, in fact, the sounds were not congruent with the motor gestures but were mounted onto the video track in an incongruous although perfectly synchronized way. The data showed that the actual years of study at the Conservatory correlated directly with the performance in the task. It was as if the more advanced students had so firmly internalized the connection between sound, gesture, and image that they automatically perceived a possible incongruity, with a percentage of error that decreased linearly as the years of practice increased. This happened thanks to the ability of multimodal neurons to create audio-visuomotor correlations that increased with the years of study and practice, regardless of the talent and age of the individual. The first effects of cerebral modification were observable after 4–6 years of intensive study and progressively continued after graduation and master's degree. Up to three years of study, the percentage of error was close to 50%, while only after obtaining the diploma (and about 10,000 h of study), the percentage fell below 10% for music teachers. This research highlighted the crucial role of exercise in shaping brain musical functions, regardless of musical talent.

The same stimuli of the study described above were shown to 12 professional musicians and 12 naïve university students to study in real-life neural mechanisms of audio-visuomotor coding of the musical gesture for their instrument and/or for an unfamiliar instrument (Proverbio et al., 2014). While the musicians watched the stimuli, they had to decide whether the note played was double or single – an easily resolvable task – not only for their instrument but even for an unfamiliar musical instrument. Throughout the task duration, their EEG was recorded in continuous mode by means of 128 sensors placed all over their scalp. Averaged ERPs indicated that audiovisual incongruity generated a prominent N400 mismatch response for the musicians' own instrument only, since it appeared almost impossible for these subjects to reach robust decisions for the unfamiliar instrument. The swLORETA applied to the N400 response identified the areas mediating multimodal motor processing: the prefrontal cortex (PFC: attention, cognitive discrepancy), the superior and middle temporal gyri (STG and MTG: auditory coding of sound), the premotor cortex (PM: motor programming, simulation), the inferior frontal and parietal areas (IF and IP, mirror system), the extrastriate region for coding of body parts (EBA), the somatosensory cortex (maps of the fingers and the hand), the cerebellum (motor coordination), and the supplementary motor area (SMA), which encodes the learned motor sequences (Fig. 5.10). In conclusion, these data indicate the existence of audio-visuomotor MNs that respond to both visual and auditory incongruent information, thus suggesting that they can encode multimodal learned motor skill representations of musical gestures and sounds.

In summary, we have reviewed a wide neuroimaging and electrophysiological literature reporting the involvement of visuomotor MNs in many mental functions including the comprehension of actions and action intentions, understanding the others' emotional and mental state, action imitation and learning, processing of visuomotor aspects of speech, vocalizations and music, developing motor or musical skills, and many others. Some criticalities still challenge the concept that the human MNs can be viewed as roughly correspondent to the monkey's MNs, for which we have direct neurophysiological recording. First of all, MNs are not always

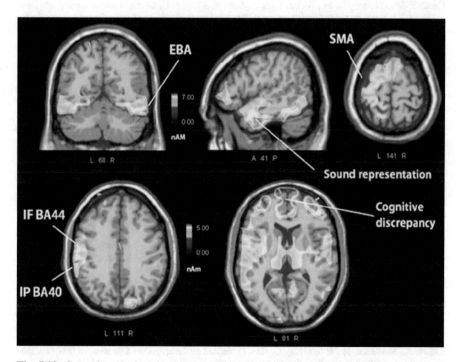

Fig. 5.10 Coronal, sagittal, and axial views of the standardized and weighted LOw REsolution electromagnetic TomogrAphy (swLORETA) applied to the N400 bioelectric response generated only for one's own musical instrument. (Taken from Proverbio et al. (2014) and redrawn)

observed while recording from the fronto/parietal areas of the monkey's brain, and their incidence can be very variable, ranging from 8.9% for ventral intra-parietal areas (VIP) to 60% for premotor dorsal areas (PMd). Other criticalities concern the fact that cell-recording studies are not very numerous (also for ethical reasons) and that in humans, evidences are relatively indirect (not based on intracranial recordings). It should be also borne in mind that MNs are only indirectly involved in social and affective processes, such as empathy, contributing for the visuomotor recognition of body language and gestures only.

References

Beauchamp, M. S., Argall, B. D., Bodurka, J., Duyn, J. H., & Martin, A. (2004a). Unraveling multisensory integration: Patchy organization within human STS multisensory cortex. *Nature Neuroscience, 7*(11), 1190–1192.

Beauchamp, M. S., Lee, K. E., Argall, B. D., & Martin, A. (2004b). Integration of auditory and visual information about objects in superior temporal sulcus. *Neuron, 41*(5), 809–823.

Calvo-Merino, B., Glaser, D. E., Grèzes, J., Passingham, R. E., & Haggard, P. (2005). Action observation and acquired motor skills: An fMRI study with expert dancers. *Cerebral Cortex, 15,* 1243–1249.

Candidi, M., Sacheli, L. M., Mega, I., & Aglioti, S. M. (2014). Somatotopic mapping of piano fingering errors in sensorimotor experts: TMS studies in pianists and visually trained musically naives. *Cerebral Cortex, 24*(2), 435–443.

Craighero, L., Fadiga, L., Umiltà, C. A., & Rizzolatti, G. (1996). Evidence for visuomotor priming effect. *Neuroreport, 8,* 347–349.

Creem-Regehr, S. H., & Lee, J. N. (2005). Neural representations of graspable objects: Are tools special? *Cognitive Brain Research, 22,* 457–469.

Dinstein, I. (2008). Human cortex: Reflections of mirror neurons. *Current Biology, 18*(20), 956–959.

Fadiga, L., Craighero, L., Buccino, G., & Rizzolatti, G. (2002). Speech listening specifically modulates the excitability of tongue muscles: A TMS study. *The European Journal of Neuroscience, 2,* 399–402.

Ghazanfar, A. A., & Schroeder, C. E. (2006). Is neocortex essentially multisensory? *Trends in Cognitive Sciences, 10*(6), 278–285.

Goldenberg, G., & Spatt, J. (2009). The neural basis of tool use. *Brain, 132*(Pt 6), 1645–1655.

Hasegawa, T., Matsuki, K., Ueno, T., Maeda, Y., Matsue, Y., Konishi, Y., & Sadato, N. (2004). Learned audio-visual cross-modal associations in observed piano playing activate the left planum temporale. An fMRI study. *Brain Res Cogn, 20,* 510–518.

Iacoboni, M. (2005). Understanding others: Imitation, language, empathy. In S. Hurley & N. Chater (Eds.), *Perspectives on imitation: From cognitive neuroscience to social science* (Vol. 1, pp. 77–99). MIT Press.

Iacoboni, M., Molnar-Szakacs, I., Gallese, V., Buccino, G., Mazziotta, J. C., et al. (2005). Grasping the intentions of others with one's own mirror neuron system. *PLoS Biology, 3*(3), e79.

Kohler, E., Keysers, C., Umiltà, M. A., Fogassi, L., Gallese, V., & Rizzolatti, G. (2002). Hearing sounds, understanding actions: Action representation in mirror neurons. *Science, 297*(5582), 846–848.

Lahav, A., Saltzman, E., & Schlaug, G. (2007). Action representation of sound: Audiomotor recognition network while listening to newly acquired actions. *The Journal of Neuroscience, 27,* 308–314.

Lelord, G., Cochin, S., Adrien, J. L., Barthelemy, C., & Martineau, J. (1998). Latent imitation of human movements presented on a videoscopic screen, disclosed by electroencephalographic mapping in the spectator. *Bulletin de L'Academie Nationale de Medecine, 182*(4), 833–42, Discussion 843-4.

Liberman, A. M., & Mattingly, I. G. (1985). The motor theory of speech perception revised. *Cognition, 21*, 1–36.

Nyström, P., Ljunghammar, T., Rosander, K., & von Hofsten, C. (2011). Using mu rhythm desynchronization to measure mirror neuron activity in infants. *Developmental Science, 14*, 327–335.

Oberman, L. M., Ramachandran, V. S., & Pineda, J. A. (2008). Modulation of mu suppression in children with autism spectrum disorders in response to familiar or unfamiliar stimuli: The mirror neuron hypothesis. *Neuropsychologia, 46*(5), 1558–1565.

Orlandi, A., Zani, A., & Proverbio, A. M. (2017). Dance expertise modulates visual sensitivity to complex biological movements. *Neuropsychologia, 104*, 168–181.

Paraskevopoulos, E., Kuchenbuch, A., Herholz, S. C., & Pantev, C. (2012). Musical expertise induces audiovisual integration of abstract congruency rules. *The Journal of Neuroscience, 32*, 18196–18203.

Pfurtscheller, G., Brunner, C., Schlögl, A., & Lopes da Silva, F. H. (2006). Mu rhythm (de)synchronization and EEG single-trial classification of different motor imagery tasks. *Neuro Image, 31*, 153–159.

Proverbio, A. M. (2012). Tool perception suppresses 10-12Hz μ rhythm of EEG over the somatosensory area. *Biological Psychology, 91*(1), 1–7.

Proverbio, A. M., Riva, F., & Zani, A. (2010). When neurons do not mirror the agent's intentions: Sex differences in neural coding of goal-directed actions. *Neuropsychologia, 48*(5), 1454–1463.

Proverbio, A. M., Adorni, R., & D'Aniello, G. E. (2011a). 250 ms to code for action affordance during observation of manipulable objects. *Neuropsychologia, 49*(9), 2711–2717.

Proverbio, A. M., D'Aniello, G. E., Adorni, R., & Zani, A. (2011b). When a photograph can be heard: Vision activates the auditory cortex within 110 ms. *Scientific Reports, 1*, 54.

Proverbio, A. M., Crotti, N., Manfredi, M., Adorni, R., & Zani, A. (2012). Who needs a referee? How incorrect basketball actions are automatically detected by basketball players' brain. *Scientific Reports. Nature, 2*, 883; https://doi.org/10.1038/srep00883.

Proverbio, A. M., Azzari, R., & Adorni, R. (2013). Is there a left hemispheric asymmetry for tool affordance processing? *Neuropsychologia, 51*(13), 2690–2701.

Proverbio, A. M., Calbi, M., Manfredi, M., & Zani, A. (2014). Audio-visuomotor processing in the musician's brain: An ERP study on professional violinists and clarinetists. *Scientific Reports, 29*(4), 5866.

Proverbio, A. M., Calbi, M., Manfredi, M., & Zani, A. (2014a). Comprehending body language and mimics: an ERP and neuroimaging study on Italian actors and viewers. *PLoS One, 9*(3), e91294.

Proverbio, A. M., Gabaro, V., Orlandi, A., & Zani, A. (2015a). Semantic brain areas are involved in gesture comprehension: An electrical neuroimaging study. *Brain and Language, 147*, 30–40.

Proverbio, A. M., Attardo, L., Cozzi, M., & Zani, A. (2015b). The effect of musical practice on gesture/sound pairing. *Frontiers in Psychology, Auditory Cognitive Neuroscience, 6*, 376.

Pulvermüller, F., & Shtyrov, Y. (2006). Language outside the focus of attention: The mismatch negativity as a tool for studying higher cognitive processes. *Progress in Neurobiology, 79*, 49–71.

Rizzolatti, G., & Craighero, L. (2004). The mirror-neuron system. *Annual Review of Neuroscience, 27*, 169–192.

Rizzolatti, G., & Luppino, G. (2001). The cortical motor system. *Neuron, 31*(6), 889–901.

Rizzolatti, G., & Sinigaglia, C. (2016). The mirror mechanism: A basic principle of brain function. *Nature Reviews. Neuroscience, 17*(12), 757–765.

Rizzolatti, G., Fadiga, L., Matelli, M., Bettinardi, V., Paulesu, E., Perani, D., & Fazio, F. (1996). Localization of grasp representations in humans by PET: 1. Observation versus execution. *Experimental Brain Research, 111*, 246–252.

Schmidt, S. N. L., Hass, J., Kirsch, P., & Mier, D. (2021). The human mirror neuron system – a common neural basis for social cognition? *Psychophysiology, 58*, e13781.

Schulz, M., Ross, B., & Pantev, C. (2003). Evidence for training-induced crossmodal reorganization of cortical functions in trumpet players. *Neuroreport, 14*(1), 157–161.

Tranel, D., Damasio, H., Eichhorn, G. R., Grabowski, T., Ponto, L. L., & Hichwa, R. D. (2003). Neural correlates of naming animals from their characteristic sounds. *Neuropsychologia, 41*(7), 847–854.

Usui, S., Senju, A., Kikuchi, Y., Akechi, H., Tojo, Y., Osanai, H., & Hasegawa, T. (2013). Presence of contagious yawning in children with autism spectrum disorder. *Autism Research and Treatment, 2013*, 971686.

van Elk, M., van Schie, H. T., Hunnius, S., Vesper, C., & Bekkering, H. (2008). You'll never crawl alone: Neurophysiological evidence for experience-dependent motor resonance in infancy. *NeuroImage, 43*(4), 808–814.

Wright, T. M., Pelphrey, K. A., Allison, T., McKeown, M. J., & McCarthy, G. (2003). Polysensory interactions along lateral temporal regions evoked by audiovisual speech. *Cerebral Cortex, 13*, 1034–1043.

Chapter 6
Sex Differences in Social Cognition

Alice Mado Proverbio

Abstract Several studies have demonstrated sex differences in empathy and social abilities. This chapter reviews studies on sex differences in the brain, with particular reference to how women and men process faces and facial expressions, social interactions, pain of others, infant faces, faces in things (*pareidolia*), living vs. non-living information, purposeful actions, biological motion, erotic vs. emotional information. Sex differences in oxytocin-based attachment response and emotional memory are also discussed. Overall, the female and male brains show some neurofunctional differences in several aspects of social cognition, with particular regard to emotional coding, face processing and response to baby schema that might be interpreted in the light of evolutionary psychobiology.

Keywords Hemispheric asymmetries · Facial expressions · Parental response · Face pareidolia · Sex hormones

Introduction

Genetic and hormonal influences are long known to affect the human brain and determine a variety of anatomical and functional differences between the two sexes (see Hines, 2020, for a review). The cerebral sexual dimorphism would support marked diversities in reproductive, parental, and social behavior. A rapidly increasing literature now documents significant sex differences in the reactivity to/efficacy of drugs and pharmaceutical molecules, as well as in the incidence of neurodegenerative, neurological, and psychiatric diseases (see the entire volume dedicated to sex differences in the brain, edited by Cahill, 2017).

Besides anatomical and physiological diversities, some functional and mental differences between men and women have been recently reported by neuroscientific

A. M. Proverbio (✉)
Department of Psychology, University of Milano-Bicocca, Milan, Italy
e-mail: mado.proverbio@unimib.it

© The Author(s) 2023
P. S. Boggio et al. (eds.), *Social and Affective Neuroscience of Everyday Human Interaction*, https://doi.org/10.1007/978-3-031-08651-9_6

studies (e.g. for the following abilities: verbal fluency (Sokolowski et al., 2020), emotion recognition (Connoly et al., 2019; Li et al., 2020), face perception (e.g., Zhou & Meng, 2020), and empathy, as shown by a recent survey examining the empathy quotients of 671,606 individuals (Greenberg et al., 2018).

Several studies have demonstrated sex differences in empathy and related capacities. This chapter reviews studies on sex differences in the brain, with particular reference to how women and men process faces and facial expressions, social interactions, pain of others, infant faces, faces in things (*pareidolia* phenomenon), opposite- vs. own-sex faces, living vs. non-living information, incongruent/inappropriate behavior, motor actions, biological motion, erotic vs. emotional information. Sex differences in oxytocin-based attachment response and emotional memory are also discussed. Overall, the female and male brains show some neuro-functional differences in several aspects of social cognition, with particular regard to emotional coding, face processing, and response to baby schema, which might be interpreted in the light of evolutionary psychobiology.

In this chapter, a recent and comprehensive review of neuroimaging, electrophysiological, and behavioral findings in the literature supporting the hypothesis of a sex difference in social cognition is provided and discussed, under the framework of cognitive neuroscience and evolutionary psychobiology theories.

The main sex differences in social brain possibly refer to:

- Hemispheric lateralization for face processing (Bourne, 2005; Proverbio et al., 2006a, b, 2010a, b, c, 2011a, b, c, 2012)
- Facial expression decoding (e.g., Orozco & Ehlers, 1988)
- Emotional response to negative affective information (empathic distress: Hofer et al., 2007; Klein et al., 2003; empathy for pain, sympathetic response: Singer et al., 2004; Han et al., 2008; Proverbio et al., 2009)
- Understanding body language and action goals (Canessa et al. 2012, Proverbio et al., 2011a, b, c)
- Interest in faces and social information (Pavlova et al., 2014, 2015; Proverbio et al., 2008)
- Parental response (Seifritz et al., 2003; Sander et al., 2007)

Hemispheric Asymmetries for Face Processing

While it is currently believed that face processing predominantly activates the right hemisphere in humans, some data reveal a lesser degree of lateralization of brain functions related to face coding in women than men. For example, a left hemispheric involvement of the occipito/temporal cortex in women for the processing of human faces has been demonstrated in two independent studies, showing a bilateral pattern of activity of the face fusiform face area (FFA) indexed by N170 ERP response in females, as opposed to the typical male right-sided hemispheric asymmetry (Proverbio et al., 2006b, 2012). In more detail, Proverbio and co-workers (2012) recorded ERPs in 50 right-handed women and men in response to 390 faces of male and female infants, children or adult, and technological objects, in a landscape detection task. Results showed no sex difference in the amplitude of N170 to objects, a much larger face-specific response over the right hemisphere in men and a bilateral response in women (see Fig. 6.1). Furthermore, a lack of the face-age coding effect was found over the left hemisphere in men, with no differences in N170 to faces as a function of age. Conversely, N170 showed to be sensitive to face age (e.g., differentiating children from adults), over both hemispheres in women.

Overall, these findings are in line with many studies that show differences between men and women in the degree of lateralization of cognitive and affective processes. Substantial data support greater hemispheric lateralization in men than women for linguistic tasks and for spatial tasks. Sex differences have also been found in the lateralization of visual-spatial processes such as object construction and mental rotation tasks, in which males are typically right hemisphere dominant and females bilaterally distributed. Consistent with this pattern of results are the data provided by Bourne (2005), who examined the lateralization of processing positive facial emotion in a group of 276 right-handed individuals. Subjects were asked to observe a series of chimeric faces with contrasting expressions and to decide which face they thought looked happier. The results showed that males were more strongly lateralized than women, showing a greater perceptual asymmetry in favor of the left visual field (RH). A similar pattern of results has been reported by Tiedt et al. (2013). Inter-hemispheric transfer-time of face-related inputs seems to be also asymmetric across sexes: N170 recorded in men have faster latencies in the left visual field (LVF)/RH → LH (170 ms) direction than in the right-visual field (RVF)/LH → RH (185 ms) direction, while it is symmetric in women (Proverbio et al., 2012). Figure 6.2 shows larger delays in N1 latency (due to callosal transfer) relative to the ipsilateral stimulation, for stimuli presented to the RVF (left hemisphere), in men.

In men, N170 was significantly earlier ($p < 0.0007$) for ipsilateral (crossed) responses over the left than right hemisphere. This effect was not found in women, who showed an IHTT of equal latency in the two directions. As for contralateral (uncrossed) responses, N170 was earlier over the RVF/LH than LVF/RH in women and of equal latency for both hemispheres in men. One potential explanation of the findings is that interhemispheric transfer time (IHTT) would be more rapid and symmetric in women than men. Notwithstanding the large electrophysiological literature in favor of this hypothesis, in neuroimaging domain, the standard and to-be-expected pattern of lateralization for face processing is still considered to be the right-sided activation of the fusiform gyrus and of the right occipital face area for both sexes (e.g., Jacques et al., 2019).

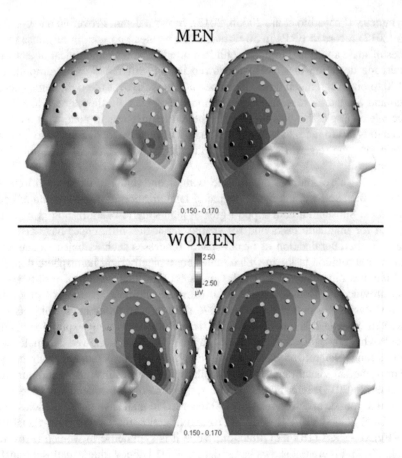

Fig. 6.1 Isocolor voltage topographical maps (left- and right-side views) showing N170 scalp distribution in female and male observers. N170 response is relative to adult face processing. The time window corresponds to its peak (150–170 ms) of maximum activation. (Taken from Proverbio et al. (2012), with permission from the authors and the editor)

Affective Facial Expressions and Emotions

Several studies have provided evidence of a woman's greater accuracy in interpreting emotional states and mind reading (Babchuk et al., 1985; Wingenbach et al., 2018). One potential explanation of these findings is that the primary role of female humans (and primates in general) in breastfeeding and rearing young offspring would have improved their ability to interact with them affectively and to understand their non-verbal behavior.

In this regard, Proverbio et al. (2007) examined the roles of sex and expertise in interpreting infant expression in a group of 34 men and women differing in their experience with infants (Fig. 6.3). The participants were subdivided into two groups (experts or non-experts) on the basis of their specific familiarity with infant facial

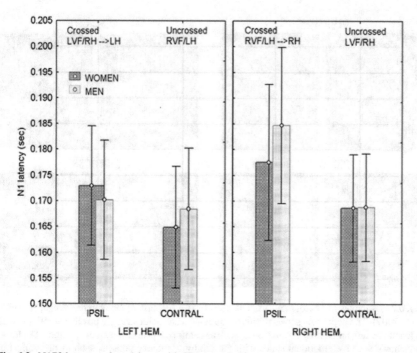

Fig. 6.2 N170 latency values (along with SD) recorded in women and men in response to lateralized faces, as a function of cerebral hemisphere and stimulus contra-laterality (collapsed across occipito/temporal electrode sites). In this study, ERPs were recorded in strictly right-handed people (16 men and 17 women) engaged in a face-sex categorization task. Occipital P1 and occipito/temporal N170 were left lateralized in women and bilateral in men. N170 to contralateral stimuli was larger over the RH in men and the LH in women. Inter-hemispheric transfer time (IHTT) was approximately 4 ms at the P1 level and approximately 8 ms at the N170 level. It was asymmetric in men, with faster latencies in the left visual field (LVF)/RH → LH (170 ms) direction than in the right-visual field (RVF)/LH → RH (185 ms) direction and symmetric in women. These findings suggest that the asymmetry in callosal transfer times might be due to faster transmission times of face-related information via fibers departing from the more efficient to the less efficient hemisphere (Proverbio et al., 2012)

expressions. In detail, individuals considered "non-expert" were those without children, nieces or nephews, and without a specific familiarity/skill with neonates or pre-school age children acquired through professional activities. In contrast, individuals with natural or adopted children, nieces or nephews under the age of 5 years old, as well as nursery school teachers or infant school teachers were considered "experts." Women showed a significantly higher level of decoding accuracy compared to men; furthermore, expertise positively affected facial expressions decoding among women only. These results suggest that in judging emotional facial expressions of infants, there is an interaction of biological (i.e., sex) and cultural factors (such as familiarity with infantile mimicry).

In an electrophysiological study performed on the same set of stimuli (Proverbio et al., 2006b), it was investigated whether viewers' sex affected the visual cortical

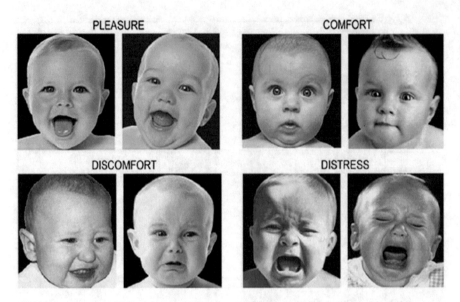

Fig. 6.3 Examples of photographs used as stimuli, as a function of facial expressions (Proverbio et al., 2007). The upper row shows positive emotional states with strongly positive emotions, such as joy on the left, and mildly positive ones, such as comfort or peacefulness, on the right. The lower row shows negative emotional states with the mildly negative emotions, such as discomfort or disappointment, on the left and strongly negative ones, such as displeasure or pain, on the right

response at various stages of perceptual processing during a judgment task of infant happy/distressed expression. All infants were unfamiliar to viewers. The lateral occipital P110 response was much larger and occurred earlier in women than in men, regardless of facial expression, thus indicating a sex difference in early visual processing. Furthermore, P110 latency was earlier in response to distressed than neutral children in women only, thus possibly showing a prioritized processing of biologically relevant information in the female brain (Fig. 6.4).

The role of viewer sex in the emotional evaluation and psychological reactivity to human faces of various age, sex, and typology has been deeply explored. Table 6.1 shows some of the main gender differences in facial expression processing.

In general, perception of aversive faces would activate an amygdala-based arousal response able to affect general stimulus processing (Phelps & LeDoux, 2005). Furthermore, stimuli inducing greater arousal in the percipient would be subjected to a prioritized processing because of their biological relevance, for example, infant faces would trigger an instinctive parental response. In this respect, it has been reported that erotic stimuli are particularly arousing for men as compared to women. Sabatinelli et al. (2004) provided fMRI evidence that perception of erotic pictures is associated with a much larger activation of the extra-striate visual cortex in men vs. women, while Huynh et al. (2012) showed the opposite effect in women, with high-intensity erotic visual stimuli de-activating the primary visual cortex as compared to low-intensity erotic movies and neutral movies. Conversely,

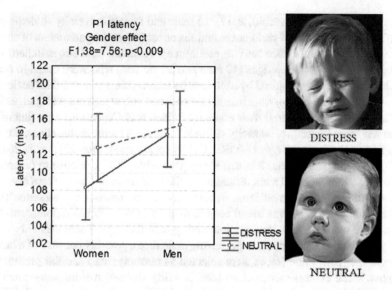

Fig. 6.4 Mean latency (in ms) of the P1 component (along with SD) recorded at the lateral occipital area (independent of hemispheric site) and analyzed according to subjects' sex and type of facial expression. (Taken and modified from Proverbio et al., 2006b study, with permission of the authors and the editor)

Table 6.1 Main gender differences related to facial expression and decoding abilities

Women
• Better decoding ability and emotion recognition (Connoly et al., 2019; Thomson and Voyer, 2014)
• More expressive (e.g., crying, smiling: frequency and intensity) (McDuff et al., 2017)
• More interested in faces and social information (Proverbio et al., 2008)

Men
• Slower response times (RTs) in recognizing expressions (Wingenbach et al., 2018)
• React stronger than females to angry faces (Sawada et al., 2014)
• Learn how to mask feelings (also use mustaches or beard to appear more aggressive, Craig et al., 2019)

perception of body mutilations (stimulating the empathic circuits) would be associated with a stronger activation of the extra-striate visual cortex in women vs. men. These findings have been interpreted by the authors of the studies according to the hypothesis that the degree of cerebral arousal and mobilization of attentional resources devoted to stimulus processing would depend on its biological relevance for the observers. It is well known that women respond differently than man to erotic information. Some women feel repulsed by muscular, erotic male photos. In general, while men are more sexually aroused by visual stimuli, women seem to be more sexually aroused by auditory, tactile, or emotionally relevant information (see Chung et al., 2013).

In a recent study (Proverbio, 2017), 15 male and female university students evaluated 400 human faces of various age and sex according to the parameters of arousal and valence. The same face set was preliminary validated (and sex-matched) by a group of 20 independent judges (10 men and 10 women) who were asked to evaluate the degree of trust inspired by each face by means of a 3-points Likert scale. The aim was to explore the possible interaction of facial characteristics with judges' sex and age. Participants shared their ethnicity (which was Caucasian) with that of the observed faces (therefore, ethnicity or "race" was not a factor in this study, nor was the so-called *"other-race effect"* (ORE; Caldara et al., 2004; Proverbio et al., 2011a).

Overall, the data collected in this study (Proverbio, 2017), relative to heterosexual young adults, showed a sex difference in the evaluation of human faces along the arousal and valence dimensions. Specifically, an opposite-sex preference (with higher valence ratings) was found only in men, in favor of female adolescents (but not mature women), thus strongly interacting with face age. There was only a tendency for women participants in preferring male faces, possibly because of a lack of specific aesthetic value (faces were selected as normotypical) and the presence of negative facial expressions (such as hate, hostility, disgust) making some faces not particularly attractive. Female subjects showed a preference for the faces of children and the elderly (as compared to other age ranges) in the arousal evaluation. The female appreciation of elderly faces might be interpreted in the light of a greater empathetic attitude for fragile persons, whereas the female preference for children faces would rely on specific neural mechanisms sensitive to child-like cues in face stimuli. Overall, women rated all human faces as more arousing and more positive than men, possibly indicating a preference, or greater interest, for faces, facial expressions, and social information in general (Proverbio et al., 2008). This piece of evidence fits with the Baron-Cohen model of sexual dimorphism in empathy and facial expression coding ability (Baron-Cohen et al., 2001; Baron-Cohen and Wheelwright, 2004). In the light of this framework, it can be proposed that the higher female ratings of valence and arousal found in the present study might reflect a greater attentional allocation to (or interest for) human faces as sensory signals (Pavlova et al., 2014, 2015).

While it seems that females generally are significantly faster and more accurate at emotion recognition, some studies failed to show consistent gender differences while varying experimental conditions (Klein & Hodges, 2001).

Parental Response

The viewers' age and the possible interaction with face age are also explored in the literature on the so-called *baby schema* effect, which predicts a preference for, and a perceptual advantage of infant vs. adult faces (Brosch et al., 2007; Glocker et al., 2009a; Luo et al., 2011; Proverbio et al., 2011a, b), as nicely reported in a review by Hahn and Perrett (2014).

The literature shows that the adult visual and the orbitofrontal cortices are specifically activated and aroused by the view of infants, also providing a pleasant sensation through the dopaminergic reward circuitry. This would happen to a greater extent in women than men, according to some authors (Hahn et al., 2013; Nitschke et al., 2004; Parsons et al., 2011, 2013). Indeed, behavioral studies showed how women might be more responsive to baby schema than men and better able to decode infant expressivity (Proverbio et al., 2007; Babchuk et al., 1985). In an electrophysiological study (Proverbio et al. 2006a, b) aimed at investigating the neural response to baby schema in female and male adult individuals, ERP results revealed a larger sensory P100 response to faces in women than in men (irrespective of whether they were parents themselves or nulliparous). These findings may possibly be interpreted as a sign of greater perceptual sensitivity (or increased arousal response) in women than men at the view of unrelated infants. Similar studies have shown that infant faces hold greater incentive salience for women than they do for men (Hahn et al., 2013; Parsons et al., 2011, 2013). Again, infant faces have been shown to capture women's attention to a greater extent that adult faces, whereas infant faces capture men's attention more so than same-sex faces, but much less than opposite-sex faces (Cárdenas et al., 2013).

Several recent neuroimaging studies (Glocker et al., 2009b; Kringelbach et al., 2008; Leibenluft et al., 2004) have investigated the neural circuits subtending the so-called "parental response" to infants and identified a set of structures predominantly involving the orbito/frontal cortex devoted to social cognition and belonging to the dopaminergic reward system. The neural correlates of "maternal love" have been investigated by recording the brain activation of mothers viewing pictures of their own children. The results showed activation of brain areas linked to affect (amygdala) and in particular positive emotions (orbitofrontal cortex and connected regions belonging to the pleasure/reward circuitry such as the periaqueductal gray matter). The possible role of oxytocin in maternal love has also been determined in an electrophysiological study (Peltola et al., 2014) testing the associations of motherhood and oxytocin receptor genetic variation with neural and behavioral responses to emotional expressions of infants and adults. It was found that mothers (vs. nonmothers) and individuals carrying the rs53576 GG variant of the *OXTR* gene (vs. A-carriers) showed enhanced ERP differentiation of infants' strong versus mild intensity facial expressions (i.e., pleasure and distress vs. comfort and discomfort).

Overall, the parental role (having own children) has been associated with a greater sensitivity to infant facial expression. In an electrophysiological study performed in parent vs. nulliparous adults, it was shown that the perceptual N160 response reflected the earliest discrimination of mild vs. strong painful facial expressions in parents (especially in mothers) but not in nulliparous individuals. These findings possibly suggest a strong interactive influence of genetic predisposition and parental status on the responsivity of visual brain areas (Proverbio et al., 2006a). Again, the data showed larger P3 responses in mothers versus all other groups (including fathers and nulliparous women), possibly indicating a greater perceptual sensitivity (or increased arousal response) in mothers, at the view of unrelated infants (Fig. 6.5).

As for the auditory modality, other studies have demonstrated a female vs. male enhanced response to the infant vocalizations (cry and laughter) (Sander et al. 2007; Seifritz et al., 2003) supporting the hypothesis of a sex difference in the parental response to infantile communicative signals.

Interest in Social Stimuli

In Proverbio's (2017) previously described study, regardless of faces' sex, women's ratings were significantly higher for both arousal and valence dimensions, thus suggesting that women might be more interested or aroused by the specific sensory stimulus (the human face). This data fits with some electrophysiological literature providing evidence of a greater female electro-cortical responsivity to faces and people than to inanimate scenarios such as landscapes.

In a study by Proverbio et al. (2008), 24 men and women viewed 220 images portraying persons or landscapes (see Fig. 6.6 for some examples of stimuli) and ERPs were recorded from 128 sites. In women, but not in men, the N2 component (210–270 ms of latency) was much larger to persons than to scenes. Inverse solution (swLORETA) showed significant bilateral activation of face-devoted areas (namely, the fusiform gyrus, BA19/37) in both sexes when viewing persons as opposed to scenes. However, only women showed a source of activity in the superior temporal gyrus (STG) and in the right middle occipital gyrus (MOG), extrastriate body area (EBA), and only men in the left parahippocampal area (PPA). This was interpreted

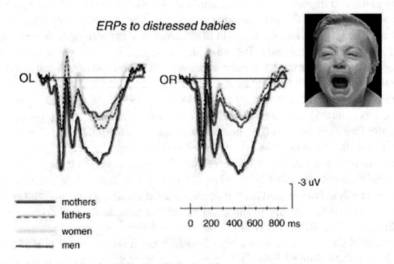

Fig. 6.5 ERPs signals recorded over left and right lateral occipital sites following presentations of infant facial expressions exhibiting strongly negative emotions, according to viewer group. Smaller P300 amplitudes were recorded in fathers vs. mothers, especially with infant expressions of suffering. (Taken from Proverbio et al., 2006a, with authors' and editors' permission)

Fig. 6.6 Examples of social and nonsocial stimuli used to evaluate the interest in social informa-tion, regardless of stimulus color richness and perceptual complexity. (Taken from Proverbio et al. (2008)'s study)

as an index of a greater female interest in, or attention to, this class of biologically relevant signals (human faces and bodies).

Whatever the cause, little neuroscientific evidence of such preference of the female brain for social stimuli has been reported, in contrast to the large body of behavioral evidence showing that females have greater social and affective compe-tence. For example, substantial literature has accumulated indicating that women are better than men at decoding facial expressions of emotion (Thomson & Voyer, 2014). Various studies have demonstrated differences between the ways in which men and women perceive (Proverbio et al., 2016), process (Canessa et al., 2012), express (McDuff et al., 2017), and experience emotions (Proverbio et al., 2009). Research generally suggests that women are more able, as well as more inclined, to express their own emotions to conspecifics (McDuff et al., 2017). Furthermore, they show greater ease in decoding non-verbal indicators connected to the expression of emotions. It has been reported that female children across various human cultures are prone to spend more time with their younger siblings, or their simulacra (baby dolls), than are their male counterparts. It is quite difficult to determine whether this socially oriented behavior is entirely due to cultural factors (such as the style of upbringing) or to a biological difference dependent on genetic factors. Since, in Proverbio's study (2008), showing a greater interest for social stimuli, no behavioral response or attention allocation to social information was required by the task (consisting in detecting rare Mondrian pictures), the stronger responsivity to persons than landscapes in women would reflect a privileged processing of images depicting conspecifics in the female brain. Consistent with this hypothesis, numerous studies (e.g., Wingenbach et al., 2018) have demonstrated that women are provided with a greater ability to decipher the emotions through facial expressions or other non-verbal communication than man and are more inclined and more competent in expressing their emotional experiences to others (Dimberg & Lundquist, 1990). Further evidence has demonstrated that women, as compared to men, react more strongly when viewing affective stimuli (such as IAPS) involving human beings, thus showing higher empathic responses (Proverbio et al., 2009). In this regard, some authors have established a link between sex, social skills, and action processing because of the strong association between the known action observation/execution properties of the motor mirror system and the theorized social functions of the human mirror system (Oberman et al., 2007).

Action and Body Language Understanding

Several sex differences have been reported in action understanding tasks. Female participants have been found to be better at understanding the action purpose as compared to men, as indexed by earlier and larger discriminative ERP responses to incongruent and purposeless behavior (Proverbio et al., 2010c). Perception of plausible and understandable actions (e.g., smiling couple clinking glasses of champagne) was contrasted with that of implausible and unintelligible actions (e.g., businesswoman balancing on one foot in the desert). ERP data showed early processing of the action's purpose in the female brain, with a larger parietal N200 to understandable behavior. Source reconstruction (swLORETA) located the neural generators of this effect in the inferior/parietal, left inferior/frontal, left and right premotor areas, right cingulate cortex, right superior/temporal and extra-striate cortex belonging to the so-called "human mirror-neuron system (MNS)." Anterior

N400 discriminative response (implausible–plausible) was greater in women than men (see Fig. 6.7). The data suggest that congruent/incongruent actions are processed differently from the two sexes, with a prevalence of limbic and cingulate activation in women, and orbito/frontal one in men, along with a right STG activation of comparable amplitude in men and women.

Consistently, the combined fMRI and ERP study by Canessa et al. (2012) and Proverbio et al. (2011c) found differences across male and female participants involving a stronger activation of the action understanding system, the STS, and the ventral premotor cortex (associated with the mirror resonance of others' actions) during the observation of cooperative (vs. affective) scenes in women. Again, other studies provided evidence of sex differences in the development of brain mechanisms for processing biological motion (Anderson et al., 2013). In an fMRI study involving the visual perception of point-light displays of coherent and scrambled biological motion, enhanced activity during coherent biological motion perception was found in females relative to males in a network of brain regions possibly implicated in social perception, including amygdala, medial temporal gyrus, and temporal pole (Anderson et al., 2013). All in all, these pieces of evidence indicate a female superiority in social skills and sex differences in action/behavior processing.

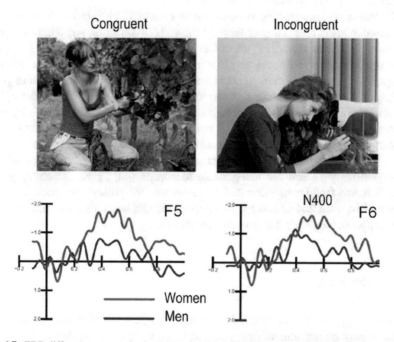

Fig. 6.7 ERP difference waves obtained by subtracting ERPs to congruent from ERPs to incongruent actions separately for men and women, over anterior scalp sites. A much larger N400 response occurred to incongruent actions in women than men. (Taken and modified from Proverbio et al., 2010c)

Face Pareidolia

Recent behavioral and electrophysiological research has shown that women are better at seeing faces, even when there are none, a perceptual illusory phenomenon called "face pareidolia" (i.e., the illusory perception of non-existent faces). Sometimes, while observing the clouds in the sky, coffee foam, or random decorative patterns, we might be struck by the impression of clearly perceiving a face that is so well defined and yet so illusory. This perceptual effect has precise neural underpinnings based on the face fusiform area.

Pavlova et al. (2015) carried out a spontaneous recognition task in which adult females and males were presented with a set of food plate images resembling faces (Arcimboldo style). Not only did women more readily recognize the images as a face (they reported images as resembling a face, on which males still did not), but gave overall more face responses. Proverbio et al. (2016) investigated the neural correlates of this sex difference, in a study in which ERPs were recorded while participants viewed pictures of animals intermixed with that of familiar objects, faces, and faces-in-things. Overall, compared to the men, the women were significantly more inclined to perceive faces in perfectly real object photographs, as shown in the preliminary face-likeness ratings assessment.

Furthermore, face-specific Vertex Positive Potential (VPP, 150–190 ms) showed a difference in the processing of faces-in-things between males and females at frontal sites; while for men VPP was of intermediate amplitude between faces and objects, for women there was no difference in VPP response to faces or faces-in-things, thus suggesting a marked anthropomorphization of objects in the latter group (Fig. 6.8). SwLORETA source reconstruction showed how in the female brain, face *pareidolia* was associated with the activation of brain areas involved in the affective processing of faces (right STS, BA22; posterior cingulate cortex, BA22; orbitofrontal cortex, BA10), which was not found in men. Normally the visual cortex separates face processing from object processing so that faces are automatically processed in ways that are inapplicable to objects (e.g., gaze detection, gender detection, and facial expression coding). However, the present data showed sexual dimorphism, with this dichotomy being much stricter in men than women because of an anthropomorphizing bias in the female brain.

Empathy for Pain

Recent findings have demonstrated that women might be more responsive than men to the sight of painful stimuli (triggering a vicarious response to pain), and therefore more empathic (Han et al., 2008). We investigated whether the two sexes differed in their cerebral responses to affective pictures portraying humans in different positive or negative contexts compared to natural or urban scenarios (Proverbio et al., 2009). Four-hundred-forty IAPS slides were presented to 24 Italian students (12 women

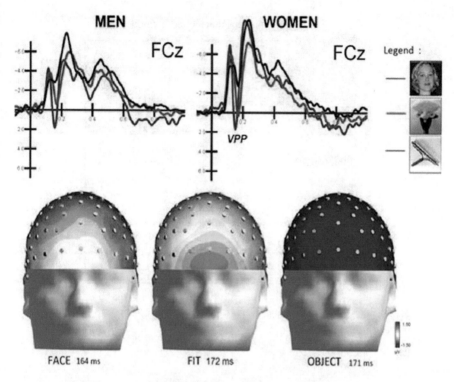

Fig. 6.8 (Top) ERP waveforms recorded in women and men as a function of stimulus type. VPP was much larger to faces and faces-in-things than objects in women. (Bottom) Mean amplitude of the N170 response recorded as a function of stimulus type and relative scalp distribution

and 12 men). An emotional impact scale was administered to all participants prior to EEG recording, showing higher emotional psychological reactions in women than men to a variety of emotional stimuli (both animated and unanimated ones), as shown in Fig. 6.9.

Occipital P115 response of ERPs was greater in response to persons than to scenes and was affected by the emotional valence of the human pictures. A possible explanation for this piece of evidence is that the processing of biologically relevant stimuli was prioritized in both sexes. A late positivity to suffering humans (visible in Fig. 6.10, blue line) far exceeded the response to negative scenes in women but not in men. Increased right amygdala and right frontal area activities were observed only in women. These data possibly indicate a sex-related difference in the brain response to humans, possibly supporting human empathy.

Previous studies have demonstrated that females show greater responsiveness in various brain areas to generically negative pictures, but to date, none has investigated the specific role of the presence of humans in determining the brain emotional response in both sexes. For example, Hofer et al. (2007) found larger activation of the right superior temporal area, right insula, right putamen, and anterior cingulate cortices during the processing of positively valenced words versus non-words for

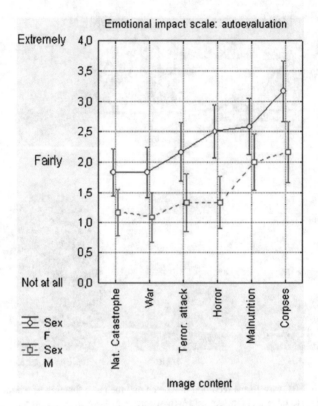

Fig. 6.9 Data obtained from the emotional impact scale (self-reporting questionnaire) administered to the 24 persons participating in the ERP experiment, separately for each image type, and according to their sex. Key: 0 = not at all; 1 = a little; 2 = fairly; 3 = very much; 4 = extremely

women versus men and interpreted these data in terms of the greater emotionality of the female sex. On the other hand, Klein et al. (2003) found increased activation of the amygdala and ACC in women in response to negative IAPS images. In our study, sex differences as a function of the affective valence of pictures were much greater for humans than scenes, thus indicating the special status of the visual image of humans for the female brain, especially in interaction with affective information. Our data are consistent with the more recent literature suggesting that women are more empathic than men are when viewing suffering humans (Han et al., 2008; Schulte-Rüther et al., 2008; Singer et al., 2004).

Sexual Hormones and Oxytocin

The literature has shown that social processes, and in particular, the neural response to opposite-sex faces, may vary as a function of hormonal phase of women. Furthermore, oral contraceptive pill use can affect cognition and alter resting state

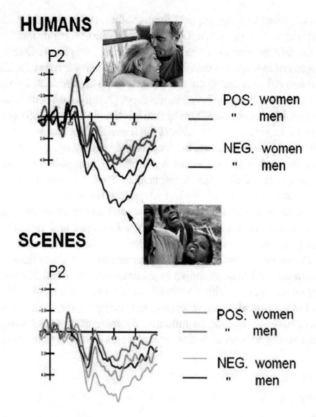

Fig. 6.10 ERPs recorded at right parietal sites as a function of stimulus content and valence and viewer's sex. A large effect of both emotional content of the stimulus is visible (evidenced by comparing ERPs to negative vs. positive unanimated scenes) and an effect of empathy for pain, especially in women (evidenced by comparing ERPs to negative scenes vs. ERPs to pictures portraying humans)

functional connectivity. Indeed, women using oral contraceptives have been shown to differ from non-pill users in memory, mental rotation, and affective memory tasks (Nielsen et al., 2011, 2014). In conclusion, the hormonal control, or lack of it, represents an important variable in determining the neurofunctional behavior of the female brain, and it should be monitored in studies on sex differences.

Several authors (Alexander & Hines, 2002) pointed out the genetic/biological nature of female preference for social stimuli. For example, evidences of toy preference in nonhuman primates (*Cercopithecus aethiops sabaeus*) have been provided, with male vervets preferring to play with unanimated fast-moving toys (e.g., cars or balls) and female vervets preferring the contact with dolls. These data suggest that sexually differentiated interest for infants/dolls arose early in human evolution, prior to the emergence of a distinct hominid lineage. Comparative studies are quite relevant at this regard since monkeys are not subject to the cultural influences proposed to explain human sex differences in social cognition.

Furthermore, other findings support the hypothesis of biological, predetermined sex differences in social interest, not dependent on cultural conditioning, but linked to the genetic role of women as primary offspring caregivers. One of the most important pieces of evidence is the observation of an early interest for infants traceable in all human cultures and historical periods in young females. Remarkably, the same phenomenon has been observed in monkeys (juvenile baboons, macaques, and rhesus monkeys: Herman et al., 2003; Maestripieri and Roney, 2006) as reflected by a higher rate of interaction with infants in females than males. The interaction includes behaviors like embracing, holding, carrying, playing, grooming, touching, staying close to, and it is unaffected by hormone manipulations. According to Maestripieri and Pelka (2002), sex differences in interest in infants across the lifespan should be interpreted as a biological adaptation for parenting. Neuro-hormonal studies carried out in humans have shown that the early interest for infants may be modulated by hormonal factors. For example, Leveroni and Berenbaum (1998) reported that girls precociously exposed to high levels of androgens (because of congenital adrenal hyperplasia) displayed less interest in infants than their normal sisters. Consistently, it has been shown in primates that maternal hormonal changes influence social interaction with unrelated infants (Ramirez et al., 2004), making adult females more empathic and receptive. In this regard, oxytocin has been shown to affect the empathic attitude in humans, by increasing social trust, and even improving the ability to infer affective mental states of others (Domes et al., 2007).

Conclusion

On the basis of a review of the relevant literature, it is concluded that many of the sex differences in social cognition may be related to the (biologically determined) role of females as primary offspring caregivers (as opposed to fighters/hunters, e.g., Kuhn and Stiner, 2006). This distinction may be associated with females' greater empathic attitude, ability to understand body language and facial expressions, attachment and responsivity to infants (Oxytocin-mediated), early interest for infants, interest for social information, emotional responsivity, lesser incidence of autistic, psychopathic and sociopathic disorders. In this way, this chapter provides a unified framework for understanding the multifaceted consequences of a sexual dimorphism in human parental behavior.

Acknowledgements We wish to thank Roberta Adorni, Valentina Brignone, Valeria De Gabriele, Marzia Del Zotto, Jessica Galli, Valentina Lozano, Eleonora Martin, Silvia Matarazzo, Roberta Mazzara, Mirella Manfredi, Laura Paganelli, Federica Riva, Laura Trestianu, and Alberto Zani for their kind contributions.

Supported by 13974 2015-ATE-0052 grant entitled "Emotional responses and gender differences in individuals with high traits of psychopathy, impulsivity and empathy" from University of Milano–Bicocca.

References

Alexander, G. M., & Hines, M. (2002). Sex differences in response to children's toys in nonhuman primates (Cercopithecus aethiops sabaeus). *Evolution and Human Behaviour, 23*, 467–479.

Anderson, L. C., Bolling, D. Z., Schelinski, S., Coffman, M. C., Pelphrey, K. A., & Kaiser, M. D. (2013). Sex differences in the development of brain mechanisms for processing biological motion. *NeuroImage, 83*, 751–760.

Babchuk, W. A., Hames, R. B., & Thompson, R. A. (1985). Sex differences in the recognition of infant facial expressions of emotion: The primary caretaker hypothesis. *Ethology & Sociobiology, 6*, 89–101.

Baron-Cohen, S., & Wheelwright, S. (2004). The empathy quotient: An investigation of adults with Asperger syndrome or high functioning autism, and normal sex differences. *Journal of Autism and Developmental Disorders, 34*, 163–175.

Baron-Cohen, S., Wheelwright, S., Hill, J., Raste, Y., & Plumb, I. (2001). The Reading the mind in the eyes' test revised version: A study with normal adults, and adults with Asperger syndrome or high-functioning autism. *Journal of Child Psychology and Psychiatry, 42*, 241–251.

Bourne, V. J. (2005). Lateralised processing of positive facial emotion: Sex differences in strength of hemispheric dominance. *Neuropsychologia, 43*, 953–956.

Brosch, T., Sander, D., & Scherer, K. R. (2007). That baby caught my eye. Attention capture by infant faces. *Emotion, 7*, 685–689.

Cahill, L. (2017). An issue whose time has come: Sex/gender influences on nervous system function. *J. Neurosci. Res, 1*, 1–791. January/February.

Caldara, R., Rossion, B., Bovet, P., & Hauert, C. A. (2004). Event-related potentials and time course of the "other-race" face classification advantage. *Neuroreport, 15*, 905–910.

Canessa, N., Alemanno, F., Riva, F., Zani, A., Proverbio, A. M., Mannara, N., Perani, D., & Cappa, S. F. (2012). The neural bases of social intention understanding: The role of interaction goals. *PLoS One, 7*(7), e42347.

Cárdenas, R. A., Harris, L. J., & Becker, M. W. (2013). Sex differences in visual attention toward infant faces. *Evolution and Human Behavior, 34*(4), 280–287.

Chung, W., Lim, S., Yoo, J., et al. (2013). Gender difference in brain activation to audio-visual sexual stimulation; do women and men experience the same level of arousal in response to the same video clip? *International Journal of Impotence Research, 25*, 138–142.

Connolly, H. L., Lefevre, C. E., Young, A. W., & Lewis, G. J. (2019). Sex differences in emotion recognition: Evidence for a small overall female superiority on facial disgust. *Emotion, 19*(3), 455–464.

Craig, B. M., Nelson, N. L., & Dixson, B. J. W. (2019). Sexual selection, agonistic signaling, and the effect of beards on recognition of Men's anger displays. *Psychological Science, 30*(5), 728–738.

Dimberg, U., & Lundquist, L. O. (1990). Gender differences in facial reactions to facial expressions. *Biological Psychology, 30*, 151–159.

Domes, G., Heinrichs, M., Michel, A., Berger, C., & Herpertz, S. C. (2007). Oxytocin improves 'mind-Reading' in humans. *Biological Psychiatry, 61*, 731–733.

Glocker, M. L., Langleben, D. D., Ruparel, K., Loughead, J. W., Gur, R. C., & Sachser, N. (2009a). Baby schema in infant faces induces cuteness perception and motivation for caretaking in adults. *Ethology, 115*, 257–263.

Glocker, M. L., Langleben, D. D., Ruparel, K., Loughead, J. W., Valdez, J. N., Griffin, M. D., Sachser, N., & Gur, R. C. (2009b). Baby schema modulates the brain reward system in nulliparous women. *PNAS, 106*, 9115–9119.

Greenberg, D. M., Warrier, V., Allison, C., & Baron-Cohen, S. (2018). Testing the empathizing-systemizing theory of sex differences and the extreme male brain theory of autism in half a million people. *Proceedings of the National Academy of Sciences of the United States of America, 115*(48), 12152–12157.

Hahn, A. C., & Perrett, D. I. (2014). Neural and behavioral responses to attractiveness in adult and infant faces. *Neuroscience & Biobehavioral Reviews, 46*(4), 591–603.

Hahn, A. C., Xiao, D., Sprengelmeyer, R., & Perrett, D. I. (2013). Gender differences in the incentive salience of adult and infant faces. *Quarterly Journal of Experimental Psychology, 66*(1), 200–208.

Han, S., Fan, Y., & Mao, L. (2008). Gender difference in empathy for pain: An electrophysiological investigation. *Brain Res, 27*(1196), 85–93.

Herman, R. A., Measday, M. A., & Wallen, K. (2003). Sex differences in interest in infants in juvenile rhesus monkeys: Relationship to prenatal androgen. *Hormones and Behavior, 43*, 573–583.

Hines, M. (2020). Neuroscience and sex/gender: Looking back and forward. *The Journal of Neuroscience, 40*(1), 37–43.

Hofer, A., Siedentopf, C. M., Ischebeck, A., Rettenbacher, M. A., Verius, M., Felber, S., & Fleischhacker, W. (2007). Sex differences in brain activation patterns during processing of positively and negatively valenced emotional words. *Psychological Medicine, 37*(1), 109–119.

Huynh, H. K., Beers, C., Willemsen, A., Lont, E., Laan, E., Dierckx, R., Jansen, M., Sand, M., Weijmar Schultz, W., & Holstege, G. (2012). High-intensity erotic visual stimuli de-activate the primary visual cortex in women. *The Journal of Sexual Medicine, 9*(6), 1579–1587.

Jacques, C., Jonas, J., Maillard, L., Colnat-Coulbois, S., Koessler, L., & Rossion, B. (2019). The inferior occipital gyrus is a major cortical source of the face-evoked N170: Evidence from simultaneous scalp and intracerebral human recordings. *Human Brain Mapping, 40*(5), 1403–1418.

Klein, K. J. K., & Hodges, S. D. (2001). Gender differences, motivation, and empathic accuracy: When it pays to understand. *Personality and Social Psychology Bulletin, 27*, 720–730.

Klein, S., Smolka, M. N., Wrase, J., Gruesser, S. M., Mann, K., Braus, D. F., et al. (2003). The influence of gender and emotional valence of visual cues on fMRI activation in humans. *Pharmacopsychiatry, 36*(3), 5191–5194.

Kringelbach, M. L., Lehtonen, A., Squire, S., Harvey, A. G., Craske, M. G., Holliday, I. E., Green, A. L., Aziz, T. Z., Hansen, P. C., Cornelissen, P. L., & Stein, A. (2008). A specific and rapid neural signature for parental instinct. *PLoS One, 3*, e1664.

Kuhn, S. L., & Stiner, M. C. (2006). What's a mother to do? A hypothesis about the division of labor and modern human origins. *Current Anthropology, 47*(6), 953–980.

Leibenluft, E., Gobbini, M. I., Harrison, T., & Haxby, J. V. (2004). Mothers' neural activation in response to pictures of their children and other children. *Biological Psychiatry, 56*, 225.

Leveroni, C., & Berenbaum, S. (1998). Early androgen effects on interest in infants: Evidence from children with congenital adrenal hyperplasia. *Developmental Neuropsychology, 14*, 321–340.

Li, G., Zhang, S., Le, T. M., Tang, X., & Li, C. R. (2020). Neural responses to negative facial emotions: Sex differences in the correlates of individual anger and fear traits. *NeuroImage, 16*(221), 117–171.

Luo, L. Z., Li, H., & Lee, K. (2011). Are children's faces really more appealing than those of adults? Testing the baby schema hypothesis beyond infancy. *Journal of Experimental Child Psychology, 110*, 115–124.

Maestripieri, D., & Pelka, S. (2002). Sex differences in interest in infants across the lifespan. *Human Nature, 13*, 327–344.

Maestripieri, D., & Roney, J. R. (2006). Evolutionary developmental psychology: Contributions from comparative research with nonhuman primates. *Developmental Review, 26*, 120–137.

McDuff, D., Kodra, E., Kaliouby, R. E., & LaFrance, M. (2017). A large-scale analysis of sex differences in facial expressions. *PLoS One, 12*(4), e0173942.

Nielsen, S. E., Ertman, N., Lakhani, Y. S., & Cahill, L. (2011). Hormonal contraception usage is associated with altered memory for an emotional story. *Neurobiology of Learning and Memory, 96*(2), 378–384.

Nielsen, S. E., Ahmed, I., & Cahill, L. (2014). Postlearning stress differentially affects memory for emotional gist and detail in naturally cycling women and women on hormonal contraceptives. *Behavioral Neuroscience, 128*(4), 482–493.

Nitschke, J. B., Nelson, E. E., Rusch, B. D., Fox, A. S., Oakes, T. R., & Davidson, R. J. (2004). Orbitofrontal cortex tracks positive mood in mothers viewing pictures of their newborn infants. *NeuroImage, 21*, 583–592.

Oberman, L. M., Pineda, J. A., & Ramachandran, V. S. (2007). The human mirror neuron system: A link between action observation and social skills. *Social Cognitive and Affective Neuroscience, 2*(1), 62–66.

Orozco, S., & Ehlers, C. L. (1988). Gender differences in electrophysiological responses to facial stimuli. *Biol Psychiatry, 44*(4), 281–289.

Parsons, C. E., Young, K. S., Kumari, N., Stein, A., & Kringelbach, M. L. (2011). The motivational salience of infant faces is similar for men and women. *PLoS One, 6*(5).

Parsons, C. E., Stark, E. A., Young, K. S., Stein, A., & Kringelbach, M. L. (2013). Understanding the human parental brain: A critical role of the orbitofrontal cortex. *Social Neuroscience, 8*, 525–543.

Pavlova, M. A., Sokolov, A. N., & Bidet-Ildei, C. (2014). Sex differences in the Neuromagnetic cortical response to biological motion. *Cereb Cortex, 25*(10), 3468–3474.

Pavlova, M. A., Scheffler, K., & Sokolov, A. N. (2015). Face-n-food: Gender differences in tuning to faces. *PLoS One, 10*(7), e0130363.

Peltola, M. P., Yrttiaho, S., Puura, K., Proverbio, A. M., Mononen, N., Lehtimäki, T., & Leppänen, J. M. (2014). Motherhood and oxytocin receptor genetic variation are associated with selective changes in electrocortical responses to infant facial expressions. *Emotion, 14*(3), 469–477.

Phelps, E. A., & LeDoux, J. E. (2005). Contributions of the amygdala to emotion processing: From animal models to human behavior. *Neuron, 48*, 175–187.

Proverbio, A. M. (2017). Sex differences in social cognition: The case of face processing. *Journal of Neuroscience Research, 95*(1–2), 222–234. https://doi.org/10.1002/jnr.23817

Proverbio, A.M. & Galli, J. (2016). Women are better at seeing faces where there are none: An ERP study of face pareidolia, *Social Cognitive and Affective Neuroscience*, first published online May 5, 2016. https://doi.org/10.1093/scan/nsw064

Proverbio, A. M., Brignone, V., Matarazzo, S., Del Zotto, M., & Zani, A. (2006a). Gender and parental status affect the visual cortical response to infant facial expression. *Neuropsychologia, 4*(14), 2987–2999.

Proverbio, A. M., Brignone, V., Matarazzo, S., Del Zotto, M., & Zani, A. (2006b). Gender differences in hemispheric asymmetry for face processing. *BMC Neuroscience, 7*, 44.

Proverbio, A. M., Matarazzo, S., Brignone, V., Del Zotto, M., & Zani, A. (2007). Processing valence and intensity of infant expressions: The roles of expertise and gender. *Scandinavian Journal of Psychology, 48*(6), 477–485.

Proverbio, A. M., Zani, A., & Adorni, R. (2008). Neural markers of a greater female responsiveness to social stimuli. *BMC Neuroscience, 9*, 56.

Proverbio, A. M., Adorni, R., Zani, A., & Trestianu, L. (2009). Sex differences in the brain response to affective scenes with or without humans. *Neuropsychologia, 47*(12), 2374–2388.

Proverbio, A. M., Riva, F., Martin, E., & Zani, A. (2010a). Neural markers of opposite-sex bias in face processing. *Frontiers in Psychology, 18*(1), 169.

Proverbio, A. M., Riva, F., Martin, E., & Zani, A. (2010b). Face coding is bilateral in the female brain. *PLoS One, 21*, 5.

Proverbio, A. M., Riva, F., & Zani, A. (2010c). When neurons do not mirror the agent's intentions: Sex differences in neural coding of goal-directed actions. *Neuropsychologia, 48*(5), 1454–1463.

Proverbio, A. M., De Gabriele, V., Manfredi, M., & Adorni, R. (2011a). No race effect (ORE) in the automatic orienting toward baby faces: When ethnic group does not matter. *Psychology, 2*(9), 931–935.

Proverbio, A. M., Riva, F., Zani, A., & Martin, E. (2011b). Is it a baby? Perceived age affects brain processing of faces differently in women and men. *Journal of Cognitive Neuroscience, 23*(11), 3197–3208.

Proverbio, A. M., Riva, F., Paganelli, L., Cappa, S. F., Canessa, N., Perani, D., & Zani, A. (2011c). Neural coding of cooperative vs. affective human interactions: 150 ms to code the action's purpose. *PLoS One, 6*(7), e22026.

Proverbio, A. M., Mazzara, R., Riva, F., & Manfredi, M. (2012). Sex differences in callosal transfer and hemispheric specialization for face coding. *Neuropsychologia, 50*(9), 2325–2332.

Ramirez, S. M., Bardi, M., French, J. A., & Brent, L. (2004). Hormonal correlates of changes in interest in unrelated infants across the peripartum period in female baboons (Papio hamadryas anubis sp.). *Hormones and Behavior, 46*, 520–528.

Sabatinelli, D., Flaisch, T., Bradley, M. M., Fitzsimmons, J. R., & Lang, P. J. (2004). Affective picture perception: Gender differences in visual cortex? *Neuroreport, 15*(7), 1109–1112.

Sander, K., Frome, Y., & Scheich, H. (2007). FMRI activations of amygdala, cingulate cortex, and auditory cortex by infant laughing and crying. *Human Brain Mapping, 28*(10), 1007–1022.

Sawada, R., Sato, W., Kochiyama, T., Uono, S., Kubota, Y., Yoshimura, S., et al. (2014). Sex differences in the rapid detection of emotional facial expressions. *PLoS One, 9*(4), e94747.

Schulte-Rüther, M., Markowitsch, H. J., Shah, N. J., Fink, G. R., Piefke, M. (2008) Gender differences in brain networks supporting empathy. *Neuroimage 42*(1), 393–403. https://doi.org/10.1016/j.neuroimage.2008.04.180.

Seifritz, E., Esposito, F., Neuhoff, J. G., et al. (2003). Differential sex-independent amygdala response to infant crying and laughing in parents versus nonparents. *Biological Psychiatry, 54*(12), 1367–1375.

Singer, T., Seymour, B., O'Doherty, J., Kaube, H., Dolan, R. J., & Frith, C. D. (2004). Empathy for pain involves the affective but not sensory components of pain. *Science, 303*(5661), 1157–1162.

Sokołowski, A., Tyburski, E., Sołtys, A., & Karabanowicz, E. (2020). Sex differences in verbal fluency among young adults. *Advances in Cognitive Psychology, 16*(2), 92–102.

Thompson, A. E., & Voyer, D. (2014). Sex differences in the ability to recognize non-verbal displays of emotion: A meta-analysis. *Cognition and Emotion, 28*(7), 1164–1195.

Tiedt, H. O., Weber, J. E., Pauls, A., Beier, K. M., & Lueschow, A. (2013). Sex-differences of face coding: Evidence from larger right hemispheric M170 in men and dipole source modelling. *PLoS One, 8*(7), e69107.

Wingenbach, T. S. H., Ashwin, C., & Brosnan, M. (2018). Sex differences in facial emotion recognition across varying expression intensity levels from videos. *PLoS One, 13*(1), e0190634.

Zhou, L., & Meng, M. (2020). Do you see the "face"? Individual differences in face pareidolia. *Journal of Pacific Rim Psychology, 14*, E2. https://doi.org/10.1017/prp.2019.27

Chapter 7
Development of Morality and Emotional Processing

Lucas Murrins Marques ⓘ, Patrícia Cabral, William Edgar Comfort ⓘ, and Paulo Sérgio Boggio ⓘ

Abstract Emotions play a very important role in moral judgments. Hume argues that morality is determined by feelings that make us define whether an attitude is virtuous or criminal. This implies that an individual relies on their past experience to make a moral judgment, so that when the mind contemplates what it knows, it may trigger emotions such as disgust, contempt, affection, admiration, anger, shame, and guilt (Hume D. An enquiry concerning the principles of morals, 1777 ed. Sec. VI, Part I, para, 196, 1777). Thus, even so-called "basic" emotions can be considered as moral emotions. As Haidt (The moral emotions. In: Handbook of affective sciences, vol 11, 852–870, Oxford University Press, 2003) points out, all emotional processing that leads to the establishment and maintenance of the integrity of human social structures can be considered as moral emotion. Consequently, the construct of "morality" is often characterized by a summation of both emotion and cognitive elaboration (Haidt J. Psychol Rev, 108(4):814, 2001).

Keywords Emotional processing · Morality · Moral psychology

Introduction

Emotions play a very important role in moral judgments. Hume argues that morality is determined by feelings that make us define whether an attitude is virtuous or criminal. This implies that an individual relies on their past experience to make a moral judgment, so that when the mind contemplates what it knows, it may trigger emotions such as disgust, contempt, affection, admiration, anger, shame, and guilt

L. M. Marques (✉)
Instituto de Medicina Fisica e Reabilitacao, Hospital das Clinicas HCFMUSP, Faculdade de Medicina, Universidade de Sao Paulo, Sao Paulo, Brazil

P. Cabral · W. E. Comfort · P. S. Boggio
Social and Cognitive Neuroscience Laboratory, Developmental Disorders Program, Center for Health and Biological Sciences, Mackenzie Presbyterian University, Sao Paulo, Brazil

© The Author(s) 2023
P. S. Boggio et al. (eds.), *Social and Affective Neuroscience of Everyday Human Interaction*, https://doi.org/10.1007/978-3-031-08651-9_7

(Hume, 1777). Thus, even so-called "basic" emotions can be considered as moral emotions. As Haidt (2003) points out, all emotional processing that leads to the establishment and maintenance of the integrity of human social structures can be considered as moral emotion. Consequently, the construct of "morality" is often characterized by a summation of both emotion and cognitive elaboration (Haidt, 2001).

According to the Social Intuitionist Model (Haidt, 2001), moral judgment is substantially influenced by "intuitions," i.e. automatic affective reactions. In turn, these intuitions appear to have evolved from physiological reactions in response to external threats and opportunities over our phylogenetic history (Bloom, 2012) and now play a role in resolving situations that threaten the integrity of human social structures (Haidt, 2003). A later hypothesis, the Moral Foundations Theory (MFT; Haidt & Joseph, 2004) based on the assumptions in the Social Intuitionist Model, posits that these intuitions emerge whenever at least one of the six universally human "moral foundations" is violated: (i) Care; (ii) Fairness; (iii) Loyalty; (iv) Authority; (v) Sanctity; and (addended by Haidt, 2012) (vi) Liberty.

In summary, the violations of these six foundations can be described and exemplified as follows (Graham et al., 2013; Graham et al., 2011): (i) Care/harm– situations that involve impairment in emotional and physical care between humans and humans in relation to animals (e.g., Physical aggression in response to an affective betrayal); (ii) Fairness/cheating–situations involving cheating (e.g., The use of public money for personal purposes); (iii) Loyalty/betrayal–situations in which an individual shows disloyalty toward a person or entity (e.g., An employee who works simultaneously for a competing company); (iv) Authority/subversion– situations involving disrespect and disregard for a figure of authority (e.g., Talking loudly during a religious ceremony); (v) Sanctity/degradation–situations involving the "degradation" of moral principles (e.g., Engaging in sexual behavior such as incest); and (vi) Liberty/oppression–situations involving the restriction of personal freedom (e.g., Forcing individual to wear a specific item of clothing).

As presented by Haidt (2008), the first three characterize foundations oriented toward the valuation of the individual (Individualizing Foundations), while the last three value the collective (Cohesive Foundations). In this sense, the recent literature on moral processing is based on the assumptions of TFM (Haidt, 2003, 2008, 2012; Graham et al., 2013), stimulating, for example, the development of instruments such as the Moral Foundations Questionnaire (Graham et al., 2011) and the Moral Foundations Vignettes (Clifford et al., 2015). On the other hand, the degree to which moral dilemmas are involved in the processing of emotions varies consistently with the influence that emotion has on moral judgment (Greene et al., 2001). However, Haidt and Greene disagreed about the role of reason in moral psychology because of Greene's belief in the relevance of thought in a manual way—which is the rational and controlled judgment system—in contrast to the automatic mode, regulated by emotion and intuition, defended by Haidt (2001), who considers emotion as the only source of moral judgment, rationalized by the manual mode (Greene, 2013). In addition, it is estimated that the moral judgment changes according to social and cultural influences (Haidt et al., 1993). However, this conception contrasts with the widespread belief in the twentieth century that a rational and deliberate process takes part in the moral decision (Kohlberg, 1969; Turiel, 1983). Although the notion that judgment is based on the emotional implications of morality is strong, the evidence is still considered insufficient and unproven by some, who argue that emotions can have little influence on moral judgment (Huebner et al., 2009). The recent literature on moral processing is based predominantly on the assumptions of the MFT (Graham et al., 2013) and forms part of the theoretical framework for the development of research instruments such as the Moral Foundations Vignettes (MFVs; Clifford et al., 2015). Furthermore, a group of researchers have recently criticized the MFT, arguing that it fails to cite specific activation modules for triggering the violation of each foundation (and an ensuing affective reaction). In the face of these criticisms, in addition to the importance of factors such as *Nativism, Cultural Learning, Intuitionism,* and *Pluralism* to account for the development of personal morality (see Graham et al., 2013, for a more in-depth analysis), a group of researchers predominantly represented by Kurt Gray have recently developed the Theory of Dyadic Morality (TDM; Schein & Gray, 2018), which suggests that morality or moral violations are represented socially through different forms of harm, but nevertheless have the same ontological basis.

As highlighted by Pizarro (2000), emotions are typically understood as processes antagonistic to moral judgments, sometimes not considering their impact on judgment processes, sometimes assuming that emotions harm judgments. However, a series of contemporary studies points out the close relationship between the two phenomena, frequently highlighting the causal role that emotional modulation plays in the impact of moral judgment (Haidt et al., 1993; Schnall et al., 2008). This impact sometimes contributes to judgment, in cases where, for example, emotional disgust related to a moral violation guides the recrimination of such a violation. On the other hand, emotions can also guide immoral behavior, for example, in cases where positive effects guide acts of injustice or corruption, such as those often observed in political contexts.

Attitudes and judgments can be taken automatically, without necessarily reasoning, based on pre-established concepts or in a complex way, using different perspectives (Van Bavel et al., 2015). As noted by Koenigs et al. (2007), some brain structures are related to more deontological moral judgments, and when these structures suffer brain injuries, the most intuitive judgments predominate, demonstrating that moral judgments are present in both situations. However, cognitive processes may be present to a greater or lesser extent. Moreover, it is worth mentioning that some studies have demonstrated that emotional intuitions can significantly impact moral judgment and reasoning both in adults and children (Danovitch & Bloom, 2009; Malti & Ongley, 2014). As such, differences in moral judgment at distinct stages of development may often be due to individual differences in the development of emotional processing and the regulation of these emotional intuitions (Eisenberg, 2000).

Emotional Processing

Several studies have identified overlapping areas in the brain responsible for both moral judgment and emotional processing, including the insula (Vicario et al., 2017; Ying et al., 2018), amygdala (Decety et al., 2012; Harenski et al., 2014), orbitofrontal cortex (OFC; Fumagalli & Priori, 2012) and ventromedial prefrontal (PFC; Shenhav & Greene, 2014; Pascual et al., 2013), and anterior cingulate cortex (ACC; Pascual et al., 2013). Moll and Oliveira-Souza (2007) suggest that this overlap may be due to the dependence of moral reasoning and judgment on the engagement of multiple emotion-related systems in the brain, citing the ventromedial PFC as one of the key nodes in this network as an interface between emotional experience and moral decision-making.

There have been frequent reviews of research into moral judgment and decision-making due to the increasing importance of moral behavior and reasoning in modern life. Several reviews have dedicated themselves to establishing a neural basis responsible for the cognitive processes underlying moral reasoning (Forbes & Grafman, 2010; Van Bavel et al., 2015). A greater understanding of the physiology of the "moral brain" has been possible by the so-called boom of functional neuroimaging studies (Greene & Haidt, 2002). Verplaetse et al. (2014) have identified some of the key nodes in a neural system subserving moral cognition, including (i) medial frontal gyrus; (ii) the superior temporal sulcus; (iii) the temporoparietal junction; (iv) orbitofrontal cortex; (v) ventromedial PFC; and (vi) dorsolateral PFC. In particular, some structures of the PFC deserve to be highlighted as they have a distinct impact on the cognitive and social processes underlying moral judgment (Forbes & Grafman, 2010).

The dorsolateral PFC has been implicated in many aspects of moral intuition; Forbes and Grafman (2010) suggest an auxiliary function of the right dorsolateral PFC in the integration of complex emotional responses that are generated by the evaluation of information from the context that is being judged, increasing the weight of emotion in this decision. However, Greene et al. (2004) find evidence that demonstrate greater involvement of the same region in more difficult personal moral dilemmas, which require greater rational cognitive processing.

On the other hand, Greene (2007) found that patients with lesions in the ventromedial PFC showed more utilitarian moral judgments, with less cognitive elaboration. More recently, another study regarding group categorization demonstrated that the ventromedial PFC showed greater activation in situations in which participant evaluated themselves as belonging to a specific group, compared to situations in which they did not belong (Molenberghs & Morrison, 2012), revealing the role of ventromedial PFC in social categorization as well. However, the studies described above only reveal correlations between different forms of moral judgment and brain activation.

Interestingly, emotions themselves may be moral in character, including such complex emotions as guilt, shame, and righteousness (Turner & Stets, 2006). These moral emotions often signal emotional arousal in response to moral violations or conformity but may have a primary role as "triggers" for more basic emotions such as anger, fear, and hatred. Similarly, our emotional reactions to moral violations of fairness and our propensity to engage in prosocial behavior have been shown to depend on similar neural substrates as reactions to situations eliciting disgust (Sanfey et al., 2003; Tabibnia et al., 2008).

Morality

Relatively few studies have been published on the development of the psychological and neural underpinnings of moral judgments. To date, the primary theories in this field continue to be those proposed by Jean Piaget and Lawrence Kohlberg, two of the most significant scholars of moral and cognitive development of the twentieth century, who saw morality primarily in terms of justice, care, and respect for authority (Bloom & Wynn, 2016).

Piaget

For Piaget et al. (1989), moral values are constructed from the interaction between the subject and the various social environments which he/she engages with, and it is through daily coexistence with others in adulthood that we build our moral values, principles, and norms. Processes of internal organization and adaptation are necessary for these interactions to occur, which Piaget's model categorizes as interactions of assimilation and accommodation. Assimilation schemas vary according to the stage of individual development and are defined as strategies for conflict resolution based on pre-existing cognitive structures and knowledge. Furthermore, Piaget argues that the development of morality is composed of three phases: (i) a "pre-moral" phase, (ii) a "heteronomous" phase, and (iii) an "autonomous" phase.

The first "pre-moral" phase, present in children of up to 5 years of age, is where the child bases their rules of conduct on their immediate needs instead of a set of moral norms which supersede behavior. When the child obeys an internally generated rule, the behavior is reinforced through habit and not by a sense of right and wrong. A baby who cries until fed is an example of moral behavior in this phase.

The second phase, that of heteronomous morality, is typically present in children aged 5–10 years. In this stage, morality corresponds to behavior, which complies with social rules and norms, with any interpretation other than this does not correspond to a correct attitude. A poor man who steals medicine to save his wife's life is committing an equal moral wrong as a man who murders his wife, according to heteronomous reasoning.

Finally, during the third phase of moral development, autonomous morality, individuals set moral codas and rules by mutual agreement.

However, as pointed out by Vozzola (2014), there are stronger points that should be considered in Piaget's classical theory, such as the interference of the environment in development and what we can structure in order to stimulate the child, but there are also other aspects that must be considered, such as the fact that Piaget underestimates the role of culture and education in fostering cognition and moral development. This important role of cognitive development in moral development is evident in a study by Smetana and Ball (2018) showed that children make distinctive moral judgments regarding physical damage and psychological damage (both from Care Foundation) because the first is concrete while the latter may have no direct and observable consequences, and therefore requires a more advanced understanding of the thoughts and feelings of others (Helwig et al., 2001; Smetana et al., 2012). In particular, understanding young children's judgment relative to psychological damage is hampered by the difficulty in coordinating moral assessments with an understanding of intentions, actions, and outcomes (Jambon & Smetana, 2014).

Kohlberg

Kohlberg (1976) divided moral development into intervals based on the responses he observed to hypothetical dilemmas presented in the form of stories, concluding that there are three main levels of moral reasoning with two stages each.

The first level is that of "preconventional morality," which is divided into an initial stage of orientation to punishment and obedience, where the child decides what is wrong on the basis of what behavior is punished, and a subsequent stage of individualism, instrumental purpose and exchange, where the child follows rules when it is in his/her immediate interest. This level is largely related to the moral foundation of authority, which values both respect for the rules established by a moral authority and punishments for moral transgressions. The role and importance of authority figures and social norms guiding the individual's principles of right and wrong are also established at this level.

The second level is that of "conventional morality" which is divided into an initial stage of mutual interpersonal expectations, relationships, and interpersonal conformism, where those actions that meet the expectations of the family or other significant social grouping are deemed to be morally right (directly related to the moral foundation of loyalty). The later stage in this level, that of social system and consciousness, emphasizes that moral actions are those defined by broader social groups (e.g., a nation or people) or by society as a whole (Kohlberg, 1976).

Finally, the third level is that of "postconventional morality," which is divided into an initial stage of orientation by the social contract, where the attitudes of the individual are directed to act in order to achieve the "greater good for the greatest number of people" (i.e. utilitarianism), and a subsequent stage of universal ethical principles, where the individual develops and follows ethical principles through reflection and personal choice to determine what is morally right (Kohlberg, 1976).

As with Piaget, Vozzola (2014) also points to the strong and weak points which can be highlighted in Kohlberg's theory. Kohlberg primarily asserts that it is through development that people can construct "a deeper understanding of particular social practices or of more specific social contexts" in qualitative divisions based on hypothetical dilemmas.

Current Perspectives

More recently, Saarni (2011) has highlighted the construction of emotional competence as a key milestone in moral development, as a set of cognitive and regulatory skills and goal-oriented behavior that emerges over time relative to the individual's social context. As discussed by Eisenberg (2000), individual factors such as cognitive development and temperament influence the development of emotional competency, which can also be influenced by social experiences and learning, including the individual's social relations history and beliefs. Also, emotion regulation habilities may mediate how emotional intuitions impacts moral judgment and reasoning. Thus, some skills of emotional competence, described above, are: (i) ability to discern and understand others' emotions based on situation and expressive clues; (ii) the capacity for empathy and sympathy involving the emotional experiences of others; and (iii) ability to soften the intensity of aversive and distressing emotions using self-regulation (Eisenberg, 2000).

Saarni (2011) also states that child's relationship with their caregivers is characterized by the initial context in which there is the unfolding of the emotional life of the child, causing this relationship to structure the child's life for the development of emotional skills and future relationships social rights (see also Graziano et al., 2010; John & Gross, 2004). The same author goes on to say that a safe bond between the caregiver and the child leaves the child free to explore the world and engage with peers, since an insecure or unstable attachment is associated with emotional and social incompetence, particularly in the areas of understanding emotions and anger regulation. Typically, in relation to the development of emotional abilities, in younger children, the expression of emotions and their regulation are less developed, requiring a greater support and reinforcement of the social environment. The development of these skills does not occur in isolation, and its progression is intricately linked with cognitive development (Eisenberg, 2000; Saarni, 2011).

In this sense, some studies have investigated the influence of emotional regulation on moral judgment (Feinberg et al., 2012; Lee et al., 2013; Li et al., 2017; Zhang et al., 2017; Helion & Ochsner, 2018). For example, one of the studies pointed out that cognitive reappraisal habit influences the rigidity of moral judgment, so that individuals who have a high frequency of cognitive reappraisal also have a more liberal moral judgment (Feinberg et al. (2012). In this same sense, another study revealed that the habit of cognitive reappraisal, in addition to being related to less conservative behaviors, is also related to less behavior in support of conservative policies, which demonstrates that this cognitive control has as much influence on moral judgment as on moral attitudes (Lee et al., 2013).

Conclusions

Several studies address the relationship between emotion and moral judgment (Pizarro, 2000; Greene et al., 2001; Haidt, 2001; Helmuth, 2001; Haidt, 2003; Koenigs et al., 2007; Moll & de Oliveira-Souza, 2007; Tangney et al., 2007; Huebner et al., 2009; Feinberg et al., 2012; Zhang et al., 2017; Wagemans et al., 2018), sometimes highlighting the duality between faster/intuitive and slower judgments/deontological, others defending the domain that emotions cause in guiding decision-making processes (Haidt, 2012; Greene, 2013). In one way or another, there is a great interest by moral psychologists in studying the relationship between these two phenomena, since this relationship affects areas such as law, politics, public health, and interpersonal relationship processes in general. In addition, emotions are currently being discussed as active processes, no longer as a mere physiological consequence of a given stimulus, highlighting the important role of cognitive processes, such as the regulation of emotion, in modulating the emotional response. In that sense, the specific assessment of different moral foundations for different ages can contribute to a better understanding of the development of moral judgment throughout the different stages of development. In addition, it is essential to highlight the importance of assessing the development of moral judgment also during adulthood, as well as in different sexes.

References

Bloom, P. (2012). Moral nativism and moral psychology. In *The social psychology of morality: Exploring the causes of good and evil* (pp. 71–89). American Psychological Association.

Bloom, P., & Wynn, K. (2016). *What develops in moral development*. D. Barner, A. S. Baron (pp. 347–364). Oxford University Press.

Clifford, S., Iyengar, V., Cabeza, R., & Sinnott-Armstrong, W. (2015). Moral foundations vignettes: A standardized stimulus database of scenarios based on moral foundations theory. *Behavior Research Methods, 47*(4), 1178–1198.

Danovitch, J., & Bloom, P. (2009). Children's extension of disgust to physical and moral events. *Emotion, 9*(1), 107.

Decety, J., Michalska, K. J., & Kinzler, K. D. (2012). The contribution of emotion and cognition to moral sensitivity: a neurodevelopmental study. *Cerebral Cortex, 22*(1):209–20. https://doi.org/10.1093/cercor/bhr111. Epub 2011 May 26. PMID: 21616985.

Eisenberg, N. (2000). Emotion, regulation, and moral development. *Annual Review of Psychology, 51*(1), 665–697.

Feinberg, M., Willer, R., Antonenko, O., & John, O. P. (2012). Liberating reason from the passions: Overriding intuitionist moral judgments through emotion reappraisal. *Psychological Science, 23*(7), 788–795.

Forbes, C. E., & Grafman, J. (2010). The role of the human prefrontal cortex in social cognition and moral judgment. *Annual Review of Neuroscience, 33*, 299–324. ISSN 0147-006X.

Fumagalli, M., & Priori, A. (2012). Functional and clinical neuroanatomy of morality. *Brain, 135* (Pt 7):2006–21. https://doi.org/10.1093/brain/awr334. Epub 2012 Feb 13. PMID: 22334584.

Graham, J., Nosek, B. A., Haidt, J., Iyer, R., Koleva, S., & Ditto, P. H. (2011). Mapping the moral domain. *Journal of Personality and Social Psychology, 101*(2), 366.

Graham, J., Haidt, J., Koleva, S., Motyl, M., Iyer, R., Wojcik, S. P., & Ditto, P. H. (2013). Moral foundations theory: The pragmatic validity of moral pluralism. In *Advances in experimental social psychology* (Vol. 47, pp. 55–130). Elsevier.

Graziano, P. A., Keane, S. P., & Calkins, S. D. (2010). Maternal behaviour and children's early emotion regulation skills differentially predict development of children's reactive control and later effortful control. *Infant and Child Development, 19*(4), 333–353.

Greene, J. D. (2007). Why are VMPFC patients more utilitarian? A dual-process theory of moral judgment explains. *Trends in Cognitive Sciences, 11*(8), 322–323. ISSN 1364-6613.

Greene, J. (2013). *Moral tribes: Emotion, reason, and the gap between us and them.* Atlantic Books.

Greene, J., & Haidt, J. (2002). How (and where) does moral judgment work? *Trends in Cognitive Sciences, 6*(12), 517–523. ISSN 1364-6613.

Greene, J. D., Sommerville, R. B., Nystrom, L. E., Darley, J. M., & Cohen, J. D. (2001). A fMRI investigation of emotional engagement in moral judgment. *Science, 293*(5537), 2105–2108.

Greene, J. D., et al. (2004). The neural bases of cognitive conflict and control in moral judgment. *Neuron, 44*(2), 389–400. ISSN 0896-6273.

Haidt, J. (2001). The emotional dog and its rational tail: A social intuitionist approach to moral judgment. *Psychological Review, 108*(4), 814.

Haidt, J. (2003). The moral emotions. In *Handbook of affective sciences* (Vol. 11, pp. 852–870). Oxford University Press.

Haidt, J. (2008). Morality. *Perspectives on Psychological Science, 3*(1), 65–72.

Haidt, J. (2012). *The righteous mind: Why good people are divided by politics and religion.* Vintage.

Haidt, J., & Joseph, C. (2004). Intuitive ethics: How innately prepared intuitions generate culturally variable virtues. *Daedalus, 133*(4), 55–66.

Haidt, J., Koller, S. H., & Dias, M. G. (1993). Affect, culture, and morality, or is it wrong to eat your dog? *Journal of Personality and Social Psychology, 65*(4), 613.

Harenski, C. L., Edwards, B. G., Harenski, K. A., & Kiehl, K. A. (2014). Neural correlates of moral and non-moral emotion in female psychopathy. *Frontiers in Human Neuroscience, 25*, 8:741. https://doi.org/10.3389/fnhum.2014.00741. PMID: 25309400; PMCID: PMC4174863.

Helion, C., & Ochsner, K. N. (2018). The role of emotion regulation in moral judgment. *Neuroethics, 11*(3), 297–308.

Helmuth, L. (2001). *Moral reasoning relies on emotion.* American Association for the Advancement of Science.

Helwig, C. C., Zelazo, P. D., & Wilson, M. (2001). Children's judgments of psychological harm in normal and noncanonical situations. *Child Development, 72*(1), 66–81.

Huebner, B., Dwyer, S., & Hauser, M. (2009). The role of emotion in moral psychology. *Trends in Cognitive Sciences, 13*(1), 1–6.

Hume, D. (1777). *An enquiry concerning the principles of morals,* 1777 ed. Sec. VI, Part I, para, 196.

Jambon, M., & Smetana, J. G. (2014). Moral complexity in middle childhood: Children's evaluations of necessary harm. *Developmental Psychology, 50*(1), 22.

John, O. P., & Gross, J. J. (2004). Healthy and unhealthy emotion regulation: Personality processes, individual differences, and life span development. *Journal of Personality, 72*(6), 1301–1334.

Koenigs, M., Young, L., Adolphs, R., Tranel, D., Cushman, F., Hauser, M., & Damasio, A. (2007). Damage to the prefrontal cortex increases utilitarian moral judgements. *Nature, 446*(7138):908–11. https://doi.org/10.1038/nature05631. Epub 2007 Mar 21. PMID: 17377536; PMCID: PMC2244801.

Kohlberg, L. (1969). *Stage and sequence: The cognitive-developmental approach to socialization.* Rand McNally.

Kohlberg, L. (1976). Moral stages and moralization: The cognitive-development approach. In *Moral development and behavior: Theory research and social issues* (pp. 31–53). Holt, Rinehart & Winston.

Lee, J. J., Sohn, Y., & Fowler, J. H. (2013). Emotion regulation as the foundation of political attitudes: Does reappraisal decrease support for conservative policies? *PLoS One, 8*(12), e83143.

Li, Z., Wu, X., Zhang, L., & Zhang, Z. (2017). Habitual cognitive reappraisal was negatively related to perceived immorality in the harm and fairness domains. *Frontiers in Psychology, 8*, 1805.

Malti, T., & Ongley, S. F. (2014). The development of moral emotions and moral reasoning. In *Handbook of moral development* (Vol. 2, pp. 163–183). Routledge.

Molenberghs, P., & Morrison, S. (2012). The role of the medial prefrontal cortex in social categorization. *Social Cognitive and Affective Neuroscience, 9*(3), 292–296. ISSN 1749-5024.

Moll, J., & de Oliveira-Souza, R. (2007). Moral judgments, emotions and the utilitarian brain. *Trends in Cognitive Science, 11*(8):319–21. https://doi.org/10.1016/j.tics.2007.06.001. Epub 2007 Jun 29. PMID: 17602852.

Pascual, L., Rodrigues, P., & Gallardo-Pujol, D. (2013). How does morality work in the brain? A functional and structural perspective of moral behavior. *Frontiers in Integrative Neuroscience, 7*, 65. https://doi.org/10.3389/fnint.2013.00065. PMID: 24062650; PMCID: PMC3770908.

Piaget, J., Garcia, R. V., Garcia, R., & Lara, J. (1989). *Psychogenesis and the history of science.* Columbia University Press.

Pizarro, D. (2000). Nothing more than feelings? The role of emotions in moral judgment. *Journal for the Theory of Social Behaviour, 30*(4), 355–375.

Saarni, C. (2011). Emotional development in childhood. In *Encyclopedia on early childhood development* (pp. 1–7). Centre of Excellence for Early Childhood Development.

Sanfey, A. G., Rilling, J. K., Aronson, J. A., Nystrom, L. E., & Cohen, J. D. (2003). The neural basis of economic decision-making in the ultimatum game. *Science, 300*(5626):1755–8. https://doi.org/10.1126/science.1082976. PMID: 12805551.

Schein, C., & Gray, K. (2018). The theory of dyadic morality: Reinventing moral judgment by redefining harm. *Personality and Social Psychology Review, 22*(1), 32–70.

Schnall, S., Haidt, J., Clore, G. L., & Jordan, A. H. (2008). Disgust as embodied moral judgment. *Personality and Social Psychology Bulletin, 34*(8), 1096–1109.

Shenhav, A., & Greene, J. D. (2014). Integrative moral judgment: dissociating the roles of the amygdala and ventromedial prefrontal cortex. *Journal of Neuroscience, 34*(13):4741–9. https://doi.org/10.1523/JNEUROSCI.3390-13.2014. PMID: 24672018; PMCID: PMC6608126.

Smetana, J. G., & Ball, C. L. (2018). Young children's moral judgments, justifications, and emotion attributions in peer relationship contexts. *Child Development, 89*(6), 2245–2263.

Smetana, J. G., Rote, W. M., Jambon, M., Tasopoulos-Chan, M., Villalobos, M., & Comer, J. (2012). Developmental changes and individual differences in young children's moral judgments. *Child Development, 83*(2), 683–696.

Tabibnia, G., Satpute, A.B., & Lieberman, M. D. (2008). The sunny side of fairness: preference for fairness activates reward circuitry (and disregarding unfairness activates self-control circuitry). *Psychological Science, 19*(4):339–47. https://doi.org/10.1111/j.1467-9280.2008.02091.x. PMID: 18399886.

Tangney, J. P., Stuewig, J., & Mashek, D. J. (2007). Moral emotions and moral behavior. *Annual Review of Psychology, 58*, 345–372.

Turiel, E. (1983). *The development of social knowledge: Morality and convention.* Cambridge University Press.

Turner, J. H., & Stets, J. E. (2006). Moral emotions. In *Handbook of the sociology of emotions* (pp. 544–566). Springer, Boston, MA

Van Bavel, J. J., Feldman-Hall, O., & Mende-Siedlecki, P. (2015). The neuroscience of moral cognition: From dual processes to dynamic systems. *Current Opinion in Psychology, 6*, 167–172.

Verplaetse, J., et al. (2014). *Moral brain.* Springer. ISBN 9400791291.

Vicario, C. M., Rafal, R. D., Martino, D., & Avenanti, A. (2017). Core, social and moral disgust are bounded: A review on behavioral and neural bases of repugnance in clinical disorders. *Neuroscience and Biobehavioral Reviews, 80*:185–200. https://doi.org/10.1016/j.neubiorev.2017.05.008. Epub 2017 May 12. PMID: 28506923.

Vozzola, E. C. (2014). *Moral development: Theory and applications.* Routledge.

Ying, X., Luo, J., Chiu, C. Y., Wu, Y., Xu, Y., & Fan, J. (2018).Functional dissociation of the posterior and anterior insula in moral disgust. *Frontiers in Psychology, 9*, 860. https://doi.org/10.3389/fpsyg.2018.00860. PMID: 29910758; PMCID: PMC5992674.

Wagemans, F., Brandt, M. J., & Zeelenberg, M. (2018). Disgust sensitivity is primarily associated with purity-based moral judgments. *Emotion, 18*(2), 277.

Zhang, L., Kong, M., & Li, Z. (2017). Emotion regulation difficulties and moral judgment in different domains: The mediation of emotional valence and arousal. *Personality and Individual Differences, 109*, 56–60.

Chapter 8
Trust in Social Interaction: From Dyads to Civilizations

Leonardo Christov-Moore, Dimitris Bolis, Jonas Kaplan, Leonhard Schilbach, and Marco Iacoboni

Abstract Human trust can be construed as a heuristic wager on the predictability and benevolence of others, within a compatible worldview. A leap of faith across gaps in information. Generally, we posit that trust constitutes a functional bridge between individual and group homeostasis, by helping minimize energy consumed in continuously monitoring the behavior of others and verifying their assertions, thus reducing group complexity and facilitating coordination. Indeed, we argue that trust is crucial to the formation and maintenance of collective entities. However, the wager that trust represents in the face of uncertainty leaves the possibility of misallocated trust, which can result in maladaptive outcomes for both individuals and groups. More specifically, trust can be thought of as a scale-invariant property of minimizing prediction error within ascending levels of social hierarchy ranging from individual brains to dyads, groups and societies, and ultimately civilizations. This framework permits us to examine trust from multiple perspectives at once,

L. Christov-Moore (✉) · J. Kaplan
Brain and Creativity Institute, University of Southern California, Los Angeles, CA, USA

D. Bolis
Independent Max Planck Research Group for Social Neuroscience, Max Planck Institute of Psychiatry, Munich-Schwabing, Germany

International Max Planck Research School for Translational Psychiatry (IMPRS-TP), Munich, Germany

Munich Medical Research School (MMRS), Dekanat der Medizinischen Fakultät, Ludwig-Maximilians-Universität München, Munich, Germany

L. Schilbach
Independent Max Planck Research Group for Social Neuroscience, Max Planck Institute of Psychiatry, Munich-Schwabing, Germany

LVR Klinikum Düsseldorf/Kliniken der Heinrich-Heine-Universität Düsseldorf, Düsseldorf, Germany

Ludwig-Maximilians-Universität, Medical Faculty, Munich, Germany

M. Iacoboni
Department of Psychiatry and Biobehavioral Sciences, Ahmanson-Lovelace Brain Mapping Center, Brain Research Institute, David Geffen School of Medicine at UCLA, Los Angeles, CA, USA

© The Author(s) 2023
P. S. Boggio et al. (eds.), *Social and Affective Neuroscience of Everyday Human Interaction*, https://doi.org/10.1007/978-3-031-08651-9_8

relating homeostasis, subjective affect and predictive processing/active inference at the individual level, with complexity and homeostasis at the collective level. We propose trust as a paradigmatic instance of an intrinsically dialectical phenomenon bridging individual and collective levels of organization, one that can be observed in daily experience and empirically studied in the real world. Here, we suggest collective psychophysiology as a promising paradigm for studying the multiscale dynamics of trust. We conclude with discussing how our integrative approach could help shine light on not only the bright but also the dark sides of trust.

Keywords Trust · Social interaction · Empathy · Homeostasis · Emergence · Scale invariance · Active inference

Introduction: A Broken Leg and a Stranger

You have fallen and broken your leg on the stairs outside your apartment. Your phone is inside and you're immobile, in agony. In that moment a stranger walks up, seemingly concerned. You look them in the eyes. They seem kind. You take a chance and ask them if they can quickly go up to your apartment, grab your phone from the coffee table, and call the emergency line. They agree to help, and you breathe a sigh of relief. In the face of your limitations in that moment, you have made a very specific wager: that they will not defy your prediction of their behavior, and particularly, that they will not do so in a way that runs counter to your interests. In other words, not only are you wagering that they will not surprise you by donning a helmet and begin singing opera, you are specifically wagering that they will not betray you by taking advantage of the situation, e.g., stealing everything and walking away. You have decided to trust them.

Trust, this bet on predictable benevolence, is a social heuristic (another word for shortcut) that enables us to navigate a world about which we have limited direct information and within which we have limited agency. Trust is a leap of faith across gaps in information, reducing the energy we would otherwise spend independently verifying others' beliefs, intentions, and actions (Braynov, 2002), and performing all the actions necessary for our survival alone. Trust allows us to plug others' hypotheses into the gaps in our information and behave in belief and in action as if those hypotheses are accurate. Doing so in no way ensures that the information is accurate, so trust is ultimately a wager, a simulation of the world that we treat as real. This reduces the energy cost associated with uncertainty, facilitating cooperation, community, and group efficiency (Lewis & Weigert, 1985; Luhmann, 1979). Due to its profound material and subjective advantages, it can be considered a form of social capital (Bachmann, 2001; Morgan & Hunt, 1994; Fukuyama, 1996; Zheng et al., 2008). By virtue of these multiple capacities, it is a foundational pillar of human interaction, ranging from pairs of people (i.e., dyads), to families and societies, all the way up to the global web of socioeconomic relations that undergirds our civilization (Misztal, 1996; Zak & Knack, 2001).

In this chapter, we will dissect the principal components of trust, the role of trust in reducing individual prediction error and group complexity (Lewis & Weigert, 1985) in and through social interaction (cf. Bolis & Schilbach, 2020a; Ramstead et al., 2018). More concretely, we will link trust to fundamental properties of predictive processing and homeostasis and, in doing so, formalize and situate our framework within the contemporary theoretical landscape. Last, we will discuss the potential maladaptive outcomes of trust formation and maintenance and possible insights on how to avoid them. This constitutes a novel framework with which to understand and study trust empirically, relating it inwards to phenomenology, cognition and affect, and outwards to informational and energetic properties of groups, within a global framework of homeostasis and free energy minimization (Table 8.1).

Conceptions and Components of Trust

Classical research on trust describes its cognitive, affective, and behavioral components, while primarily approaching it via two core accounts: from the psychological perspective, the disposition to trust is conceived as a trait difference dependent on properties of the other, such as honesty, status, benevolence, etc. From the behavioral perspective, trust is modeled as a risky but advantageous wager on future reciprocity, primarily studied through a small set of paradigms such as the prisoner's dilemma and the trust game (reviewed in Lewis & Weigert, 1985). Trust can reduced to the following: one agent (the truster) engages in a belief about a future outcome that relies on the behavior of the other (the trustee). This may be voluntary (as when you decide to give the keys to your apartment to a friend) or compelled (as in the case of your broken leg) (Bamberger, 2010; Mayer et al., 1995; McKnight & Chervany, 1996).

From the individual perspective, trust is essentially regarded as a belief in the predictability of a future outcome, whether in the cognitive and social sciences (Lewis & Weigert, 1985), in management science and business (Cui et al., 2018; da Rosa Pulga et al., 2019), or in law. Predictability here means you have well estimated the parts you can "model" and the parts that are chaotic or unpredictable, e.g. the explanatory variables and random error. Predictions about others constitute a cardinal process in social interaction on both an intrapersonal and interpersonal level (Frith & Frith, 2012; Timmermans et al., 2012). Conversely, making oneself predictable, and thus facilitating trust formation, one can help increase the chances of continuing to interact with others (Coan, 2015), while collectively decreasing metabolic cost (Theriault et al., 2021). We don't just trust those we find predictable, we also seek to make ourselves more predictable to those we trust.

However, predictability alone is not sufficient to capture what we mean when we say "I trust you." I can trust that person X will betray me if I let down my guard. Person X in this account is predictable, but I cannot say that I trust them if I am sure

Table 8.1 Glossary of terms

Active inference	An account of action according to which (biological) systems sample the environment in accordance with prior beliefs for minimizing free energy.
Dialectics	The dialectical method asserts that phenomena cannot be meaningfully understood by reducing them into single levels of description (cf. reductionism) or assuming a metaphysical independence between levels (cf. dualism) but should be rather studied in their wholeness, inner contradiction, and movement.
Dialectical misattunement hypothesis	This rethinks a psychiatric condition, such as autism spectrum conditions (ASC), not merely as a disorder of the individual brain but also as cumulative misattunement between persons, which can be thought of as disturbances in the dynamic and reciprocal unfolding of an interaction across multiple time scales, resulting in increasingly divergent prediction and (inter-)action styles.
Emergence	Emergent entities (properties or substances) "arise" out of collective (inter-) actions of more fundamental entities and yet are "novel" or "irreducible" with respect to them.
Free energy	An information theory measure that bounds or limits (by being greater than) the surprise on sampling some data, given a generative model. Put simply, with regard to an organism, free energy minimization can be thought of as a process of maintaining current living form by being restricted to a limited number of possible states.
Homeostasis	An organism's dynamic process of maintaining optimal function within the internal milieu and within its relation to the external environment.
Precision	A statistical term defined as inverse variance that can be thought of as the confidence a (biological) system places upon its beliefs.
Prediction error	The discrepancy between incoming information and (biological) system-generated hypotheses.
Predictive processing	Theory that states that (biological) systems are constantly generating and updating hypotheses about the causal structure of the environment and the self along different levels of abstraction with the ultimate goal of minimizing free energy
Prior	One's existing hypotheses about a given quantity (the distribution of expected states) before new evidence is taken into account.
Self-other resonance	A phenomenon in which neural systems involved in one's own affect, interoception, somatosensation, and behavior also process that of others in a similar way, a.k.a. mirroring, shared-circuits, experience-sharing.
Superorganism	A putative form of organization in which individual organisms form a sufficiently coordinated whole such that individuals are no longer informationally or materially independent, allowing group-level homeostatic mechanisms to emerge, e.g., ant colonies, beehives. Some authors have speculated that human societies present superorganismic properties (Kesebir, 2012).

Adapted from Bolis et al. (2017)

that they will betray me given the opportunity. To say that I trust them (rather than trust that they will do as expected) conveys a belief that they are predictably benevolent. Philosophers have addressed this by distinguishing trust from reliance, where trust can be betrayed while reliance can only be disappointed (Baier, 1986, 235). I may rely on a clock to give the time, but I do not generally feel betrayed when it breaks. Trust is thus different from reliance in that a truster accepts the risk of being betrayed. In contrast, I can feel confident in the competence of another without being invested in their benevolence (Nooteboom, 2017).

In addition to predictability and benevolence, we argue that trust requires the trustor's implicit belief that the trustee's conception of benevolence and perception of reality are compatible with their own. I can find a person's behavior both predictable and a clear expression of their benevolence according to their worldview but still not trust them because that expression of benevolence is incompatible with mine. Take the case of a fanatic who offers to ceremoniously sacrifice me in order to reunite me with my creator. I may believe that they are sincerely benevolent, but I cannot say that I trust them, because human sacrifice is not consistent with my concept of benevolence. In addition, benevolence is often expressed through the transfer of accurate information. Trust requires that we be able to generally believe another's assertions about the world. Take the case of a person or group of persons significantly deviating from an intersubjective consensus in a given spatiotemporal and, thus, sociocultural context, e.g., a schizophrenic or a person under the influence of strong hallucinogens. In addition to qualms about their benevolence, I cannot use the information they provide me about the world, unless I have access to their perspective of reality. Thus, trust in our estimation entails predictability of benevolence within an interpersonally compatible conception of reality, including one's values within that reality. In this light, trust is fundamentally a dialectical relationship between individual properties (cf. predictable benevolence) and collective properties (cf. shared reality and values).

The formation of social bonds whether they be romantic, friendly, or transactional relies on the formation and building of trust. Not surprisingly, evidence suggests that the ability to trust and the state of trusting are not only beneficial to the survival of an organism, they are also important for subjective well-being (DeNeve & Harris, 1998). To feel trusted and to trust others feels good, fosters calm, and is important for healthy, warm relationships (DeNeve, 1999). At the other extreme, the betrayal of trust is deeply traumatic and difficult to recover from, whether in relationships or societies (Lewis & Weigert, 1985). In Dante's Inferno, the lowest circle of hell was reserved for traitors. For what greater sin exists within the human mind and human mythology than betrayal? The emotional experience of being betrayed is deeply traumatic and difficult to recover from, as it constitutes a massive prediction error in the human psyche, a defiance not only of current predictions but of an entire history of belief. Indeed, the powerful subjective experience of trust and its betrayal, and this experience's ubiquity and power within human art and mythology, points to its pivotal role as a group-level homeostatic function.

Despite manifesting as a compelling component of individual experience, trust is an inherently social construct (Lewis & Weigert, 1985; Searle, 1995), much like language, power, surveillance, and accountability (Gerck, 1998a, b). Note here that this extends to trust in one's concept of self as the third person, e.g., "I don't trust myself to drive drunk." Trust acts as a group-level adaptive mechanism that makes social life more predictable and less dangerous, thereby facilitating coordination (Lewis & Weigert, 1985; Tomasello, 2014). Inversely, the absence of trust inhibits the formation of community, impedes cooperation and in doing so increases insta-bility and entropy within the group, reducing its efficiency (Braynov & Sandholm, 2002; Lewis and Wiegert, 1985; Zak & Knack, 2001). Along similar lines of thought, modern economics considers trust as an economic lubricant, reducing the cost of transactions between parties, minimizing risk and uncertainty, as well as enabling new forms of cooperation and, at the macro level, generally facilitating business, with hypothesized macroeconomic effects even on indices such as GDP and infla-tion (Morgan & Hunt, 1994; Singh, 2012; Zheng et al., 2008).

Let us imagine an illustrative example of trust's role in groups (cf. Tomasello, 2014): In a primitive world where relatively small prey was abundant, each indi-vidual anthropoid could hunt and take care of their satiation needs independently. Imagine now an abrupt ecological shift after which small prey (e.g., chickens) has been substituted by significantly bigger prey (e.g., bears), which is impossible to hunt individually. In this case, individuals have two options: either they continue going independently for the limited small prey or learn to coordinate in order to effectively catch the bigger prey. One such strategy could have been found in an individual scaring a bear which, trying to escape, falls in the trap of several other anthropoids. Here, crucially, the first anthropoid in our hypothetical scenario should trust the group of the others will provide a fair share of the food at the end of the day.

At a certain point of evolutionary history, humanoids were potentially presented with a fundamental dilemma of trust: I either act independently, risking the unavoid-able case of running out of suitable prey; or trust and coordinate with others in order to survive both as an individual and as a group, risking being potentially betrayed. However, the heuristics we use to establish trust, such as perceived similarity, group affiliation, or charismatic persuasion, can result in emergent, maladaptive outcomes. We may trust groups whose behavior, while coordinated, may be ultimately detri-mental to ourselves and others or groups whose own internal dynamics may be ultimately self-destructive. Indeed, throughout the evolution of organisms and superorganisms (Kesebir, 2012) of greater and greater complexity, there must have existed a dynamic balance between the opposing needs to minimize prediction error by trust and, on the other hand, to withhold and restrict trust because of the high risk it implies.

The Bayesian perspective construes the brain as an organ that calculates and maintains expectations about subsequent events in the environment or within the body, by combining prior experience (priors for short) and newly sensed or poste-rior information. Crucially, the more confidence (i.e., precision) is placed on the validity of existing prior expectations/beliefs; the less these are updated in the face of new incoming information (i.e., evidence). Trust in this light operates primarily upon gaps in information or points of uncertainty, allowing us to place high confi-dence in one's (or trusted others') priors in the absence of data. After all, if one possessed absolute knowledge, trust would be unnecessary.

Trust reduces the energy used in mitigating uncertainty around incomplete infor-mation and lessens the impact of conflicting data. Trust can thus be viewed as a complementary mechanism of attention. Attention putatively allows for reallocation of monitoring through selectively keeping precision (confidence) of incoming

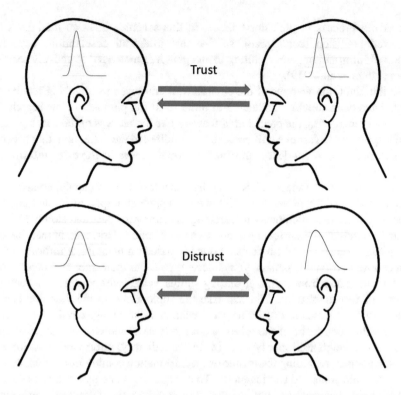

Fig. 8.1 Trust minimizes apparent prediction error and facilitates the interpersonal sharing of priors

information high or low, in order to attend or not, respectively (cf. Mirza et al., 2019; Friston, 2009). Trust acts by reallocating attention as if prediction error was low (Fig. 8.1), through the selective tuning of precision/confidence toward the trustee, resulting in a reduction of energy consumption. Trust is a wager driving informational confidence in prior beliefs about blind or occluded spots in the social and material world—the unseen priors driving others' observed behavior and our priors about the unseen world that trusted others provide.

Let us unpack this core idea intuitively. When introducing their Bayesian model of selective attention, Birza and colleagues describe the example of the lost red pen (Mirza et al., 2019). Imagine you are at your office and you have lost a red pen. How will you search for it? Or to put it more technically, how will you sample the environment? A naive robot with high luxury of time might choose to deploy a serial search, sequentially scanning all possible positions in the room until the red pen is found. However, in real life, humans deploy certain heuristics in order to optimize their sensorimotor processes in space and time and eventually the chances of maintaining their own existence. Searching for a red pen, arguably, might not be crucial for one's own life these days, yet quickly spotting the red apple on a tree and, thus, avoiding eating a poisonous fruit or encountering a wild predator might have been a life-saving process in prehistoric times. In this scheme, attention modulates the expected precision (confidence), so that task-irrelevant observations have less expected information gain, resulting in an agent less motivated to actively seek for them (Mirza et al., 2019).

In this light, the features of the environment employed in searching for the red apple have been suitably modulated in order to fit the purposes of the search. By selectively increasing the precision of features like the redness or the curving shape of the apple, these features will become more salient in the belief updating process, as high precision (confidence) information weighs more in Bayesian information integration.

Now imagine a stranger offers you a fruit that looks like an apple, assuring you it is delicious. Will you eat that fruit? A naive approach implying ample luxury of time may include meticulously researching the stranger's past and slowly forming an understanding of their feelings, intentions, and beliefs. However, in real life, one is typically forced to make a decision quickly based on incomplete information. If you choose to trust this person, you choose to consider their offer as a predictable, benevolent action from a person sharing similar values with you, and as such not worthy of further scrutiny. You will take the apple and skip the time and energy necessary for verification of its nature. Technically speaking, one is choosing to increase the confidence of the priors about the trustee, allowing one to selectively disregard or weigh less other types of (even contradictory) information in a context-specific manner, regarding the trustee or the information coming from them, as well as the specific goals and contingencies. This wager saves energy but carries the risk of both being incorrect and not learning from it, with potentially disastrous consequences. In summary, we construe trust as a selective increase of prior beliefs' precision/confidence about the trustee predicated on a wager on the predictability, benevolence and interpersonal similarity of the trustee.

The Experience and Function of Trust: Affect and Homeostasis

Think of someone you trust. What is trusting them like? What does it mean for your relationship with them? Who has not experienced the deep comfort of feeling trust in another human being? Or the heartbreaking, scarring pain that comes from experiencing trust betrayed? Trust's affective content is undeniable in daily life (Lewis & Weigert, 1985). Classical accounts of trust conceive of trust in terms of individual predispositions toward trusting, judgments based on qualities of the trustee, and as a rational cost-benefit analysis, in which trust constitutes a risky but advantageous wager on future benevolence (Lewis & Weigert, 1985). These accounts, however, largely overlook the subjective experience of trust and do not analyze what the functional role of this affective component may be, or from where it may arise.

Contemporary theories of affect posit that feelings are the extrusion into conscious experience of homeostatic processes that arise from the interaction of the body, brain, and exterior milieu (Damasio, 2018; Damasio & Carvalho, 2013). Homeostasis is here not taken to mean a stable unchanging state but rather a dynamic process aimed at minimizing the prediction error or disconnect between one's expectations of the world and one's body, and information derived from the world and one's body (cf. Seth, Suzuki, & Critchley, 2012). The feelings associated with a process such as trust (especially their valence and intensity) are informative in understanding the importance and role of that process. The intensity of trust's formation and betrayal suggests from a homeostatic perspective that there is something so profoundly useful about trusting relationships that we are evolved to seek, enjoy, foster, and preserve them. Indeed, the powerful feelings associated with trust (cf. Lewis & Weigert, 1985) drive correspondingly striking patterns of behavior and belief: if you trust someone in the extreme, you assume that they will act in your interests despite ample motivation to do otherwise. You assume that they are not misleading you even if the content of their words is unbelievable and assume that there is a rhyme and reason to their actions even if they appear nonsensical or deeply unethical.

Correspondingly, the neural correlates of trust appear related to affective processing in general. Judgments of trustworthiness involve a broad array of affective decision-making regions such as the anterior cingulate, frontal lobe, caudate, insula, and amygdala (Todorov et al., 2008; Watabe et al., 2011; Winston et al., 2002). The amygdala has been thought to be hyperactive in response to specifically untrustworthy faces (Adolphs et al., 1998; Baron et al., 2011; Santos et al., 2016) although some data suggests the amygdala may actually have a non-linear response to the strength of the trustworthiness judgment (Freeman et al., 2014; Said et al., 2009). The assessment of trustworthiness is associated with amygdalar activity (signaling trust or mistrust), as well as connectivity with and activation of regions associated with top-down control of affect like dorsolateral prefrontal cortex, temporoparietal junction, and ventromedial prefrontal cortex (Bellucci et al., 2019).

The maintenance and formation of unconditional trust relationships is associated with activation in mesolimbic reward systems associated with encouraging and reinforcing optimal behaviors (Krueger et al., 2007). When choosing to trust and interacting with a trustworthy individual, activity heightens in the orbitofrontal cortex and caudate, though this activation decreases with age suggesting that trust and cooperation become more of a given than a novel reward (Decety et al., 2004; Fett et al., 2014; Gromann et al., 2014). It is also thought that affective regions such as the insula and anterior cingulate may be hyper-responsive when there is a possibility of betrayal, reflecting the strong importance of avoiding misplacing trust (Aimone et al., 2014; Fett et al., 2014).

As social animals, we need trust because it is simply impossible to experience the whole of our social and material world directly. Much of our perception of the world relies on others (cf. participatory sense making De Jaegher & Di Paolo, 2007) and dialectical attunement (Bolis & Schilbach, 2020a, Vygotsky, 1978). Hence, from the perspective of a homeostatic organism that seeks to minimize prediction error about the world, what could be better than having an affective marker of the validity of one's models of others? Individuals within trusting (and hence coordinated) groups expend radically less energy in monitoring the world, getting information and verifying it, predicting the behavior of others, and ensuring their benevolence. Conversely, high unpredictability and a high possibility of malevolence or betrayal within our immediate environment is energy-intensive to monitor and manage, detrimental to homeostasis, highly suboptimal for group efficiency, and correspondingly regarded as both stressful and unpleasant (DeNeve, 1999; DeNeve & Harris, 1998; Singh, 2012).

The affective nature of trust supports a crucial role for it in the formation and maintenance of efficient, coordinated groups, from dyads to civilizations (Lewis & Weigert, 1985; Zak & Knack, 2001). Much as the theory of embodied cognition speaks of the brain's cognitive processes as situated within the body (Wilson & Foglia, 2011), we cannot disregard that as the brain exists within the body, the body exists and grows within a society in virtually every instance in our evolution. It has been theorized that we "extend" our cognition via external tools (Clark & Chalmers, 1998), such as our phones. Perhaps our cognition, knowledge, and affect are also extended within our trusted groups. Our knowledge of the world, our reactions to it, and our motivations within it are informed by others we trust, as such trusted conspecifics are able to compensate for incomplete data and uncertainty, thus achieving coordinated and potentially optimal knowledge, motivation, and affective states that emerge between individuals. The wager of trust consists in our choice of who to include in this collective entity, and thereby our window to the unseen and occluded world.

If our approach to trust is correct, then the social pain of betrayal may reflect more than individual homeostasis: it reflects "damage" to the collective, a reflection into the individual of group homeostatic signals: When I trust you, I'm in a sense saying "Let's form a superorganism, a collective entity, together." This may be why violations of trust are deeply traumatic and have such an enduring impact on subsequent behavior and inference. When betrayed we are saying: "We were a superorganism together. I did not second guess your inferences about the world, or your future behavior towards me, and it was great. As it turns out, I was wrong. My choice of superorganism and the investment it implied was incorrect. What a surprising disappointment." The intensity and valence of the affective response when trust is betrayed suggests that it constitutes a massive prediction error that must be incorporated into behavior and decision making, and remembered about that person, group or institution, lest we risk future negative consequences. Via this bridge between phenomenology and individual and collective forms of homeostasis, we can draw novel links between the subjective experience of trust and predictive theories surrounding cognition.

Trust as Dialectical Bridge Between Individual Prediction Error and Group Complexity

We posit that trust is a characteristic example of a phenomenon bridging individual accounts of cognition that emphasize the minimization of prediction error (Clark, 2013) and group accounts of efficiency and coordination that emphasize the reduction of complexity (Lewis & Weigert, 1985), unified by the principle of free energy (cf. Bolis & schilbach, 2020a; Ramstead et al., 2018) (Fig. 8.2). Leaning on dialectical (e.g., Vygotsky, 1978) and Bayesian (e.g., Clark, 2013; Friston, 2013) accounts of cognition and action, we also regard trust as a core process of dialectical attunement. Dialectical attunement construes human becoming as the dynamic interplay between (social) internalization and (collective) externalization in and through culturally mediated social interaction (Bolis, 2020; Bolis & Schilbach, 2020a). Here, internalization is thought of as the co-construction of bodily hierarchical models of the (social) world and the organism. Externalization is taken as the collective transformation of the world.

Our notion of internalization is largely based on predictive processing conceptualizations (Bolis, 2020; cf. Friston, 2013; Clark, 2013). Predictive processing has been defined as a hierarchical bidirectional process through which an organism adjusts itself in order to "optimally" predict environmental and bodily regularities. In brain function, predictions are continuously generated and propagated from higher levels of the neural hierarchy to lower ones in an attempt to explain away prediction errors, i.e., the discrepancy between incoming information and generated predictions. On the other hand, prediction errors are propagated from lower levels

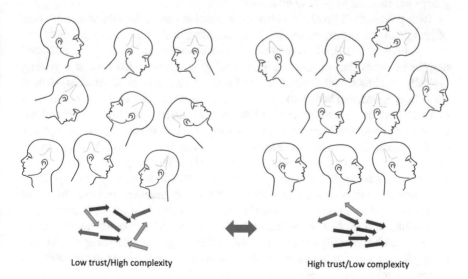

Low trust/High complexity   High trust/Low complexity

Fig. 8.2 At a collective level, trust facilitates efficient group behavior, reducing group complexity and overall energy consumption

of the hierarchy to higher ones in order to suitably readjust the organism. The ultimate goal of such a process is to minimize prediction error as precisely as possible, through processes such as perception and learning. Such hierarchical structures should be considered as collectively shaped. First, we dynamically embody others in and through social interaction, shaping each other's hierarchical structure (Bolis, 2020; Bolis & Schilbach, 2020a), and second, such structures might even be socially extended into interbodily configurations (Ramstead et al., 2018). The structure and culture of social groups are two possible avenues to achieve such configurations.

Yet, organisms such as humans are not passive observers of reality who merely try to adapt to it (cf. Bolis, 2020; Bolis & Schilbach, 2020a). Organisms continuously interact with their world (including their own body), adjusting it according to their prior expectations (cf. Friston, 2013; Clark, 2013). For instance, the body temperature tends to fall behind expected values in extremely cold environments. Bodily tremor, lighting a fire, or choosing to go in a warm space typically reverses such a bodily temperature decrease, helping keep it within well-defined bounds. Processes of actively controlling the body and the environment in order to minimize prediction error have been described as active inference (Friston, 2013; see also Clark, 2013). However, such processes should not be exclusively attributed to the individual. For instance, "architecture and technology can be viewed as a collective effort for reducing overall uncertainty by transforming the environment according to bodily and interpersonal expectations" (Bolis, 2020). In a nutshell, humans actively co-construct and co-regulate—in interaction with other organisms—their ecosocial niches, with the ultimate aim to facilitate survival of not only the individual but also the social group and the species as a whole (cf. dialectical attunement). Here we suggest that multiscale processes of trust between individuals and groups are cardinal to such an endeavor.

Such processes of prediction error minimization can be thought of as processes of complexity reduction subserving life. According to the free energy principle theory (cf. Friston, 2013), life is thought of as a natural process leading to a restricted number of states. For instance, a human being, conceptualized here as a system, typically inhabits a well-defined range of states across several dimensions, such as temperature, size, and body structure. A human corporeal system maintains an order of a certain extent. Such a process of life implies a tendency to resist the second law of thermodynamics of keeping disorder (entropy) as low as possible. As entropy can be mathematically defined as the mean value of surprise over time, a living system needs to also keep surprise as low as possible. However, the precise calculation of surprise is not accessible to a living system, as it should be aware of the dynamics of all possible states of a given world. Therefore, in practice, an upper bound of surprise (i.e., free energy), as opposed to the exact value of surprise, is kept as low as possible. In turn, free energy minimization can be cast as prediction error minimization under simplifying assumptions. Taken together, according to the free energy theory, a living organism, such as a human being, achieves staying alive via effectively minimizing overall prediction error.

Similarly, trust can be viewed as a multiscale process of dialectical attunement via interpersonal prediction error minimization. Put simply, the trustor outsources a part of their active inference and predictive processing to the trustee. By assuming predictability and benevolence of another, within a compatible set of values and worldview, the trustor gives up control over the actions of the trustee. In so doing, it inextricably links interindividual processes of making sense and controlling the world, allowing for potential complexity reduction in broader scales, as well as within each individual (Fig. 8.2). Of course, trust also entails the risk of severe increase of complexity across various scales (from harm of the trustor to disorder of the group) in the case of betrayal or breakdown of trust.

Trust does not solely allow us to minimize uncertainty about conceptions and models of other people, groups, and institutions, it also mediates the extent to which we reduce uncertainty about the world at large and ourselves, by mediating the extent to which we internalize the predictive models and world views of other individuals, groups, and institutions. Trust mediates not only our view of others but also the extent to which we accept them as windows into a world that we can never see and understand in its entirety for ourselves. In other words, within a group, trust operates as a gate on the intergroup sharing of priors as well as a gate on the level of precision applied to those priors. We wish to emphasize that the increased coordination and reduced complexity afforded by intergroup trust lies separate or orthogonal to the potential maladaptive outcomes of that group coordination. Take for example, the collective suicides of doomsday cults. You may well argue that they are capable of incredibly effective coordinated behavior, but the outcome of that coordination and integration is clearly maladaptive. This has applications to understanding both the phenomenon of groups centered around charismatic leaders as well as emergent networks of trust like those that emerge in social media echo chambers.

The Formation and Maintenance of Trust: Bottom-Up and Top-Down Interactions

Despite the risk to material and psychological benefits it entails (Lewis & Weigert, 1985), trust is often established quickly and without thorough verification due to the frequent interactions necessary for social life (particularly in large societies). As trust ultimately lies in a probabilistic belief about the other, not in certainty, there is a question of how much we can "trust trust" (Gambetta, 2000). We often have to rely on explicit and implicit heuristics that allow us to quickly attribute trust that emerge in the intersection of top-down and bottom-up processes, all of which hinge on perceived similarity.

From the top-down, explicit direction, the predominant marker of trust is group identity as manifested in markers of group affiliation (Platow et al., 2012) and stereotypes (Foddy et al., 2009). In practice, these provide the quickest markers and indicator of trustworthiness. From the bottom-up perspective, there are both active and passive forms of implicit trust formation. On the passive end, people who are viewed as more similar to oneself are more likely to be empathized with, trusted, and vice versa (DeBruine, 2002, 2005). Conversely, similar neural responses to naturalistic stimuli in audiences predict affiliation (Parkinson et al., 2018). On the active end, enforced or created similarity, either through mimicry or through joint action and cooperation, can also create a sense of affinity that translates into trust, empathy, and subsequent benevolence (Chartrand & Bargh, 1999). Both of these paths (top-down and bottom-up) point to trust as a bridge between individual phenomenology and homeostasis, as well as group formation and maintenance.

This can be observed in the bidirectional interaction between implicit and explicit forms of coordination, whether affective, somatomotor, or behavioral: Studies have found that mimicry of behavior can create a sense of affiliation and trust, and these are in turn associated with increased behavioral, affective, and cognitive coordination (cf. "The Chameleon Effect", Chartrand & Bargh, 1999). Conversely, neuroscience has found that subjects' pain centers become active when observing another in pain, a phenomenon which is considered a marker of empathic concern (Reviewed in Lamm, 2011). In fact, not only my perception of the pain of the other, but even my very feeling of pain is socially modulated, being dependent on both embodied social factors and personal attachment styles (Fotopoulou & Tsakiris, 2017). Curiously, when subjects are induced to distrust a mock subject (after observing them engage in traitorous behavior), this vicarious pain response is diminished (Hein & Singer, 2008). This diminished response is also observed when observing members of a perceived out-group experience pain (Hein & Singer, 2008), suggesting that trust is a mediator of coordination, even at a basic somatomotor level, modulating the extent to which the perceived internal states of another will impact my own. Coordination is the key word here: when people trust each other and begin to resonate with each other, they do not necessarily solely mimic/imitate each other. Rather they become part of a coordinated group; they plug into a collective entity linked by common priors and shared mutual confidence, with downstream effects on their internal states. Put simply, you internalize the entities around you, and they internalize you (cf. Bolis & Schilbach, 2020a). Our brains become less independent from one another when we trust each other.

An illustrative example of extremes of trust formation and its manifestation in coordination and optimality is in the training of groups of soldiers into platoons. From the start, they are stripped of other affiliations, allegiances, and markers, and generally without explanation induced to engage in coordinated motor behavior, such as simultaneous and repetitive behavioral exertions and utterances. They also undergo similarly harrowing painful experiences and suffering. These early difficult experiences are often cited as the foundation of the later sense of affiliation and trust to an extent rarely observed in the civilian world. From the top-down, soldiers are encouraged to strip themselves of group affiliations and identities and submit to a common identity and a common goal and unifying mythology, as well as a belief in their complete interdependence and complete reliance on one another for their very survival. The result, when successful, is a degree of coordination, efficiency, and ability to quickly align and pursue common goals that are virtually unparalleled in daily life. In all instances, there is a convergence of shared experience, coordinated behavior, repeated interaction, as well as a shared worldview and identity.

Whether emerging from explicit or implicit sources, trust either consist of—or at least uses as—its prime heuristic, similarity. Perceived similarity is correlated with trust when controlling for other variables, and engaging in behavioral synchrony or coordination with others can foster a sense of affiliation and trust (DeBruine, 2002, 2005; Chartrand & Bargh, 1999). Similarity here acts as a heuristic element. By its virtue, I can make fewer extraneous assumptions about behavior, as well as develop models of others using my own priors. It should be clear that similarity in this case does not simply refer to the static, superficial identification of similarity, as in a photograph, but to the whole action set of a person, their behavioral cadence and the idiosyncrasies of their movements and expressions, which are reflective of their internal drives, preferences, and values. Perhaps through the practices of institutionalized as well as spontaneous mimicry, coordination, and joint attention, there is also fostered a sense of a shared worldview. This manifests at the cognitive level but more importantly, at the level of values, underlying drives, points of attention, reaction patterns, affective trajectories, and also less independent internal states by virtue of self-other resonance (e.g., mirroring, contagion, mimicry).

Theories on attachment revolve around trust. Whether the caregivers are mostly predictable in meeting needs leads to either trusting relationships later on or something maladaptive like the need to seek absolute security, etc. (Mikulincer, 1998). Deficits of trustfulness and the ability to form trusting relationships imply deficits in forming parts of coordinated groups, because one's somatomotor and affective integration into group state might be inefficient. The possible exception to this trend is in affiliation with large-scale group identity in which there is less implicit face to face, bottom-up somatomotor/affective interaction, e.g., ideological camps from a computer screen. This implies that people with emotional, affective problems and trust issues interpersonally may still be able to form part of mass movements. Hence, the authoritarian loner archetype described in works such as Eric Hoffers' *The True Believer*: disaffected loners or self-perceived losers/exiles who are fertile terrain for recruitment into fanatical ideologies and associated tribal identities.

Interpersonal traumas and breakdowns in relationships can be reduced to or at least related to a violation or erosion of trust (Lewis & Weigert, 1985). A current danger relevant to our subject is the breakdown in shared realities in large-scale societies, which has occurred in large part through social media algorithms through which people are "siloed" into only viewing content that already conforms and encourages their pre-existing beliefs, encouraging disparate but more importantly incommensurable realities, thus irrevocably destroying the capacity for trust. If I perceive someone to have an incompatible concept of benevolence—one so incompatible that it affects my perception of their predictability—then I am incapable of trusting them and hence incapable of forming part of coordinating groups with them, a phenomenon which you can see crystallized in the disruption of work forces, in teams, on ideological grounds and at a larger scale in the fragmentation of societies in periods of civil war.

We should not solely focus on the loss of the positive effects of trust within groups: the loss of coordination, the loss of empathy, or the loss of affiliation. We must also examine the gain of negative properties and outcomes, beyond dislike or disagreement, down to a dysfunction in empathy and associated coordination at the somatic motor and affective level. The loss of trust and affiliation may modulate our subconscious affective and somatomotor processing such that when confronted with members of a distrusted group, their suffering has a reduced power to move us and motivate sympathy and compassion. This could hypothetically make it more likely for harm to be inflicted by otherwise typical people. For instance, dehumanization of the other (an ultimate form of trust disruption) through language coupled with reinforcement of in-group coordination (through rallies, synchronized behaviors, and symbols) has been deliberately deployed for enabling subsequent mass atrocities (cf. Scarry, 1985).

Empirical Implications and Future Experiments

Due to conceptual and methodological constraints, research of trust has largely focused on either the individual or the sociological (reviewed in Lewis & Weigert, 1985). Here, we emphasize the importance of studying intrapersonal and interpersonal processes in their inherent interrelation, as they unfold during social interactions and beyond (Bolis & Schilbach, 2017, 2020a, b; Bolis, 2020; De Jeagher & Di Paolo, 2007; Dumas et al., 2014, 2020). The concept of trust as a bridge between individual prediction error and group complexity (Fig. 8.2) is intended to spur novel work. To this end, we suggest empirically studying the links between the phenomenology of trust, the behavioral and neural correlates of minimization of prediction error, as well as the complexity and efficiency at the group level. As we discussed throughout this chapter, trust can be thought of as lying at the dynamic intersection of the individual and the collective, entailing both bottom-up (somatomotor and affective forms of coordination, on one hand) and top-down (contextual and reputation-based factors) processes. Combined, trust can thus be studied as a single interconnected construct. Here, we envision a research line which will elucidate the links between trust processes across scales, extending from implicit behavior and contagion, all the way up to conscious phenomenology and further up to patterns of collective coordination in brain and behavior. In what follows, we describe an experimental framework, namely collective psychophysiology, as well as an analysis scheme, namely multi-level analysis of intersubjectivity that could help us do so (cf. Bolis & Schilbach, 2020a).

Traditionally, psychophysiology as a research paradigm has enabled the empirical investigation of the interrelation between physiological and psychological processes, offering important insights about the mechanisms at the level of the individual. However informative this kind of endeavor may have been, the inherent dynamics of social constructs, such as trust, will remain largely unexplored until dynamic interpersonal processes are systematically considered, as, for instance, (social) cognition might be fundamentally different when we really interact with others (Schilbach et al., 2013). In fact, it has been argued that the most important experience of the other stems from the archetypal situation of face-to-face social interaction, while all other cases remain mere products of it (Berger & Luckmann, 1967).

Building upon empirical frameworks of interpersonal research (e.g., Bolis & Schilbach, 2020a; Dumas, 2011; Froese et al., 2015; Koike et al., 2016; Montague et al., 2001; Schilbach et al., 2013), the paradigm of collective psychophysiology allows for the empirical investigation and systematic manipulation of real-time social interaction, across various modalities and temporal scales. To give a concrete example of such a framework, in the special case of two-person psychophysiology (Bolis & Schilbach, 2018), study participants sit opposite each other, working on tasks either individually or collectively, while being able to interact via a micro-camera communication system. Such a two-person framework allows for the monitoring and systematic manipulation of processes that lie in different levels of organization, from psychophysiology to culture. In fact, via systematically controlling the diversity of the interacting individuals across various dimensions, such as age, culture, social class, or even psychological condition, core interpersonal processes of trust can be put to the test: emerging contextual and interpersonal differences and similarities in social interactions might prove equally, or even more important than individual traits in building and maintenance relationships of trust (cf. the dialectical misattunement hypothesis; Bolis et al., 2017). For instance, it has been shown that it is the interpersonal similarity of (autistic) traits that primarily predicts friendship quality in the general population and not the traits per se (Bolis et al., 2020).

While laboratory studies typically offer excellent experimental control, collective psychophysiology of trust can (and should) be eventually examined beyond the laboratory walls, where it manifests itself, in real-world social life. Two paradigmatic cases of such scenarios could be found in pedagogy and psychotherapy. In fact, in line with pedagogical and clinical insights (cf. Bolis, 2020; Bolis & Schilbach, 2020a; Hendren & Kumagai, 2019; Koole & Tschacher, 2016; Lee, 2007; Ramseyer & Tschacher, 2011; Terrell & Terrell, 1984; Thompson et al., 2004), we suggest that the formation of trust and interpersonal attunement between the educator and the student, as well as the therapist and the patient, might be a first step of pivotal importance to an eventual educational and therapeutic success. Here, multipersonal neuroimaging and motion tracking could be deployed in order to capture the interpersonal mechanisms of real-time social interactions in classrooms and psychotherapeutic settings (cf. Bolis et al., 2017; Dikker et al., 2017; Lahnakoski et al., 2020; Tschacher et al., 2014), being complemented by digital phenotyping and interactive self-reports (cf. Bolis et al., 2020; Insel, 2017).

It may be fruitful as well to explore phenomena that exemplify our framework and its predictions in action: one possible example is placebo, in which individuals seemingly internalize not just the abstract beliefs about the properties of a drug or procedure/ritual but also the internal states that would be implied by the belief. One could possibly interpret this as the internalization of affective and homeostatic priors, which powerfully suggests that the mitigating factors in the placebo effect are the degree of trust in the physician/clinician/healer/belief system surrounding the treatment, as well as trustfulness as a trait in the patient. This merits study.

Parts of this chapter discussed trust in terms of interpersonal predictive processing and active inference. Here, we suggest moving from exclusively focusing on the isolated individual, toward a multilevel understanding of intersubjectivity and psychopathology (cf. Bolis & Schilbach, 2017, 2020a). It would be also interesting to observe the ways and extent to which the dynamic formation of trust within a group and its neural and behavioral correlates, as well as group-level measures such as efficiency and complexity. Not only observing formed groups but rather observing the formation of groups and their maintenance, and how the success or failure of this process is modulated by bottom-up and top-down processes, what their respective contributions are, and also how these factors co-vary with the emergence of group schisms and reformations, induced and spontaneous. This could be fruitfully applied in specific instances where trust mediates information flow, such as in the classroom setting. Taken together, collective psychophysiology, we suggest, appears as a promising empirical framework for studying trust, enabling both great experimental control and ecological validity (cf. Bolis, 2020; Bolis & Schilbach, 2020a).

Conclusion

To conclude, we define trust at its core as a belief and behavior in accordance with predictable benevolence in another within a compatible worldview. Classical theories and studies on trust still bear the mark of behaviorism, with little regard for social interaction dynamics and the informational depth provided by subjective experience, particularly affect, and lacking any computational account of how these relate to brain function and cognition, due also to limitations in experimental methods and conceptual commitments at the time.

Our primary contribution may be to tie the group level reduction of complexity not only to extant concepts of trust at the individual level, mainly individual differences in trustfulness, or game theoretical conceptions of trust, but deeper, to fundamental theories about cognition, specifically predictive processing and active inference. In relating levels of analysis and inquiry that encompass a conceptual space bridging cortical hierarchies, upwards/outwards to hierarchies of individuals within groups and groups within civilization, we have a tentative framework with which to potentially examine trust at all these levels separately, in relation, and simultaneously. Thus, we provide a novel synthesis and a framework which is capable of empirical implementation and moreover carries with it novel domains in which to study the phenomenon of trust and its manifestations in daily life.

How does the novel synthesis presented in this framework help us? What does it add to our knowledge beyond description? How can we perhaps mitigate the maladaptive outcomes of the heuristics we use for the formation of trust using this model and understand them better? Broadly, our model provides a unified framework and common language with which to address a massive swath of human experience, with putative scale-free properties. More specifically, it may focus awareness on the extent to which the perception of similarity or dissimilarity among people can create coordination that extends far beyond superficial measures of worldview or declarations of goodwill, down to coordinations between somatomotor and affective states.

This highlights the tremendous importance of trust for maintaining and facilitating the wellbeing of society, and the tremendous risk that comes when, due to short-sighted ambitions, leaders, and institutions undergo actions that create distrust between the components of society (such as local population and immigrants), as this goes beyond damaging goodwill to possibly forming a material, concrete antecedent to schisms in relationships, groups, societies, and civilizations. Unveiling the mechanisms of trust formation and breakdown across various domains of human life, ranging from relationships and pedagogy to marketing and politics, may help facilitate group coordination and individual well-being, but also boost immunity against the misuses of trust. Taken together, we hope this work ultimately makes starkly apparent that trust's role and importance is difficult to overestimate.

From a broken leg and a kind stranger, we have attempted to take the reader inwards to the predictive and homeostatic processes underlying cognition and affect, and outwards, to the formation and maintenance of collective entities. In doing so, we hope we have conveyed that trust is a root fundament of the structure of human life in every form of interaction, from dyads to civilizations.

References

Adolphs, R., Tranel, D., & Damasio, A. R. (1998). The human amygdala in social judgment. *Nature, 393*(6684), 470–474.

Aimone, J. A., Houser, D., & Weber, B. (2014). Neural signatures of betrayal aversion: An fMRI study of trust. *Proceedings of the Royal Society B: Biological Sciences, 281*(1782), 20132127.

Bachmann, R. (2001). Trust, power and control in transorganizational relations. *Organization Studies, 22*(2), 337–365. https://doi.org/10.1177/0170840601222007

Baier, A. (1986). *'Trust and antitrust'. Ethics* (Vol. 96, pp. 231–260). Reprinted in: Moral Prejudices. Cambridge University Press.

Bellucci, G., Molter, F., Park, S. Q., (2019) Neural representations of honesty predict future trust behavior. *Nature Communications 10*(1), 5184 https://doi.org/10.1038/s41467-019-13261-8

Bamberger, W. (2010). *Interpersonal trust – Attempt of a definition.* Scientific report, Technische Universität München. Retrieved August 16, 2011.

Baron, S. G., Gobbini, M. I., Engell, A. D., & Todorov, A. (2011). Amygdala and dorsomedial prefrontal cortex responses to appearance-based and behavior-based person impressions. *Social Cognitive and Affective Neuroscience, 6*(5), 572–581.

Berger, P. L., & Luckmann, T. (1967). *The social construction of reality: A treatise in the sociology of knowledge.* Doubleday.

Bolis, D. (2020). *'I interact therefore I am': Human becoming in and through social interaction.* Doctoral dissertation. Max Planck Institute of Psychiatry and LMU.

Bolis, D., & Schilbach, L. (2017). Beyond one Bayesian brain: Modeling intra-and interpersonal processes during social interaction: Commentary on "Mentalizing homeostasis: The social origins of interoceptive inference" by Fotopoulou & Tsakiris. *Neuropsychoanalysis, 19*(1), 35–38.

Bolis, D., & Schilbach, L. (2018). Observing and participating in social interactions: Action perception and action control across the autistic spectrum. *Developmental Cognitive Neuroscience, 29*, 168–175. https://doi.org/10.1016/j.dcn.2017.01.009

Bolis, D., & Schilbach, L. (2020a). 'I interact therefore I am': The self as a historical product of dialectical attunement. *Topoi, 39*(3), 521–534. https://doi.org/10.1007/s11245-018-9574-0

Bolis, D., & Schilbach, L. (2020b). 'Through others we become ourselves': The dialectics of predictive coding and active inference. *Behavioral and Brain Sciences, 43*, e93.

Bolis, D., Balsters, J., Wenderoth, N., et al. (2017). Beyond autism: Introducing the dialectical misattunement hypothesis and a bayesian account of intersubjectivity. *Psychopathology.* https://doi.org/10.1159/000484353

Bolis, D., Lahnakoski, J., Seidel, D., Tamm, J., & Schilbach, L. (2020). Interpersonal similarity of autistic traits predicts friendship quality. *Social Cognitive and Affective Neuroscience.*

Braynov, S. (2002). Contracting with uncertain level of Trust. *Computational Intelligence, 4*, 501–514. https://doi.org/10.1111/1467-8640.00200

Chartrand, T. L., & Bargh, J. A. (1999). The chameleon effect: The perception–behavior link and social interaction. *Journal of Personality and Social Psychology 76*(6), 893–910. https://doi.org/10.1037/0022-3514.76.6.893

Clark, A. (2013). Whatever next? Predictive brains, situated agents, and the future of cognitive science. *Behavioral and Brain Sciences, 36*(03), 181–204.

Clark, A., & Chalmers, D. J. (1998). The extended mind. *Analysis, 58*(1), 7–19. https://doi.org/10.1093/analys/58.1.7. JSTOR 3328150.

Coan, J. A., & Sbarra D. A. (2015). Social Baseline Theory: the social regulation of risk and effort. Current Opinion in Psychology 187-91 S2352250X14000396 https://doi.org/10.1016/j.copsyc.2014.12.021

Cui, V., Vertinsky, I., Robinson, S., & Branzei, O. (2018). Trust in the workplace: The role of social interaction diversity in the community and in the workplace. *Business and Society, 57*(2), 378–412. https://doi.org/10.1177/0007650315611724

da Rosa Pulga, A. A., Basso, K., Viacava, K. R., Pacheco, N. A., Ladeira, W. J., & Dalla Corte, V. F. (2019). The link between social interactions and trust recovery in customer–business relationships. *Journal of Consumer Behaviour, 18*(6), 496–504. https://doi.org/10.1002/cb.1788

Damasio, A. (2018). *The strange order of things: Life, feeling, and the making of cultures.* Pantheon.

Damasio, A., & Carvalho, G. B. (2013). The nature of feelings: Evolutionary and neurobiological origins. *Nature Reviews Neuroscience, 14*(2), 143–152. https://doi.org/10.1038/nrn3403

De Jaegher, H., & Di Paolo, E. (2007). Participatory sense-making. *Phenomenology and the Cognitive Sciences, 6*(4), 485–507.

DeBruine, L. M. (2002). Facial resemblance enhances trust. *Proceedings of the Royal Society of London, Series B: Biological Sciences, 269*(1498), 1307–1312. PMC 1691034. 7 July 2002. https://doi.org/10.1098/rspb.2002.2034

DeBruine, L. M. (2005). Trustworthy but not lust-worthy: context-specific effects of facial resemblance. *Proceedings of the Royal Society B, 272*(1566), 919–922. https://doi.org/10.1098/rspb.2004.3003. JSTOR 30047623. PMC 1564091. 3 November 2005.

Decety, J., Jackson, P. L., Sommerville, J. A., Chaminade, T., & Meltzoff, A. N. (2004). The neural bases of cooperation and competition: An fMRI investigation. *NeuroImage, 23*(2), 744–751.

DeNeve, K. M. (1999). Happy as an extraverted clam? The role of personality for subjective well-being. *Current Directions in Psychological Science, 8*(5), 141–144. https://doi.org/10.1111/1467-8721.00033

DeNeve, K. M., & Cooper, H. (1998). The happy personality: A meta-analysis of 137 personality traits and subjective Well-being (PDF). *Psychological Bulletin, 124*(2), 197–229. https://doi.org/10.1037/0033-2909.124.2.197

Dikker, S., Wan, L., Davidesco, I., Kaggen, L., Oostrik, M., McClintock, J., Rowland, J., Michalareas, G., Van Bavel, J. J., Ding, M., & Poeppel, D. (2017). Brain-to-Brain Synchrony Tracks Real-World Dynamic Group Interactions in the Classroom. *Current Biology 27*(9) 1375–1380 S0960982217304116 https://doi.org/10.1016/j.cub.2017.04.002

Dumas, G. (2011). Towards a two-body neuroscience. *Communicative & Integrative Biology, 4*(3), 349–352.

Dumas, G., Kelso, J. A., & Nadel, J. (2014). Tackling the social cognition paradox through multiscale approaches. *Frontiers in Psychology, 5*, 882.

Dumas, G., Gozé, T., & Micoulaud-Franchi, J. A. (2020). "Social physiology" for psychiatric semiology: How TTOM can initiate an interactive turn for computational psychiatry?.

Fett, A. K. J., Gromann, P. M., Giampietro, V., Shergill, S. S., & Krabbendam, L. (2014). Default distrust? An fMRI investigation of the neural development of trust and cooperation. *Social Cognitive and Affective Neuroscience, 9*(4), 395–402.

Foddy, M., Platow, M. J., & Yamagishi, T. (2009). Group-based trust in strangers: The role of stereotypes and expectations. *Psychological Science, 20*(4), 419–422. https://doi. org/10.1111/j.1467-9280.2009.02312.x

Fotopoulou, A., & Tsakiris, M. (2017). Mentalizing homeostasis: The social origins of interoceptive inference. *Neuropsychoanalysis, 19*(1), 3–28.

Freeman, J. B., Stolier, R. M., Ingbretsen, Z. A., & Hehman, E. A. (2014). Amygdala responsivity to high-level social information from unseen faces. *Journal of Neuroscience, 34*(32), 10573–10581.

Friston K. (2009). The free-energy principle: a rough guide to the brain?. *Trends in Cognitive Sciences 13*(7) 293–301 S136466130900117X https://doi.org/10.1016/j.tics.2009.04.005

Friston K. (2013). Life as we know it. *Journal of The Royal Society Interface 10*(86) 20130475 https://doi.org/10.1098/rsif.2013.0475

Frith C. D., & Frith U. (2012). Mechanisms of Social Cognition. *Annual Review of Psychology 63*(1), 287–313 10.1146/annurev-psych-120710-100449

Froese, T., Iizuka, H., & Ikegami, T. (2015). Embodied social interaction constitutes social cognition in pairs of humans: A minimalist virtual reality experiment. *Scientific Reports, 4*, 3672. https://doi.org/10.1038/srep03672

Fukuyama, F. (1996). *Trust: The social virtues and the creation of prosperity.* Touchstone Books.

Gambetta, D. (2000). Can we trust trust? In D. Gambetta (Ed.), *Trust: Making and breaking cooperative relations, electronic edition, Department of Sociology* (pp. 213–237). University of Oxford, chapter 13.

Gerck, E. (1998a). Trust points, digital certificates: Applied internet security by J. Feghhi, J. Feghhi and P. Williams, Addison-Wesley, ISBN 0-201-30980-7.

Gerck, E. (1998b Jan 23). 'Definition of trust'. Mcwg.org. Retrieved January 4, 2013.

Gromann, P. M., Shergill, S. S., De Haan, L., Meewis, D. G. J., Fett, A. J., Korver-Nieberg, N., & Krabbendam, L. (2014). Reduced brain reward response during cooperation in first-degree relatives of patients with psychosis: An fMRI study. *Psychological Medicine, 44*(16), 3445.

Hendren, E. M., & Kumagai, A. K. (2019). A matter of trust. *Academic Medicine, 94*(9), 1270–1272.

Hein G., & Singer T. (2008). I feel how you feel but not always: the empathic brain and its modulation. *Current Opinion in Neurobiology 18*(2), 153–158 S0959438808000706 https://doi. org/10.1016/j.conb.2008.07.012

Insel, T. R. (2017). Digital phenotyping: Technology for a new science of behavior. *JAMA, 318*(13), 1215–1216.

Kesebir, S. (2012). The superorganism account of human sociality: How and when human groups are like beehives. *Personality and Social Psychology Review, 16*(3), 233–261. https://doi. org/10.1177/1088868311430834

Krueger, F., McCabe, K., Moll, J., Kriegeskorte, N., Zahn, R., Strenziok, M., Heinecke, A., & Grafman, J. (2007). Neural correlates of trust. *Proceedings of the National Academy of Sciences 104*(50), 20084–20089. https://doi.org/10.1073/pnas.0710103104

Koike, T., Tanabe, H. C., Okazaki, S., et al. (2016). Neural substrates of shared attention as social memory: A hyperscanning functional magnetic resonance imaging study. *NeuroImage, 125*, 401–412. https://doi.org/10.1016/j.neuroimage.2015.09.076

Koole, S. L., & Tschacher, W. (2016). Synchrony in psychotherapy: A review and an integrative framework for the therapeutic alliance. *Frontiers in Psychology, 7*, 862.

Lahnakoski J. M., Forbes, P. A.G., McCall, C., & Schilbach, L. (2020). Unobtrusive tracking of interpersonal orienting and distance predicts the subjective quality of social interactions. *Royal Society Open Science 7*(8) 191815 https://doi.org/10.1098/rsos.191815

Lamm C., Decety J., & Singer, T. (2011). Meta-analytic evidence for common and distinct neural networks associated with directly experienced pain and empathy for pain. *NeuroImage 54*(3), 2492–2502. S1053811910013066. https://doi.org/10.1016/j.neuroimage.2010.10.014

Lee, S.-J., (2007) The relations between the student–teacher trust relationship and school success in the case of Korean middle schools. *Educational Studies 33*(2) 209–216 8 10.1080/03055690601068477

Lewis, J. D., & Weigert, A. (1985). Trust as a social reality. *Social Forces, 63*(4), 967–985. https://doi.org/10.1093/sf/63.4.967

Luhmann, N. (1979). *Trust and power*. John Wiley & Sons.

Mayer, R. C., Davis, J. H., & Schoorman, F. D. (1995). An integrative model of organizational trust. *Academy of Management Review, 20*(3), 709–734. CiteSeerX 10.1.1.457.8429. https://doi.org/10.5465/amr.1995.9508080335

McKnight, D. H., & Chervany, N. L. (1996). The meanings of trust. Scientific report, University of Minnesota. Archived 2011-09-30 at the Wayback machine.

Mikulincer, M. (1998). Attachment working models and the sense of trust: An exploration of interaction goals and affect regulation. *Journal of Personality and Social Psychology, 74*(5), 1209.

Misztal, B. (1996). *Trust in Modern Societies: The search for the bases of social order*. Polity Press. ISBN 0-7456-1634-8.

Mirza, B. M., Adams R. A., Friston, K., & Parr T. (2019). Introducing a Bayesian model of selective attention based on active inference. *Scientific Reports 9*(1), 13915. https://doi.org/10.1038/s41598-019-50138-8

Montague, P. R., Berns, G. S., Cohen, J. D., et al. (2001). Hyperscanning: Simultaneous fMRI during linked social interactions. *NeuroImage*. https://doi.org/10.1006/nimg.2002.1150

Morgan, R., & Hunt, S. (1994). The commitment-trust theory of relationship marketing. *The Journal of Marketing, 58*(3), 20–38. https://doi.org/10.2307/1252308. JSTOR 1252308.

Nooteboom, B. (2017). *Trust: Forms, foundations, functions, failures and figures*. Edward Elgar Publishing. ISBN 9781781950883. Retrieved 29 October 2017 – via Google Books.

Parkinson, C., Kleinbaum, A. M., & Wheatley, T. (2018). Similar neural responses predict friendship. *Nature Communications, 9*, 332. https://doi.org/10.1038/s41467-017-02722-7

Platow, M. J., Foddy, M., Yamagishi, T., Lim, L., & Chow, A. (2012). Two experimental tests of trust in in-group strangers: The moderating role of common knowledge of group membership. *European Journal of Social Psychology, 42*, 30–35. https://doi.org/10.1002/ejsp.852

Ramstead, M. J. D., Benjamin, P., Badcock, J., & Friston, K. (2018). Answering Schrödinger's question: A free-energy formulation. *Physics of Life Reviews* 241–216. https://doi.org/10.1016/j.plrev.2017.09.001

Ramseyer, F., & Tschacher, W. (2011). Nonverbal synchrony in psychotherapy: Coordinated body movement reflects relationship quality and outcome. *Journal of Consulting and Clinical Psychology, 79*(3), 284.

Said, C. P., Baron, S. G., & Todorov, A. (2009). Nonlinear amygdala response to face trustworthiness: Contributions of high and low spatial frequency information. *Journal of Cognitive Neuroscience, 21*(3), 519–528.

Santos, S., Almeida, I., Oliveiros, B., & Castelo-Branco, M. (2016). The role of the amygdala in facial trustworthiness processing: A systematic review and meta-analyses of fMRI studies. *PLoS One, 11*(11), e0167276.

Scarry, E. (1985). *The body in pain: The making and unmaking of the world*. Oxford University Press.

Schilbach, L., Timmermans, B., Reddy, V., et al. (2013). Toward a second-person neuroscience. *The Behavioral and Brain Sciences, 36*, 393–414. https://doi.org/10.1017/S0140525X12000660

Searle, J. R. (1995). *The construction of social reality*. The Free Press.

Seth, A. K., Suzuki, K., & Critchley, H. D. (2012). An interoceptive predictive coding model of conscious presence. *Frontiers in Psychology, 2*, 395. https://doi.org/10.3389/fpsyg.2011.00395

Singh, T. B. (2012). A social interactions perspective on trust and its determinants. *Journal of Trust Research, 2*(2), 107–135. https://doi.org/10.1080/21515581.2012.708496

Terrell, F., & Terrell, S. (1984). Race of counselor, client sex, cultural mistrust level, and premature termination from counseling among Black clients. *Journal of Counseling Psychology, 31*(3), 371.

Theriault, J. E., Young L., & Barrett L. F. (2021). The sense of should: A biologically-based framework for modeling social pressure. *Physics of Life Reviews* 36 100-136 S157106452030004X https://doi.org/10.1016/j.plrev.2020.01.004

Thompson, V. L. S., Bazile, A., & Akbar, M. (2004). African Americans' perceptions of psycho-therapy and psychotherapists. *Professional Psychology: Research and Practice, 35*(1), 19.
Timmermans, B., Schilbach, L., Pasquali, A., & Cleeremans, A. (2012). Higher order thoughts in action: consciousness as an unconscious re-description process. *Philosophical Transactions of the Royal Society B: Biological Sciences 367*(1594), 1412–1423. https://doi.org/10.1098/rstb.2011.0421
Todorov, A., Baron, S. G., & Oosterhof, N. N. (2008). Evaluating face trustworthiness: A model based approach. *Social Cognitive and Affective Neuroscience, 3*(2), 119–127.
Tomasello, M. (2014). *A natural history of human thinking*. Harvard University Press.
Tschacher, W., Rees, G. M., & Ramseyer, F. (2014). Nonverbal synchrony and affect in dyadic interactions. *Frontiers in Psychology, 5*, 1323.
Vygotsky, L.S. (1978). Mind in society: the development of higher psychological processes. Harvard University Press. Original work 1930–1935. Translated in Greek, (1997) by A. Bibou and S Vosniadou, Gutenberg.
Watabe, M., Ban, H., & Yamamoto, H. (2011). Judgments about others' trustworthiness: An fMRI study. *Letters on Evolutionary Behavioral Science, 2*(2), 28–32.
Wilson, R. A., & Foglia, L. (2011). In E. N. Zalta (Ed.), *Embodied Cognition*. The Stanford Encyclopedia of Philosophy (Fall 2011 Edition). 25 July 2011.
Winston, J. S., Strange, B. A., O'Doherty, J., & Dolan, R. J. (2002). Automatic and intentional brain responses during evaluation of trustworthiness of faces. *Nature Neuroscience, 5*(3), 277–283.
Zak, P. J., & Knack, S. (2001). Trust and growth. *Economic Journal, 111*, 295–321.
Zheng, J., Roehrich, J. K., & Lewis, M. A. (2008). The dynamics of contractual and rela-tional governance: Evidence from long-term public-private procurement arrangements. *Journal of Purchasing and Supply Management, 14*(1), 43–54. https://doi.org/10.1016/j.pursup.2008.01.004

Part III
Clinical Neuroscience

Chapter 9
The Time Has Come to Be Mindwanderful: Mind Wandering and the Intuitive Psychology Mode

Óscar F. Gonçalves and Mariana Rachel Dias da Silva

Abstract No matter how hard you try—pinching different parts of your body, slapping your face, or moving restlessly in your seat—you cannot prevent your mind from occasionally escaping from the present experience as you enter into a mental navigation mode. Sometimes spontaneously, others deliberately, your mind may move to a different time—you may see yourself running an experiment inspired by the chapter you just finished reading or you may imagine yourself on a quantum leap into the future as you fantasize about the delivery of your Nobel Prize acceptance speech. Your mind may move to a distinct space, for example, as you replay last weekend's party or anticipate a most desirable date, and may even venture into the mind of another (e.g., as you embody the mind of the author you are currently reading). Our minds can accomplish all this mental navigation in fractions of a second, allowing us to see ourselves or even impersonate different people across space and time. While teleportation and time travel may never be physically possible, our wandering minds are indeed very accomplished "time machines" (Suddendorf T, Corballis MC, Behav Brain Sci 30(3), 2007).

Keywords Mind wandering · Perceptual decoupling · Mental improvisation · Mental navigation

Ó. F. Gonçalves (✉)
Proaction Lab, CINEICC – Faculty of Psychology and Educational Sciences, University of Coimbra, Coimbra, Portugal
e-mail: oscar@fpce.uc.pt

M. R. D. da Silva
Tilburg University Cognitive Science and Artificial Intelligence Department, Tilburg, The Netherlands

© The Author(s) 2023
P. S. Boggio et al. (eds.), *Social and Affective Neuroscience of Everyday Human Interaction*, https://doi.org/10.1007/978-3-031-08651-9_9

Introduction

No matter how hard you try—pinching different parts of your body, slapping your face, or moving restlessly in your seat—you cannot prevent your mind from occasionally escaping from the present experience as you enter into a mental navigation mode. Sometimes spontaneously, others deliberately, your mind may move to a different time—you may see yourself running an experiment inspired by the chapter you just finished reading or you may imagine yourself on a quantum leap into the future as you fantasize about the delivery of your Nobel Prize acceptance speech. Your mind may move to a distinct space, for example, as you replay last weekend's party or anticipate a most desirable date, and may even venture into the mind of another (e.g., as you embody the mind of the author you are currently reading). Our minds can accomplish all this mental navigation in fractions of a second, allowing us to see ourselves or even impersonate different people across space and time. While teleportation and time travel may never be physically possible, our wandering minds are indeed very accomplished "time machines" (Suddendorf & Corballis, 2007).

The concept of mind wandering is still very fuzzy and heterogeneous. As such, distinct authors seldom agree on a common definition (Christoff et al., 2016; Seli et al., 2018). Despite this lack of agreement, the adoption of a family resemblances view of mind wandering, which embraces the heterogeneity of the phenomenon, is key to further advancing the field. Here, I define *mind wandering* as the process by which the mind decenters from the current task and stimulus conditions (Stawarczyk et al., 2011a), moving freely (Christoff et al., 2016) toward multiple space, time, and/or mind positions (Corballis, 2013).

In what follows, and as summarized in Fig. 9.1, we will maintain that first this wandering process represents our mind/brain's default mode. Second, we describe three distinct but interrelated psychological mechanisms involved in mind wandering—perceptual decoupling, mental improvisation, and mental navigation. Third, we argue that mind wandering has the core function of priming our minds into a psychosocial mode (i.e., a folk/intuitive psychology). Finally, we conclude by suggesting that maybe the time has come to move beyond what Corballis (2015) refers to as the "bad press" that mind wandering has been facing and start acknowledging the benefits of mind wandering.

Minds Wandering by Default

Let us begin by substantiating the claim that mind wandering constitutes our mind's default mode. Recently, Killingsworth and Gilbert (2010) published in *Science* the results of a real-time large-scale thought sampling report. Thought probes were sent to participants randomly throughout the day by means of a smartphone application, requiring participants to report on the content and nature of their thoughts. An

Minds Wander by Default

Mind's Default Mode	Brain's Default Mode

Mind Wandering Processes

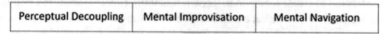

Perceptual Decoupling	Mental Improvisation	Mental Navigation

Mind Wandering Function

Priming Psychosocial Mode

Fig. 9.1 Nature and functions of mind wandering

analysis of responses from 2250 adults confirmed that, for about half of the day (i.e., 47%), individuals reported to be mind wandering (i.e., "thinking about something other than what they were currently doing"). Interestingly, mind wandering was transversal to most of their daily activities. In fact, the nature of people's activities explained no more than 3.5% of the between-person variance in mind wandering.

The ubiquity of mind wandering is even more impressive if we move beyond daily wakeful activity. When we add dreaming to the equation, the prevalence of mind wandering increases dramatically. As defended by Fox et al. (2013), dreams, particularly during REM (Rapid Eye Movement) sleep (Mutz & Javadi, 2017), may be considered an extreme form of mind wandering (Andrews-Hanna et al., 2018; Domhoff, 2018), sharing common audio-visual, fantasy, and spontaneous activity (Christoff et al., 2016). This is likely the reason why mind wandering is often taken as synonymous with "daydreaming" (Regis, 2013; Stawarczyk, 2018). As in REM dreaming, waking mind wandering entails a process of spontaneous activity eluding the frontiers of space and time. Curiously, reports of impersonation have been reported in dreams as well in dreaming phenomenology (Schredl, 2019).

By default, both day and night, our minds wander. It is now widely acknowledged that the brain remains highly active during states of mind wandering. Marcus E. Raichle et al. (2001) coined the term *Default Mode Network* (DMN) to refer to a network connecting the medial frontal cortex with the posterior cingulate, precuneus and inferior parietal cortex, shown to be particularly active when individuals are not requested to perform a specific task in an fMRI (functional magnetic resonance imaging) environment (Raichle, 2010). In such task-negative (default) states, our brains sustain high levels of activity. Metabolically speaking, our brain is a very expensive organ. It spends about 10 times more energy than what would be expected from its volume and mass, such that the majority of its metabolism is associated with "off-task" and "stimulus independent activity" (70–80%). Conversely, it is estimated that "task-evoked activity" accounts for no more than 5% of the brain's total energy consumption (Raichle, 2010).

There is now abundant evidence that our mind's default mode (i.e., mind wandering) is supported by the brain's DMN (Christoff et al., 2009; Kirschner et al., 2012; Mason et al., 2007). To illustrate, a recent study by Scheibner et al. (2017) confirms such evidence of DMN activity during mind wandering. In their study, participants were instructed to either focus on their own breathing (internal attention condition) or on tones (external attention condition) while fixating on a white cross. Once the cross turned red, participants were requested to report if they were either focused or mind wandering. Core regions of the DMN (medial prefrontal cortex, posterior cingulate cortex, and left temporoparietal junction) were significantly more active during instances of mind wandering than when participants reported being focused (either externally or internally). Relatedly, Stawarczyk et al. (2011b) also found that, when contrasted with being on-task, mind wandering was associated with clusters of increased activity in core DMN nodes (e.g., medial prefrontal, posterior cingulate, inferior parietal lobe). However, in addition, this activity was also evident in extended nodes of the DMN (e.g., parahippocampal cortex; inferior and medial temporal gyrus), indicating that core regions of the DMN interact with subnetworks, including the medial temporal lobe subsystem and the dorsal medial subsystem. Meta-analyses using Neurosynth (http://neurosynth.org) indicate that, while the core DMN nodes are engaged in self-referential processes, the medial temporal and dorsal medial subsystems are engaged in episodic memory and social cognition, respectively (Andrews-Hanna et al., 2014). As such, different DMN subsystems seem to be supporting mental processes that are prevalent during mind wandering (i.e., self-referential, episodic, and social cognitive processes). A recent study by Poerio et al. (2017) confirmed that the connectivity between and within different DMN subsystems supports the multicomponent nature of mind wandering, particularly with regard to perceptual decoupling and memory retrieval. Importantly, nodes of this DMN are often anti-correlated with nodes active during tasks requiring focused attention (Fox et al., 2005). However, we also note that executive networks are also known to be a neural correlate of off-task thinking (Dixon et al., 2017), including mind wandering (Christoff et al., 2009; Kam & Handy, 2014). Activity in these networks may seem counterintuitive, considering their recruitment during task-positive, goal-directed thought. However, recent studies indicate that executive networks also serve to regulate attention back and forth between the external environment and internal thoughts and are similarly recruited during mind wandering in order to sustain an internal train of thought (Christoff et al., 2016). Also an indirect confirmation of the role played by the DMN in mind wandering is the research confirming that extended regions of the DMN are involved during REM dreaming (Fox et al., 2013; Sämann et al., 2011). In sum, mind wandering constitutes the mind as well as the brain's default mode. Different DMN subsystems seem to cooperate, allowing the mind to perceptually decouple and to venture into a mode of mental improvisation and mental navigation.

Mind Wandering Processes

As illustrated in Fig. 9.2, mind wandering depends on three interconnected processes: perceptual decoupling, mental improvisation, and mental navigation. The process can be triggered either by bottom-up (e.g., perceptual fatigue–Boksem et al., 2006) or top-down mechanisms (e.g., memory retrieval–Baird et al., 2011).

In our lab, we are currently studying the contribution of these three different mechanisms to the mind wandering process. We administered a large-scale questionnaire study in which we asked participants to answer questions concerning individuals' tendency to disengage from the environment as they mind wander, concerning the dynamics and variability of mind wandering thoughts, and concerning the general tendency to mind wander across space and time. Specifically, items from our *Mind Wandering Inventory* (Gonçalves et al., 2020) were intended to capture the following dimensions:

1. Perceptual decoupling (e.g., My mind often disconnects from what surrounds me)
2. Mental improvisation (e.g., My thoughts jump easily from one subject to another)
3. Mental navigation across time (e.g., My thoughts travel frequently through time–past or future), space (e.g., I often imagine that I'm somewhere else), and minds (e.g., I often imagine what others are thinking or feeling)

Moreover, we examined the relationship between trait levels of mind wandering assessed with the Mind Wandering Inventory and state mind wandering probed during a vigilance task (Dias da Silva et al., 2020) and have validated the questionnaire with neurophysiological electroencephalogram (EEG) data (Dias Da Silva, Gonçalves, & Postma, 2022).

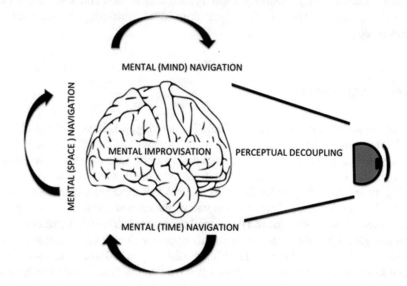

Fig. 9.2 Mind wandering processes

Perceptual Decoupling

Mind wandering entails at least some degree of perceptual decoupling. While decoupling from the immediate perceptual experience, the individual switches to an internal processing mode. This is illustrated by Smilek et al. (2010), who found that during a reading task, individuals tend to blink significantly more when reporting to be mind wandering. In addition, they found that during mind wandering periods, there were a smaller number of ocular fixations, suggesting that direct eye avoidance is associated with the elimination of external stimulation sources and priming of internal processing. Moreover, Bristow et al. (2005) demonstrate that eye blinking is associated with the deactivation of a fronto-parietal network responsible for visual attention. Together, these findings indicate that perceptual decoupling from the immediate experience represents an important component of mind wandering.

Also supporting the evidence for perceptual decoupling are studies which show that the amplitude of early (P1 and N1) and late (P3) perceptual evoked potentials are attenuated during mind wandering. Since both the P1 and N1 are early ERP components indexing processing during the sensory input stage, their reduction is taken as evidence for an inhibitory effect of mind wandering on external processing (Kam & Handy, 2013; Schooler et al., 2011). However, some recent findings show that, for low demanding attention tasks, individuals are able to maintain appropriate levels of alert, orienting and executive attention during mind wandering (Gonçalves et al., 2017a, b) without impacting early and late perceptual evoked potentials (Gonçalves et al., 2018b). As such, the effects of perceptual decoupling are dictated by task demands. These findings indicate that perceptual decoupling is an important, but not an absolute, condition for mind wandering. Individuals with higher executive resources (e.g., working memory) are able to maintain some degree of external processing while at the same time mind wandering (Smallwood & Schooler, 2015).

Mental Improvisation

In order to understand the function of mind wandering, it is important not only to look at the content but also to characterize the dynamics of thought. By investigating mind wandering over time, we can see that thoughts tend to evolve freely from one topic to the next, sometimes coming back to a core theme. For example, while writing an article, it may suddenly come to mind that there are a couple of emails that need to be answered. That reminds you of the current status of the computer you ordered a few weeks ago. You then recall the conversation with the salesperson. Then you get back to the emails and think about the email with the invitation to visit a foreign lab. You remember your last visit, the dinner you had with your friends and all the fun you had. This memory brings you back to the thought that you have to respond right away to that email. In sum, mind wandering dynamics seems to entail a process of free, but not necessarily random, thought movement. This dynamic of free movement is responsible for the heightened variability of thought content (Mills et al., 2018). Mills et al. (2018) coined the "default variability hypothesis" to refer to the process by which the dynamics of free movement between thoughts favors the encoding of separate memory episodes and the consolidation of episodic memories into semantic knowledge. It is precisely this variability that distinguishes mind wandering from ruminative thoughts (i.e., the persistence of sticky and recurrent thoughts). In fact, there is evidence that these ruminative thoughts are associated with task-related interference (Dias da Silva et al., 2018). The misclassification of rumination as mind wandering has been in large part responsible for the widespread misattribution of negative costs in terms of attention (Hu et al., 2012), executive functioning (McVay et al., 2013) and mood (Wilson et al., 2014) to mind wandering. In contrast, mind wandering defined as a process of mental improvisation has consistent benefits in terms of creativity (Baird et al., 2011), memory consolidation (Mills et al., 2018), and mental simulation (O'Callaghan et al., 2015).

Recently Marron et al. (2018) found that the degree of mental improvisation, as expressed in terms of free association fluency, flexibility, and semantic remoteness during a free association task, is related to an increase in activation of the DMN and a decrease in activation of core nodes from the executive network (e.g., Inferior Frontal Gyrus). Notably, these free association markers were positively correlated with creativity measures but not with intelligence scores. As in theater or music improvisation, our thoughts seem to have a mind of their own, moving freely but often recurring around a specific theme before departing into a new associative dynamic. Curiously, studies on music improvisation consistently report a deactivation in brain regions responsible for executive functioning (i.e., dorsolateral prefrontal cortex) and the concurrent activation of core nodes of the DMN (e.g., anterior cingulate; Beaty, 2015; Landau & Limb, 2017).

Mental Navigation

Mind wandering is also characterized by a process of mental navigation. While time travel is most often acknowledged in mind wandering, a mental navigation mode is better illustrated by the existence of a triple de-centration: time de-centration, space de-centration, and mind de-centration. Next, we will briefly address each of these mental navigation components.

Several studies have shown that mind wandering entails a time-travel process. Mind wandering has a remarkable temporal orientation (90%), allowing the individual to navigate between the past (~29%), present (~12%) and, above all, the future (~48%, Smallwood & Schooler, 2015). Notably, about half of the time, the mind wanders to some time in the future. This suggests that, along with an eventual consolidation of past memories, mind wandering frees the individual from the here and now, simulating the future and potentiating autobiographical planning (Stawarczyk et al., 2013).

This time travel is undissociated from the correlative process of space navigation—mentally moving from the "here and now" to "there and then". Similarly to
what happens with episodic memory, the mental evocation of past or foreseen contexts is hippocampal dependent (Tulving, 2002). A recent study by McCormick
et al. (2018) found that patients with hippocampal amnesia were no different than
healthy controls in reporting high levels of mind wandering during quiet restful
moments. However, their instances of mind wandering were found to be mostly
dependent on semantic knowledge (i.e., closer to ruminative thoughts than mental
improvisation), in contrast with the episodic content more typically found for
healthy controls. In support of this space navigation process, healthy controls
reported having mind wandering thoughts of an intense sensorial quality, particularly concerning the experience of visual scenes.

However, mind wandering is not an uninhabited scenario. That is, during mind
wandering, time and space de-centration goes often together with mind navigation.
This third type of mind de-centration is characterized by the ability to be able to
tune in with others' experiences and move into their minds and imagine how they
are thinking, feeling, or behaving. Although there are some cultural and individual
differences, individuals often adopt (~50%) a third-person perspective when mind
wandering; that is, they see the world from the viewpoint of an outside observer
(Christian et al., 2013). Building on evidence from this third-person perspective
along with findings indicating the enrolment of core DMN nodes responsible for
social cognitive processes (Davey et al., 2016; Li et al., 2014; Mars et al., 2012;
Poerio & Smallwood, 2016), we can assert along with Corballis (2015) that mind
wandering may be central for developing a theory of mind (i.e., the ability to identify or attribute mental states in ourselves and others).

Mind Wandering Function: Priming the Psychosocial Mode

We will now be maintaining that the processes of perceptual decoupling and mental
improvisation, with a triple—time, space, mind—de-centration, promote a reorien-
tation of the mind from the current physical reality (i.e., intuitive/folk physics) to a
predominantly psychosocial mode (i.e., intuitive/folk psychology). While the
understanding of the physical reality predominantly requires systematic thinking,
the comprehension of psychosocial phenomena relies mostly on reflective, creative,
and empathic processes. As a result of daily dealing with physical and psychosocial
phenomena, individuals develop a sort of intuitive physics (i.e., folk physics) and
intuitive psychology (i.e., folk psychology). On the one hand, this intuitive physics
translates the nature and degree of individual understanding of physical phenomena
into an individual theory of the world. On the other hand, this intuitive psychology
translates our personal understanding of the psychosocial reality into an individual
theory of the mind (Baron-Cohen, 1997; Kamps et al., 2017).

As stated before, the DMN supports core socio-cognitive processes involved in
developing an individual theory of the mind. The DMN, as a network supporting

mind wandering, is also central in our orientation to the psychosocial domain. Curiously, the DMN is anti-correlated with the fronto-parietal (attention/executive) network predominantly involved in processing physical phenomena. This is illustrated in an interesting study by Jack et al. (2013). In their experiment, participants were presented with several problem-solving vignettes, some portraying tasks requiring reasoning about mental states and others requiring reasoning about causal/mechanical issues. The results indicate that not only the DMN was associated with the psychosocial domain and the fronto-parietal network with the physical domain but also that the activity of these regions was reciprocally inhibited.

This type of folk psychology (versus folk physics) orientation sustained by activating or deactivating the mind's default mode is also illustrated in Simon Baron-Cohen's *Systemizing–Empathizing Theory* (Greenberg et al., 2018). According to his view, people can be allocated to a dispositional continuum ranging from empathizing (i.e., drive to identify another person's emotions and thoughts and to respond to these with an appropriate emotion) to systemizing (i.e., drive to analyze, understand, predict, control, and construct rule-based systems).

Confirming the relationship between an intuitive psychology/DMN association, Takeuchi et al. (2014) demonstrate that empathizing is positively correlated with resting state functional connectivity between different DMN nodes, particularly, the medial prefrontal cortex, the dorsal anterior cingulate, the precuneus, and the left superior temporal sulcus. In contrast, systemizing positively correlates with resting state functional connectivity in an "external attention network" between the dorsolateral prefrontal cortex and the dorsal anterior cingulate cortex.

Indeed, the DMN is active during a range of tasks related to both mind wandering (Christoff et al., 2009; Kirschner et al., 2012; Mason et al., 2007; Scheibner et al., 2017) and theory of mind (Jack et al., 2013; Oliveira Silva et al., 2018; Takeuchi et al., 2014). However, DMN activations alone do not guarantee that both are equivalent. Moreover, correlations do not imply causation. Nevertheless, correlations do provide ground for future research for investigating the manner in which mind wandering states may support such a theory of the mind. As recently shown in a series of studies, it seems that we are by default in some sort of empathizing mode (Oliveira Silva et al., 2018), with DMN activity and connectivity being central to maintaining this state of mind (Esménio et al., 2019a, b). Although research is still underway, it may very well be the case that our default mind wandering state significantly contributes to prime this similarly default empathizing/psychosocial mode, helping individuals navigate the socio-emotional world around them (Poerio & Smallwood, 2016).

Concluding Remarks: A Time to Be *Mindwanderful*

During the last decades, the concept of mindfulness has witnessed a growing popularity (Tomlinson et al., 2018). Even though different definitions are available for mindfulness (Allen et al., 2012; Kabat-Zinn, 2003; Keng et al., 2011; Moore &

Malinowski, 2009), most of the researchers see it as a process of directing the attentional focus to the individual's current experience in the present moment while avoiding thought escape into the past/future. As such, a mindful mind seems to be the opposite of a wandering mind (Schooler et al., 2014). For example, Mrazek et al. (2012) demonstrated that people with high levels of mindfulness report fewer instances of mind wandering and perform better on an attention focus task (i.e., mindful breathing).

Despite some controversy regarding conceptual and methodological aspects in mindfulness research (Van Dam et al., 2018), there is evidence for the benefits of mindfulness in terms of orientating attention to the current physical reality (Posner et al., 2015). In contrast to orienting attention to a physical reality, mind wandering optimizes an orientation to the psychosocial domain. In addition to being aware of the present moment, mindfulness can also refer to the act of being aware of one's own internal thoughts and not just stimuli in the external environment (Ellamil et al., 2016). As such, it could also be that mind wandering and mindfulness represent two ends of the same construct. Therefore, it is necessary to find an ideal balance between attention to the external world and our internal thoughts, while also mindfully being aware of the wandering mind and the benefits that might come with it. Now the question remains: How can individuals take full advantage of the benefits of mind wandering in order to facilitate navigation in the psychosocial domain?

Several studies are currently underway, testing whether we can impact mind wandering by modulating specific neural correlates. For example, building on EEG markers of mind wandering instances, we recently launched a series of studies that explore the viability of different real-time EEG protocols (e.g., SMR⇑Theta⇓; Theta⇑SMR⇓) in improving mind wandering during an attention task (Gonçalves et al., 2018a, b). Other authors are using different strategies to neuromodulate processes associated with mind wandering by using transcranial direct current simulation (Axelrod et al., 2015; Axelrod et al., 2018; Boayue et al., 2020), or even real time MEG and fMRI (Garrison et al., 2013). Although these studies are in their beginning stages, as we evolve in the identification of reliable brain predictors of mind wandering, we hope to come up with more reliable methods for detecting and impacting mind wandering (Hosseini & Guo, 2019; Jin et al., 2019).

The ideal balance between mindfulness and mind wandering is still not known (Schooler et al., 2014). However, it seems that in order to facilitate both navigation across the physical and psychosocial domains, individuals may gain an advantage by adopting a *mindwanderfulness* position—a process of strategically switching between mindfulness and mind wandering, in order to respond adaptively to the demands of physical and psychosocial domains (Gonçalves, 2019; Hasenkamp, 2018).

References

Allen, M., Dietz, M., Blair, K. S., van Beek, M., Rees, G., Vestergaard-Poulsen, P., ... Roepstorff, A. (2012). Cognitive-affective neural plasticity following active-controlled mindfulness intervention. *Journal of Neuroscience, 32*(44), 15601–15610. https://doi.org/10.1523/JNEUROSCI.2957-12.2012

Andrews-Hanna, J. R., Smallwood, J., & Spreng, R. N. (2014). The default network and self-generated thought: Component processes, dynamic control, and clinical relevance. *Annals of the New York Academy of Sciences, 1316*(1), 29–52. https://doi.org/10.1111/nyas.12360

Andrews-Hanna, J. R., Irving, Z. C., Fox, K. C. R., Spreng, R. N., & Christoff, K. (2018). The neuroscience of spontaneous thought: An evolving interdisciplinary field. In *The Oxford handbook of spontaneous thought: Mind-wandering, creativity, and dreaming*. https://doi.org/10.1093/oxfordhb/9780190464745.013.33

Axelrod, V., Rees, G., Lavidor, M., & Bar, M. (2015). Increasing propensity to mind- wander with transcranial direct current stimulation. *Proceedings of the National Academy of Sciences of the United States of America, 112*(11), 3314–3319. https://doi.org/10.1073/pnas.1421435112

Axelrod, V., Zhu, X., & Qiu, J. (2018). Transcranial stimulation of the frontal lobes increases propensity of mind-wandering without changing meta-awareness. *Scientific Reports, 8*(1), 1–14. https://doi.org/10.1038/s41598-018-34098-z

Baird, B., Smallwood, J., & Schooler, J. W. (2011). Back to the future: Autobiographical planning and the functionality of mind-wandering. *Consciousness and Cognition, 20*(4), 1604–1611. https://doi.org/10.1016/j.concog.2011.08.007

Baron-Cohen, S. (1997). Are children with autism superior at folk physics? *New Directions for Child Development, 1997*(75), 45–54. https://doi.org/10.1002/cd.23219977504

Beaty, R. E. (2015). The neuroscience of musical improvisation. *Neuroscience and Biobehavioral Reviews, 51*, 108–117. https://doi.org/10.1016/j.neubiorev.2015.01.004

Boayue, N. M., Csifcsák, G., Aslaksen, P., Turi, Z., Antal, A., Groot, J., ... Mittner, M. (2020). Increasing propensity to mind-wander by transcranial direct current stimulation? A registered report. *European Journal of Neuroscience, 51*(3), 755–780. https://doi.org/10.1111/ejn.14347

Boksem, M. A. S., Meijman, T. F., & Lorist, M. M. (2006). Mental fatigue, motivation and action monitoring. *Biological Psychology, 72*(2), 123–132. https://doi.org/10.1016/j.biopsycho.2005.08.007

Bristow, D., Haynes, J. D., Sylvester, R., Frith, C. D., & Rees, G. (2005). Blinking suppresses the neural response to unchanging retinal stimulation. *Current Biology, 15*(14), 1296–1300. https://doi.org/10.1016/j.cub.2005.06.025

Christian, B. M., Miles, L. K., Parkinson, C., & Macrae, C. N. (2013). Visual perspective and the characteristics of mind wandering. *Frontiers in Psychology, 4*, 699. https://doi.org/10.3389/fpsyg.2013.00699

Christoff, K., Gordon, A. M., Smallwood, J., Smith, R., & Schooler, J. W. (2009). Experience sampling during fMRI reveals default network and executive system contributions to mind wandering. *Proceedings of the National Academy of Sciences of the United States of America, 106*(21), 8719–8724. https://doi.org/10.1073/pnas.0900234106

Christoff, K., Irving, Z. C., Fox, K. C. R., Spreng, R. N., & Andrews-Hanna, J. R. (2016). Mind-wandering as spontaneous thought: A dynamic framework. *Nature Reviews Neuroscience, 17*(11), 718–731. https://doi.org/10.1038/nrn.2016.113

Corballis, M. C. (2013). Mental time travel: A case for evolutionary continuity. *Trends in Cognitive Sciences, 17*(1), 5–6. https://doi.org/10.1016/j.tics.2012.10.009

Corballis, M. C. (2015). *The wandering mind: What the brain does when you're not looking*. University of Chicago Press.

Davey, C. G., Pujol, J., & Harrison, B. J. (2016). Mapping the self in the brain's default mode network. *NeuroImage, 132*, 390–397. https://doi.org/10.1016/j.neuroimage.2016.02.022

Dias da Silva, M. R., Rusz, D., & Postma-Nilsenová, M. (2018). Ruminative minds, wandering minds: Effects of rumination and mind wandering on lexical associations, pitch imitation and eye behaviour. *PLoS One, 13*(11), 1–20. https://doi.org/10.1371/journal.pone.0207578

Dias da Silva, M. R., Gonçalves, Ó. F., & Postma, M. (2020). Assessing the relationship between trait and state levels of mind wandering during a tracing task. In *Annual meeting of the Cognitive Science Society 2020: Developing a mind: Learning in humans, animals, and machines. Toronto, Canada.*

Dias da Silva, M., Gonçalves, Ó. F., & Postma, M. (2022). Revisiting consciousness: Distinguishing between states of conscious focused attention and mind wandering with EEG. *Consciousness & Cognition, 101*, 1–18. https://doi.org/10.1016/j.concog.2022.103332

Dixon, M.L., Girn, M., Christoff, K. (2017). Hierarchical Organization of Frontoparietal Control Networks Underlying Goal-Directed Behavior. In: Watanabe, M. (eds) The Prefrontal Cortex as an Executive, Emotional, and Social Brain. Springer, Tokyo. https://doi.org/10.1007/978-4-431-56508-6_7

Domhoff, G. W. (2018). Dreaming is an intensified form of mind-wandering, based in an augmented portion of the default network. In K. C. R. Fox & K. Christoff (Eds.), *The Oxford handbook of spontaneous thought: Mind-wandering, creativity, and dreaming* (pp. 355–370). https://doi.org/10.1093/oxfordhb/9780190464745.013.7

Ellamil, M., Fox, K. C. R., Dixon, M. L., Pritchard, S., Todd, R. M., Thompson, E., & Christoff, K. (2016). Dynamics of neural recruitment surrounding the spontaneous arising of thoughts in experienced mindfulness practitioners. *NeuroImage, 136*, 186–196. https://doi.org/10.1016/j.neuroimage.2016.04.034

Esménio, S., Soares, J. M., Oliveira-Silva, P., Gonçalves, Ó. F., Decety, J., & Coutinho, J. (2019a). Brain circuits involved in understanding our own and other's internal states in the context of romantic relationships. *Social Neuroscience, 14*(6), 729–738. https://doi.org/10.1080/1747091 9.2019.1586758

Esménio, S., Soares, J. M., Oliveira-Silva, P., Zeidman, P., Razi, A., Gonçalves, Ó. F., … Coutinho, J. (2019b). Using resting-state DMN effective connectivity to characterize the neurofunctional architecture of empathy. *Scientific Reports, 9*(1), 1–9. https://doi.org/10.1038/s41598-019-38801-6

Fox, M. D., Snyder, A. Z., Vincent, J. L., Corbetta, M., Van Essen, D. C., & Raichle, M. E. (2005). The human brain is intrinsically organized into dynamic, anticorrelated functional networks. *Proceedings of the National Academy of Sciences of the United States of America, 102*(27), 9673–9678. https://doi.org/10.1073/pnas.0504136102

Fox, K. C. R., Nijeboer, S., Solomonova, E., Domhoff, G. W., & Christoff, K. (2013). Dreaming as mind wandering: Evidence from functional neuroimaging and first-person content reports. *Frontiers in Human Neuroscience, 7*, 1–18. https://doi.org/10.3389/fnhum.2013.00412

Garrison, K. A., Scheinost, D., Worhunsky, P. D., Elwafi, H. M., Thornhill, T. A., Thompson, E., … Brewer, J. A. (2013). Real-time fMRI links subjective experience with brain activity during focused attention. *NeuroImage, 81,* 110–118. https://doi.org/10.1016/j.neuroimage.2013.05.030

Gonçalves, Ó. F. (2019). *Cérebro Errante.* Pearson Clinical.

Gonçalves, Ó. F., Oliveira-Silva, P., de Souza-Queiroz, J., Amaro, E., Rêgo, G., Leite, J., … Boggio, P. S. (2017a). Is the relationship between mind wandering and attention culture-specific? *Psychology and Neuroscience, 10*(2), 132–143. https://doi.org/10.1037/pne0000083

Gonçalves, Ó. F., Rêgo, G., Oliveira-Silva, P., Leite, J., Carvalho, S., Fregni, F., … Boggio, P. S. (2017b). Mind wandering and the attention network system. *Acta Psychologica, 172,* 49–54. https://doi.org/10.1016/j.actpsy.2016.11.008

Gonçalves, Ó. F., Carvalho, S., Mendes, A. J., Leite, J., & Boggio, P. S. (2018a). Neuromodulating attention and mind-wandering processes with a single session real time EEG. *Applied Psychophysiology and Biofeedback, 43*(2), 143–151. https://doi.org/10.1007/s10484-018-9394-4

Gonçalves, Ó. F., Carvalho, S., Mendes, A. J., Lema, A., Leite, J., & Boggio, P. S. (2018b). Neuromodulating attention and mind-wandering processes with multi- session real-time electroencephalogram. *Porto Biomedical Journal, 3*(2), e17. https://doi.org/10.1016/j.pbj.0000000000000017

Gonçalves, Ó. F., da Silva, M. R. D., Carvalho, S., Coelho, P., Lema, A., Mendes, A. J., ... Leite, J. (2020). Mind wandering: Tracking perceptual decoupling, mental improvisation, and mental navigation. *Psychology and Neuroscience, 13*(4), 493–502. https://doi.org/10.1037/pne0000237

Greenberg, D. M., Warrier, V., Allison, C., & Baron-Cohen, S. (2018). Testing the empathizing–systemizing theory of sex differences and the extreme male brain theory of autism in half a million people. *Proceedings of the National Academy of Sciences of the United States of America, 115*(48), 12152–12157. https://doi.org/10.1073/pnas.1811032115

Hasenkamp, W. (2018). Catching the wandering mind: Meditation as a window into spontaneous thought. In K. C. R. Fox & K. Christoff (Eds.), *The Oxford handbook of spontaneous thought: Mind-wandering, creativity, and dreaming.* https://doi.org/10.1093/oxfordhb/9780190464745.013.12

Hosseini, S., & Guo, X. (2019). Deep convolutional neural network for automated detection of mind wandering using EEG signals. In *ACM-BCB 2019 – Proceedings of the 10th ACM international conference on bioinformatics, computational biology and health informatics* (pp. 314–319). https://doi.org/10.1145/3307339.3342176

Hu, N., He, S., & Xu, B. (2012). Different efficiencies of attentional orienting in different wandering minds. *Consciousness and Cognition, 21*(1), 139–148. https://doi.org/10.1016/j.concog.2011.12.007

Jack, A. I., Dawson, A. J., Begany, K. L., Leckie, R. L., Barry, K. P., Ciccia, A. H., & Snyder, A. Z. (2013). FMRI reveals reciprocal inhibition between social and physical cognitive domains. *NeuroImage, 66,* 385–401. https://doi.org/10.1016/j.neuroimage.2012.10.061

Jin, C. Y., Borst, J. P., & van Vugt, M. K. (2019). Predicting task-general mind- wandering with EEG. *Cognitive, Affective, & Behavioral Neuroscience, 19*(4), 1059–1073. https://doi.org/10.3758/s13415-019-00707-1

Kabat-Zinn, J. (2003). Mindfulness-based interventions in context: Past, present, and future. *Clinical Psychology: Science and Practice, 10,* 144–156. https://doi.org/10.1093/clipsy/bpg016

Kam, J. W. Y., & Handy, T. C. (2013). The neurocognitive consequences of the wandering mind: A mechanistic account of sensory-motor decoupling. *Frontiers in Psychology, 4,* 1–13. https://doi.org/10.3389/fpsyg.2013.00725

Kam, J. W. Y., & Handy, T. C. (2014). Differential recruitment of executive resources during mind wandering. *Consciousness and Cognition, 26*(1). https://doi.org/10.1016/j.concog.2014.03.002

Kamps, F. S., Julian, J. B., Battaglia, P., Landau, B., Kanwisher, N., & Dilks, D. D. (2017). Dissociating intuitive physics from intuitive psychology: Evidence from Williams syndrome, *168,* 146–153. *Cognition.* https://doi.org/10.1016/j.cognition.2017.06.027

Keng, S. L., Smoski, M. J., & Robins, C. J. (2011). Effects of mindfulness on psychological health: A review of empirical studies. *Clinical Psychology Review, 31*(6), 1041–1056. https://doi.org/10.1016/j.cpr.2011.04.006

Killingsworth, M. A., & Gilbert, D. T. (2010). A wandering mind is an unhappy mind. *Science, 330*(606), 932. https://doi.org/10.1126/science.1192439

Kirschner, A., Kam, J. W. Y., Handy, T. C., & Ward, L. M. (2012). Differential synchronization in default and task-specific networks of the human brain. *Frontiers in Human Neuroscience, 6,* 1–10. https://doi.org/10.3389/fnhum.2012.00139

Landau, A. T., & Limb, C. J. (2017). The neuroscience of improvisation. *Music Educators Journal, 103*(3), 27–33. https://doi.org/10.1177/0027432116687373

Li, W., Mai, X., & Liu, C. (2014). The default mode network and social understanding of others: What do brain connectivity studies tell us. *Frontiers in Human Neuroscience, 74,* 1–15. https://doi.org/10.3389/fnhum.2014.00074

Marron, T. R., Lerner, Y., Berant, E., Kinreich, S., Shapira-Lichter, I., Hendler, T., & Faust, M. (2018). Chain free association, creativity, and the default mode network. *Neuropsychologia, 118,* 40–58. https://doi.org/10.1016/j.neuropsychologia.2018.03.018

Mars, R. B., Neubert, F. X., Noonan, M. A. P., Sallet, J., Toni, I., & Rushworth, M. F. S. (2012). On the relationship between the "default mode network" and the "social brain". *Frontiers in Human Neuroscience, 6,* 1–9. https://doi.org/10.3389/fnhum.2012.00189

Mason, M. F., Norton, M. I., Van Horn, J. D., Wegner, D. M., Grafton, S. T., & Macrae, C. N. (2007). Wandering minds: The default network and stimulus- independent thought. *Science, 315*(5810), 393–395. https://doi.org/10.1126/science.1131295

McCormick, C., Rosenthal, C. R., Miller, T. D., & Maguire, E. A. (2018). Mind-wandering in people with hippocampal damage. *Journal of Neuroscience, 38*(11), 2745–2754. https://doi.org/10.1523/JNEUROSCI.1812-17.2018

McVay, J. C., Meier, M. E., Touron, D. R., & Kane, M. J. (2013). Aging ebbs the flow of thought: Adult age differences in mind wandering, executive control, and self-evaluation. *Acta Psychologica, 142*(1), 136–147. https://doi.org/10.1016/j.actpsy.2012.11.006

Mills, C., Herrera-Bennett, A., Faber, M., & Christoff, K. (2018). Why the mind wanders: How spontaneous thought's default variability may support episodic efficiency and semantic optimization. In K. C. R. Fox & K. Christoff (Eds.), *The Oxford handbook of spontaneous thought: Mind-wandering, creativity, and dreaming.* Oxford University Press.

Moore, A., & Malinowski, P. (2009). Meditation, mindfulness and cognitive flexibility. *Consciousness and Cognition, 18*(1), 176–186. https://doi.org/10.1016/j.concog.2008.12.008

Mrazek, M. D., Smallwood, J., & Schooler, J. W. (2012). Mindfulness and mind- wandering: Finding convergence through opposing constructs. *Emotion, 12*(3), 442–448. https://doi.org/10.1037/a0026678

Mutz, J., & Javadi, A.-H. (2017). Exploring the neural correlates of dream phenomenology and altered states of consciousness during sleep. *Neuroscience of Consciousness, 2017*(1), 1–12. https://doi.org/10.1093/nc/nix009

O'Callaghan, C., Shine, J. M., Lewis, S. J. G., Andrews-Hanna, J. R., & Irish, M. (2015). Shaped by our thoughts – A new task to assess spontaneous cognition and its associated neural correlates in the default network. *Brain and Cognition, 93,* 1–10. https://doi.org/10.1016/j.bandc.2014.11.001

Oliveira Silva, P., Maia, L., Coutinho, J., Frank, B., Soares, J. M., Sampaio, A., & Gonçalves, Ó. (2018). Empathy by default: Correlates in the brain at rest. *Psicothema, 30*(1), 97–103. https://doi.org/10.7334/psicothema2016.366

Poerio, G. L., & Smallwood, J. (2016). Daydreaming to navigate the social world: What we know, what we don't know, and why it matters. *Social and Personality Psychology Compass, 10*(11), 605–618. https://doi.org/10.1111/spc3.12288

Poerio, G. L., Sormaz, M., Wang, H. T., Margulies, D., Jefferies, E., & Smallwood, J. (2017). The role of the default mode network in component processes underlying the wandering mind. *Social Cognitive and Affective Neuroscience, 12*(7), 1047–1062. https://doi.org/10.1093/scan/nsx041

Posner, M. I., Rothbart, M. K., & Tang, Y. Y. (2015). Enhancing attention through training. *Current Opinion in Behavioral Sciences, 4,* 1–5. https://doi.org/10.1016/j.cobeha.2014.12.008

Raichle, M. E. (2010). Two views of brain function. *Trends in Cognitive Sciences, 14,* 180–190. https://doi.org/10.1016/j.tics.2010.01.008

Raichle, M. E., MacLeod, A. M., Snyder, A. Z., Powers, W. J., Gusnard, D. A., & Shulman, G. L. (2001). A default mode of brain function. *Proceedings of the National Academy of Sciences of the United States of America, 98*(2), 676–682. https://doi.org/10.1073/pnas.98.2.676

Regis, M. (2013). *Daydreams and the function of fantasy.* New York: Palgrave Macmillan. https://doi.org/10.1057/9781137300775

Sämann, P. G., Wehrle, R., Hoehn, D., Spoormaker, V. I., Peters, H., Tully, C., ... Czisch, M. (2011). Development of the brain's default mode network from wakefulness to slow wave sleep. *Cerebral Cortex, 21*(9), 2082–2093. https://doi.org/10.1093/cercor/bhq295

Scheibner, H. J., Bogler, C., Gleich, T., Haynes, J. D., & Bermpohl, F. (2017). Internal and external attention and the default mode network. *NeuroImage, 148,* 381–389. https://doi.org/10.1016/j.neuroimage.2017.01.044

Schooler, J. W., Smallwood, J., Christoff, K., Handy, T. C., Reichle, E. D., & Sayette, M. A. (2011). Meta-awareness, perceptual decoupling and the wandering mind. *Trends in Cognitive Sciences, 15,* 319–326. https://doi.org/10.1016/j.tics.2011.05.006

Schooler, J. W., Mrazek, M. D., Franklin, M. S., Baird, B., Mooneyham, B. W., Zedelius, C., & Broadway, J. M. (2014). The middle way. Finding the balance between mindfulness and mind-wandering. In *Psychology of learning and motivation – Advances in research and theory, 60,* 1–33. https://doi.org/10.1016/B978-0-12-800090-8.00001-9

Schredl, M. (2019). Being someone or something else in the dream: Relationship to thin boundaries. *Imagination, Cognition and Personality, 40*(1), 43–51. https://doi.org/10.1177/0276236619896272

Seli, P., Kane, M. J., Smallwood, J., Schacter, D. L., Maillet, D., Schooler, J. W., & Smilek, D. (2018). Mind-wandering as a natural kind: A family- resemblances view. *Trends in Cognitive Sciences, 22*(6), 479–490. https://doi.org/10.1016/j.tics.2018.03.010

Smallwood, J., & Schooler, J. W. (2015). The science of mind wandering: Empirically navigating the stream of consciousness. *Annual Review of Psychology, 66*(1), 487–518. https://doi.org/10.1146/annurev-psych-010814-015331

Smilek, D., Carriere, J. S. A., & Cheyne, J. A. (2010). Out of mind, out of sight: Eye blinking as indicator and embodiment of mind wandering. *Psychological Science, 21*(6), 786–789. https://doi.org/10.1177/0956797610368063

Stawarczyk, D. (2018). Phenomenological properties of mind-wandering and daydreaming: A historical overview and functional correlates. In K. C. R. Fox & K. Christoff (Eds.), *The Oxford handbook of spontaneous thought: Mind-wandering, creativity, and dreaming.* https://doi.org/10.1093/oxfordhb/9780190464745.013.18

Stawarczyk, D., Majerus, S., Maj, M., Van der Linden, M., & D'Argembeau, A. (2011a). Mind-wandering: Phenomenology and function as assessed with a novel experience sampling method. *Acta Psychologica, 136*(3), 370–381. https://doi.org/10.1016/j.actpsy.2011.01.002

Stawarczyk, D., Majerus, S., Maquet, P., & D'Argembeau, A. (2011b). Neural correlates of ongoing conscious experience: Both task-unrelatedness and stimulus-independence are related to default network activity. *PLoS One, 62*(2), 1–14. https://doi.org/10.1371/journal.pone.0016997

Stawarczyk, D., Cassol, H., & D'Argembeau, A. (2013). Phenomenology of future- oriented mind-wandering episodes. *Frontiers in Psychology, 4,* 425. https://doi.org/10.3389/fpsyg.2013.00425

Suddendorf, T., & Corballis, M. C. (2007). The evolution of foresight: What is mental time travel, and is it unique to humans? *Behavioral and Brain Sciences, 30*(3), 299–313. https://doi.org/10.1017/S0140525X07001975

Takeuchi, H., Taki, Y., Nouchi, R., Sekiguchi, A., Hashizume, H., Sassa, Y., … Kawashima, R. (2014). Association between resting-state functional connectivity and empathizing/systemizing. *NeuroImage, 99,* 312–322. https://doi.org/10.1016/j.neuroimage.2014.05.031

Tomlinson, E. R., Yousaf, O., Vittersø, A. D., & Jones, L. (2018). Dispositional mindfulness and psychological health: A systematic review. *Mindfulness. 9*(1), 23–43. https://doi.org/10.1007/s12671-017-0762-6

Tulving, E. (2002). Episodic memory: From mind to brain. *Annual Review of Psychology. 53*(1), 1–25 https://doi.org/10.1146/annurev.psych.53.100901.135114

Van Dam, N. T., van Vugt, M. K., Vago, D. R., Schmalzl, L., Saron, C. D., Olendzki, A., … Meyer, D. E. (2018). Mind the hype: A critical evaluation and prescriptive agenda for research on mindfulness and meditation. *Perspectives on Psychological Science, 13*(1), 36–61. https://doi.org/10.1177/1745691617709589

Wilson, T. D., Reinhard, D. A., Westgate, E. C., Gilbert, D. T., Ellerbeck, N., Hahn, C., … Shaked, A. (2014). Just think: The challenges of the disengaged mind. *Science, 345*(6192), 75–77. https://doi.org/10.1126/science.1250830

Chapter 10
Social Cognition Development and Socioaffective Dysfunction in Childhood and Adolescence

Claudia Berlim de Mello, Thiago da Silva Gusmão Cardoso, and Marcus Vinicius C. Alves (iD)

Abstract Social cognition refers to a wide range of cognitive abilities that allow individuals to understand themselves and others and also communicate in social interaction contexts (Adolphs, Curr Opin Neurobiol 11(2):231–239, 2001). According to Adolphs (Annu Rev Psychol 60(1):693–716, 2009), social cognition deals with psychological processes that allow us to make inferences about what is happening inside other people—their intentions, feelings, and thoughts. Although the term can be defined in many ways, it is clear that it must be safeguarded for the mental operations underlying social interactions. The most investigated cognitive processes of social cognition are emotion recognition and theory of mind (ToM), given that a whole range of socio-affective and interpersonal skills, such as empathy, derive from them (Mitchell RL, Phillips LH, Neuropsychologia, 70:1–10, 2015). Theory of mind is an intuitive ability to attribute thoughts and feelings to other people, and this ability usually matures in children in preschool age (Wellman HM, The child's theory of mind. Bradford Books/MIT, 1990), whereas emotional recognition refers to an individual's ability to identify others' emotions and affective states, usually based on their facial or vocal expressions, it is a critical skill that develops early and supports the development of other social skills (Mitchell RL, Phillips LH, Neuropsychologia, 70:1–10, 2015).

Keywords Social interactions · Theory of mind · Emotional recognition · Adolescence and childhood

C. B. de Mello
Department of Psychobiology, Universidade Federal de São Paulo, São Paulo, Brazil

T. da Silva Gusmão Cardoso
Centro Adventista Universitário de São Paulo, São Paulo, Brazil

M. V. C. Alves (✉)
Faculty of Health Sciences of Trairi, Universidade Federal do Rio Grande do Norte, Santa Cruz, Brazil

© The Author(s) 2023
P. S. Boggio et al. (eds.), *Social and Affective Neuroscience of Everyday Human Interaction*, https://doi.org/10.1007/978-3-031-08651-9_10

Introduction

Social cognition refers to a wide range of cognitive abilities that allow individuals to understand themselves and others and also communicate in social interaction contexts (Adolphs, 2001). According to Adolphs (2009), social cognition deals with psychological processes that allow us to make inferences about what is happening inside other people—their intentions, feelings, and thoughts. Although the term can be defined in many ways, it is clear that it must be safeguarded for the mental operations underlying social interactions.

The most investigated cognitive processes of social cognition are emotion recognition and theory of mind (ToM), given that a whole range of socio-affective and interpersonal skills, such as empathy, derive from them (Mitchell & Phillips, 2015). Theory of mind is an intuitive ability to attribute thoughts and feelings to other people, and this ability usually matures in children in preschool age (Wellman, 1990), whereas emotional recognition refers to an individual's ability to identify others' emotions and affective states, usually based on their facial or vocal expressions, it is a critical skill that develops early and supports the development of other social skills (Mitchell & Phillips, 2015).

Adverse Childhood Experiences (ACEs), such as parental neglect or physical, sexual or psychological abuse, especially in the early stages of development, can have particularly harmful long-term consequences for the consolidation of cognitive, affective, and emotional skills (Herzog & Schmahl, 2018). A systematic review on the associations between early social environment, early-life adversity, and social cognition in major psychiatric disorders found that emotional and physical abuse, neglect, and avoidant attachment styles were the strongest predictors of ToM and emotion recognition deficits, as well as emotional dysregulation (Rokita et al., 2018). Prolonged exposure to events of this nature can lead to brain changes, particularly in circuits involved in regulating responses to stress, configuring the concept of toxic stress (Shonkoff & Garner, 2012).

In clinical or psychopathological contexts, social and affective impairments worsen social disadvantages that many patients face. As an example, in Autism Spectrum Disorder (ASD), ToM seems to be the most impaired social cognition domain (Baron-Cohen et al., 1985), whereas in schizophrenia, social impairments appear as negative symptoms and tend to predict patients' lower mental capacity. Several evidences indicate that core social difficulties in ASD are best explained by deficits in controlled processes, such as ToM, rather than automatic ones, such as those dependent on emotional contagion (Hamilton, 2013). This discrepancy between the two seems to be in line with the hypothesis of empathy imbalance (Smith, 2009). As an example, the avoidance of eye or physical contact frequently observed in individuals diagnosed with this disorder can be explained by an exacerbated affective empathy, which is dependent on emotional contagion. On the other hand, a deficit of cognitive empathy, being associated with ToM, could justify the low performance in tasks with greater demand for mental state assignment.

Social cognition deficits are expressed, for instance, in difficulties to interpret social cues and regulate behavior accordingly, which turns engagement in social relationships especially challenging. In ADHD, there are also failures to recognize emotions, especially anger and fear (Bora & Pantelis, 2016). Contrary to what is observed in autism, however, the difficulties tend to be lighter and get better with age. Thus, deficits in social cognition are early and prominent features of many neuropsychiatric, neurodevelopment, and neurodegenerative disorders (Agnew-Blais & Seidman, 2013). In the most recent edition of the American Psychiatric Association's Diagnostic and Statistical Manual for Mental Disorders (DSM-5, 2013), social cognition has been added as one of the six main components of neuro-cognitive function, alongside memory and executive control. In addition, the DSM-5 points out that impairments in social cognition often arise as deficient ToM, reduced affective empathy, impaired social perception, or abnormal social behavior.

In this chapter, we discuss social cognition processes in typical and atypical development and its relevance for the understanding of socio-affective disorders in childhood and adolescence. We expect therefore that it will be useful for clinicians, teachers, and perhaps also for parents.

Social Interactions and Theory of Mind in Childhood and Adolescence

Social interactions depend essentially on the exchange of social and affective cues, which can be verbal and non-verbal (Frith & Frith, 2007). Verbal cues include vocalizations, tone of voice, and speech content, while the most important non-verbal cues that humans use are facial expression, body posture, and eye monitoring. Most of these cues are processed automatically and unconsciously (Frith & Frith, 2007). In this way, regardless of an unambiguous interpretation, we automatically decipher people's emotions through their facial expressions, while also looking for something interesting in the environment from clues such as the direction of others' gaze. Moreover, curiously, we tend to imitate the behavior of people with whom we have a good relationship (Frith & Frith, 2007). In an eloquent way, social cues guide our behavior through an ambiguous and complex world.

In childhood (especially in the first two years of life), human beings have limited attention and limited working memory. Given these restrictions, one possible question to ask is if babies are able to process complex social information. Sensitivity to social clues seems to develop early, considering that even in early childhood, it is already possible to locate its clues from the orientation of attention in human babies (Michel et al., 2017). In Wu and Kirkham's study (2010), 8-month-old babies were presented with two identical audiovisual events simultaneously in two different locations on one computer screen. The babies' attention was focused on a stimulus that contained a social suggestion (a face saying "Hi, baby, look at this!" and that turned to a target event) or a non-social suggestion (a red square surrounding the target event). Both stimuli directed attention equally, as measured by the time the babies stayed looking at the events. However, only babies exposed to social cues predicted the location of the signaled events, suggesting that social attention hints shape the likelihood and content of learning about events during childhood.

Reid and Striano (2007) propose that for a baby to react successfully to a social situation, four stages of cognitive processing of the task must occur: (1) the detection of socially relevant organisms; (2) the identification of socially relevant organisms; (3) the evaluation of the place of attention and direction of the individual's gaze observed in relation to the child; and (4) the detection of any attention directed to objects or involvement of objects by the observed individual. As a consequence, if the previous stage were successful, then another stage is reached; in the fifth stage, the baby is able to infer the observed objective and/or prepare an appropriate response (e.g., establishing contact).

Reid and Striano (2007) suggest that the detection of biological movement, an early cognitive ability in babies, plays a key role in the detection of co-specifics and, therefore, in identifying the specifics of social interaction in stage 1. Stage 2 is possible because babies can identify idiosyncrasies in the observed organisms, for example, discriminating between a familiar and unfamiliar individual. In relation to stage 3, these abilities are supported because human babies are sensitive to the elements of the human face, especially to follow the gaze of the observed subject. In addition, they are able to distinguish whether an adult interacts in a salient manner, providing contingent feedback on social interaction, such as smiles and vocalizations, or interacts in an irregular manner, with a delay in social return (Striano et al., 2005).

Finally, babies are competent in discerning a relationship between a person and an object (Stage 4). In another study, Reid and Striano (2005) found that the direction of a person's gaze affects the coding of new objects in 4-month-old babies. In this study, in a first experimental condition, the babies saw the face of a woman with new toys being presented on her right and left sides. The woman then directed her gaze to one side, thus capturing a certain object, and as a consequence, she averted her gaze from the other toy. In a second experimental condition, the 4-month-old babies were presented to the two toys again and looked more at the toy that had not been looked at in the first condition, probably because of the little attention paid to it, and only after that to the toy looked at in the first condition. This suggests that 4-month-old babies not only follow the adult's gaze in the first condition but also acquire more information about the object that was the focus of adult attention. It is quite convincing that human babies respond early to social cues, however, an idea valued for social and affective neuroscience is that in order to adequately manage the complex levels of social interaction that characterize our social life, human beings need to develop specific social-cognitive mechanisms, such as Theory of Mind.

Since Premack and Woodruff's (1978) seminal article "Does the chimpanzees have a theory of the mind?" raised the question of chimpanzees' ability to attribute states of mind to themselves and others, the subject of ToM has become part of human development studies. One of the first studies of ToM development as a child was the work of Wimmer and Perner (1983), who pointed to the understanding of false belief as an indicator of preschoolers' ToM. One way to test this understanding is to place the child in a task where he or she has to predict the behavior of a character who has a belief that does not correspond to reality. An example would be to present the child with a common box of chocolates, then ask him/her to open the box to check its contents; After checking, to their surprise, that the box doesn't have chocolates, but colored pencils, some questions are asked to the children, such as if you show a friend the same box of chocolates, and ask him/her what is the content, what do you think he/she would say? Why would he/she say that?

It is expected that children between the ages of 3 and 5 will be able to respond correctly to this type of task, thus revealing an understanding of beliefs about reality which, being personal representations of reality, maybe true or false. However, the ability to understand false beliefs seems to be more the result of an ongoing process of developing skills to assign mental and affective states to others. Therefore, some studies analyzing children's discourse have found that references to desires precede references to cognition (Bartsch & Wellman, 1995; Peterson & Slaughter, 2006).

According to Wellman et al. (2011), the theory of mind develops progressively and sequentially from the child's ability to understand different levels of representation of mental states: (1) various desires (people may have different desires for the same thing), (2) various beliefs (people may have different beliefs about the same situation), (3) access to knowledge (something may be true, but someone may not know it), (4) false belief (something may be true, but someone may believe in something different), and (5) hidden emotion (someone may feel one way, but show different emotions).

ToM's reasoning required by different social situations may involve the assimilation of complex levels of intentionality. For example, Ygor believes (ToM of first order of intentionality) that Larissa thinks (ToM of second order) that her aunt Marcia wants (ToM of third order) that Ygor supposes (ToM of fourth order) that Larissa wants (ToM of fifth order) that her aunt Marcia believes (ToM of fifth order). This ability of first- and third-order ToM seems to improve with age (Dumontheil et al., 2010).

Dumontheil et al. (2010) showed that the ability to adopt the point of view of another agent grows from childhood, passes through adolescence, and improves even more in adulthood. Meinhardt-Injac et al. (2020) tested a two-component model of social cognition, social perception, and cognitive social, in a sample of 267 participants between 11 and 25 years of age. In addition, they measured language, reasoning, and inhibitory control as covariables. In the study, adolescents showed a substantial improvement in ToM (social perception and false belief) and covariable measures. An interesting finding is that the social perception component increased with age, while the socio-cognitive component (false belief) increased with age and covariables measures.

The tasks of false belief can be further developed between adolescence and adult life. In the study by Valle et al. (2015), adolescents performed significantly worse than young adults in tasks of false belief involving third-order ToM, but an equal result for second-order ToM. Other components of ToM develop in adolescence as the social knowledge required in tests involving soft lies, forgeries, and strange stories (Maylor et al., 2002; Bosco et al., 2014). Happé (1994) developed a test called The Strange Stories to evaluate advanced mental capacity, which is suitable for adolescents and adults with superior functioning. In the test, participants read short vignettes and were asked to explain why a character said something that is literally not true. Therefore, successful performance requires the assignment of mental states, such as desires, beliefs, or intentions, and sometimes higher-order mental states, such as one character's beliefs in what another character knows.

Using a subset of Happé test stories (1994), Maylor et al. (2002) investigated individuals aged 16–29 when performing advanced ToM tasks in the first person and noted that participants scored an average of 4 out of an available maximum of 7, without any ceiling effect being achieved in the age group. In the study of Bosco et al. (2014), teen performance improved with age in all ToM scales, which investigate first-person and third-person ToM, first- and second-order allocating ToM, and egocentric third-person ToM. However, age differences were consistent between 11 and 13 years and then tended to stabilize between 13 and 15 years. The findings of these different studies suggest that some, but not all, components of ToM continue to develop into adulthood.

Development of Emotional Recognition and Understanding in Childhood and Adolescence

Children's emotional knowledge comprises two distinct dimensions: recognition of emotion and knowledge of the emotional situation. Recognizing emotions means that the child can label facial expressions using expressive knowledge of emotions as well as identifying emotions when expressed with verbal labels.

Nine-month-old babies are already able to discriminate between positive and negative emotions (Otte et al., 2015). In the study, 84 babies received emotional vocalizations (fearful or happy) preceded by the same facial expression or a different expression (i.e., fearful vocalization with a happy expression). The data processing of emotional information (event-related potential, or ERPs) revealed that the potentials were distinct for positive and negative emotions, and that the babies dedicated more process capacity to potentially threatening stimuli than to non-threatening ones. Between 18 and 24 months, children are already able to acquire the necessary terms to label basic emotions, both positive and negative (Widen & Russell, 2003, 2010).

According to Pons et al. (2004), children develop emotional understanding from three levels (external, mental, and reflective) and use at least nine components for this. The first level (already found in children of 5 years) is characterized by the understanding of public aspects of emotions, such as situational causality, external expression, and cues that reactivate an emotion. The second level (developed at age 7) is characterized by an understanding of the mental states of emotions, their connection with desires and beliefs, between expressed and felt emotion. The third level (between the ages of 9 and 11) is characterized by the understanding that we can feel different feelings and that they can be contradictory and even morally charged.

Regarding the components, approximately between 3 and 4 years old, children begin to recognize, and name emotions based on expressive clues (recognition component). Thus, most children in this age group can recognize basic emotions (happiness, sadness, fear, and anger) when presented in images. Also at this age, they begin to understand how external causes affect other children's emotions (external causality component). For example, they can anticipate the sadness someone feels when losing a favorite toy. Already around the age of 3 to 5, they begin to understand that the emotional reactions of people depend on their desires (component of desire). They are able to understand that two people can feel a different emotion about the same situation, because they have different desires.

Still, according to Pons et al. (2004), children between 4 and 6 years old begin to understand that a person's beliefs—being false or true—will determine their emotional reaction to a situation (component of beliefs). Also, around this age, between 3 and 6 years, children start to understand the relationship between memory and emotion (memory component). For example, children can recognize that the intensity of an emotion decreases with time and that some aspects of the current situation can reactivate past emotions. Still, in this age group, 4 and 6 years old, children already understand that the emotion expressed can be different from the emotion felt (component of the cover-up). In sequence, between the ages of 6 and 7, they are able to use different strategies to regulate emotions (component of regulation), the younger ones use behavioral strategies, and those over 8 use psychological strategies such as denial and distraction. It is also from the age of 8 that they begin to understand that a person can have various and contradictory emotions (mixed component) to a given situation. Finally, 8-year-old children are able to understand negative feelings resulting from a reprehensible moral action (morality component) (like lying, stealing, and hiding), as well as being morally dignified (like a sacrifice, or resisting a temptation, or even confessing a mistake).

Emotional recognition improves with adolescence. Functional brain imaging studies during facial emotion recognition tasks have demonstrated an increase in activation and connectivity of frontal and temporal regions from childhood to adolescence (Cohen Kadosh et al., 2011, 2013). This increase in emotional recognition tends to continue into adulthood. The study by Tousignant et al. (2017), showed that adolescents perform worse than young adults in this area of social cognition.

Social and Affective Dysfunctions

To allow a better understanding of how social cognition impacts psychosocial functioning in children and adolescents, we will focus on the findings of research that has investigated this domain of cognition in some psychopathological conditions with the presence of social and affective dysfunctions, such as schizophrenia, ASD, ADHD, and impulse control disorders.

In the last two decades, the domains of social cognition have been intensively studied in individuals with neurological conditions, with genetic syndromes, and in population groups at risk for developing the first episode of psychosis, such as children of parents with schizophrenia. There is also broad evidence of these neurological conditions such as epilepsy (Besag & Vasey, 2019). Social cognitive deficits are part of the cognitive phenotype of genetic diseases, including deletion syndromes, such as SD22q11.2, and those related to X-Chromosome numerical alterations, such as Turner and Klinefelter (Morel et al., 2018). In a review study, Agnew-Blais and Seidman (2013) found that young people belonging to families at high biological risk for developing schizophrenia, especially siblings and children of individuals with schizophrenia, show deficits in aspects of social cognition, such as ToM and emotion recognition. However, it is not yet known whether deficits in the domain of social cognition follow a pattern of delays over time or are static. A population-based, prospective cohort study evaluated 7-year-old children at high familial risk of developing schizophrenia or bipolar disorder (Christiani et al., 2019). The authors found significant impairments in social responsiveness in children at risk for schizophrenia compared to controls, but not for children at risk for bipolar disorder compared to controls (Christiani et al., 2019). In several population-based cohort studies, children and adolescents who later developed schizophrenia showed premorbid social impairments (Tarbox & Pogue-Geile, 2008; Agnew-Blais & Seidman, 2013). These findings reinforce the view of schizophrenia as a neurodevelopmental disorder and the importance of identifying in clinical practice differences in the domain of social cognition in children and adolescents belonging to groups at risk for developing schizophrenia.

In ADHD, in addition to classical assessment of executive functions, deficits in the domain of social cognition are increasingly investigated (Mohammadzadeh et al., 2016). In a meta-analysis of studies investigating social cognition in ADHD, it was reported that facial and vocal recognition skills and ToM were significantly impaired in ADHD (Bora & Pantelis, 2016). In addition, the authors rated the performance of individuals with ADHD as intermediate between ASD and healthy controls (Bora & Pantelis, 2016). An interesting finding is that deficits in social cognition appear to occur later in ADHD than in ASD and appear to depend on social interactions with family members and peers at school (Bora & Pantelis, 2016). Not surprisingly, social dysfunction is one of the most impactful aspects for the psychosocial development of children and adolescents with ADHD, as individuals with ADHD often report significant interpersonal problems, including conflict with parents, siblings, peers, and teachers (Ros & Graziano, 2018). Social dysfunction in ADHD appears to depend on a number of factors, including social skills, ability to process information and modulate social responses (social cognition), and contexts of social interaction with peers (Ros & Graziano, 2018). Although these reviews seemingly converge on a view of deficits in social cognition in children and adolescents with ADHD, most clinicians seem to ignore that ADHD patients tend to experience not only limitations in social skills but also impairments in social information processing. Furthermore, ToM skills are associated with different patterns of prosocial behavior such as helping, cooperating, and comforting (Imuta et al., 2016), and since in ADHD ToM may be significantly impaired, these patients would be less likely to develop positive social interactions with their peers, teachers, and family members.

In impulse control disorders, social cognition is an aspect of patient functioning that has gained relevance. Impulse control disorders (ICDs) are grouped as a heterogeneous group of mental disorders related to the failure to resist impulses to perform dangerous, troublesome, or disturbing behaviors. We can include in this category, pathological gambling (PG), kleptomania, pyromania, trichotillomania, internet gaming disorder (IGD), intermittent explosive disorder (IED), among others. Aspects of the social cognition domain have been investigated in at least these last two clinical conditions (Coccaro et al., 2016). In IED, the biopsychosocial model of impulsive aggression holds that the individual with the disorder usually explodes in response to social threat, and one of the main dysfunctions would be in social information processing (Coccaro et al., 2011). Thus, the individual with IED, when encoding and interpreting social cues, often performs an attribution of hostile intent in social interactions (Coccaro et al., 2011). In a study investigating an aspect of social cognition of patients with IED, attribution style, demonstrated that participants with IED have higher hostile attribution in ambiguous social situations than healthy controls and patients with other psychiatric disorders, and also that attribution style was directly related to negative emotional response (Coccaro et al., 2016).

In the case of IGD, patients are observed to have significant social impairments, with a predilection for social interactions only in the online environment than in real life (Caplan, 2010). In the cognitive model of IGD proposed by Caplan (2010), pathological internet use could be defined by two cognitive features: preference for online social interaction and worry. Preference for online social interaction can be defined as an individual's tendency to develop beliefs that online interactions and relationships are better, safer, and more comfortable than face to face (Caplan, 2010). Worry or cognitive salience is defined as obsessive thought patterns about internet use (Caplan, 2010). Some studies have investigated that a component of social cognition, emotion recognition, is related to problematic internet use in adolescents (Spada & Marino, 2017; Yavuz et al., 2019). Specifically, in relation to IGD, it is suggested that adolescents engage in gaming as a strategy to alleviate affective dysfunction caused by poor skill in recognizing negative emotions (Yavuz et al., 2019). Another study revealed that low negative emotion recognition skills were able to predict cognitive salience, tolerance, and relapse associated with IGD in adolescents (Aydın et al., 2020). These findings analyzed together allow us to think that negative emotion recognition, attribution style, and consequently the strategies used for cognitive-affective regulation used by adolescents aroused by negative affect may increase their risk of developing impulse control mental disorders, specifically IED and IGD.

Treatment

Behavioral interventions focusing on the stimulation of socio-emotional skills have been developed and their effectiveness tested in a more remedial or preventive perspective. For the follow-up of children and adolescents with a history of exposure to maltreatment, there are several approaches such as cognitive-behavioral therapy (CBT), Eye Movement Desensitization and Reprocessing (EMDR); therapies based on artistic activities or contact with animals, and family-focused interventions (e.g., systemic family therapy) (Macdonald et al., 2016).

In clinical conditions, such as autism, the most well-known interventions are those based on behavior analysis models. For children with high functioning autism, virtual reality-based training has proven to be a complementary strategy that can stimulate engagement in the therapeutic process (Didehbani et al., 2016). Simmons et al. (2019) proposed an integrative model for intervention programs considering social cognition and executive functions simultaneously. In fact, impairments in these cognitive domains have been associated with the symptomatology of both ASD and ADHD (Van der Meer et al., 2012). According to the model, for a target social and affective problem, intervention strategies should consider social cognitive (e.g. practices to recognition of emotional cues) or executive (behavioral rehearsal) components.

Other programs have a more preventive character, generally organized for implementation in schools. Emphasis on programs of this nature has been strengthened by evidence of the importance of developing socio-emotional skills, such as understanding one's point of view or self-regulation, for academic performance (Blair, 2002; Fantuzzo et al., 2007). An example is Social and Emotional Learning, or SEL (Durlak et al., 2011). The program aims to stimulate the management of emotions and the development of empathy in order to strengthen positive relationships and more responsible decision-making. In a meta-analysis, Durlak et al. (2011) reviewed 82 school-based SEL interventions implemented inside and outside the United States, covering 97,406 kindergarten to high-school students. The programs were selected according to recommended practice criteria including a clear alignment of goals and curriculum, involvement of students in all steps of implementation, attention to community needs, and reflective activities such as class discussions. Results provided evidence of positive effects on students' attitudes toward themselves and toward school and learning, on social behavior, and academic performance, independently of students' race or socioeconomic background.

Conclusion

In this chapter, we describe how social cognitive processes, such as emotion recognition and ToM, develop. These processes are of paramount importance for the understanding of normal and pathological processes in childhood and adolescence.

The continuous processing of social information and the perception of emotional states combine to build up a child's emotional knowledge and to develop social competence.

The process by which we understand our own emotions and those of others, and assign intentions and desires to others, helps to explain the different challenges we face throughout human development. The abilities to recognize emotions and assign mental states, important domains of social cognition, are crucial to assessing someone's immediate social environment, providing valuable information about the inner emotional state of others, and influencing adaptive social behavior and social interactions. The observation of social and affective dysfunction in children and adolescents raises the question of whether these deficits are risk factors for neuropsychiatric disorders, the untoward consequences of neuropsychiatric disorders, or both. Socio-affective neuroscience offers ways to elucidate this issue. In addition, in health sciences, there has been a focus on the early identification of developmental conditions in which a fast and rational intervention can favorably reach positive outcomes. In most neuropsychiatric disorders, difficulties in social functioning are evident, and if we can focus on the identification and remediation of social and affective dysfunction at an earlier age, the better.

We address how these aspects are related to healthy development and also in neuropsychiatric disorders. Since the skills involved in social cognition are extremely important for future interactions, diligent investigations of this phenomenon are useful for clinical and applied neuroscience. Bearing in mind that the understanding of the social and affective factors involved in children's development is fundamental for the understanding of the adults that will emerge from it.

References

Adolphs, R. (2001). The neurobiology of social cognition. *Current Opinion in Neurobiology, 11*(2), 231–239. https://doi.org/10.1016/S0959-4388(00)00202-6

Adolphs, R. (2009). The social brain: Neural basis of social knowledge. *Annual Review of Psychology, 60*(1), 693–716. https://doi.org/10.1146/annurev.psych.60.110707.163514

Agnew-Blais, J., & Seidman, L. J. (2013). Neurocognition in youth and young adults under age 30 at familial risk for schizophrenia: A quantitative and qualitative review. *Cognitive Neuropsychiatry, 18*, 44–82.

American Psychiatric Association (2013). Diagnostic and statistical manual of mental disorders (5th ed). Washington, DC.

Aydın, O., Güçlü, M., Ünal-Aydın, P., & Spada, M. M. (2020). Metacognitions and emotion recognition in Internet Gaming Disorder among adolescents. *Addictive Behaviors Reports, 12*, 100296. https://doi.org/10.1016/j.abrep.2020.100296

Bartsch, K., & Wellman, H. M. (1995). Children talk about the mind. New York: Oxford University Press, 234 pp

Baron-Cohen, S., Leslie, M. A., & Frith, U. (1985). Does the autistic child have a theory of mind? *Cognition, 21*, 37–46. https://doi.org/10.1016/0010-0277(85)90022-8

Besag, F. M. C., & Vasey, M. J. (2019). Social cognition and psychopathology in childhood and adolescence. *Epilepsy & Behavior, 100*(Pt B), 106210. https://doi.org/10.1016/j.yebeh.2019.03.015

Blair, C. (2002). School readiness. Integrating cognition and emotion in a neurobiological conceptualization of children's functioning at school entry. *American Psychologist, 57*(2), 111–127.

Bora, E., & Pantelis, C. (2016). Meta-analysis of social cognition in attention- deficit/hyperactivity disorder (ADHD): Comparison with healthy controls and autistic spectrum disorder. *Psychological Medicine, 46*(4), 699–716. https://doi.org/10.1017/S0033291715002573

Bosco, F. M., Gabbatore, I., & Tirassa, M. (2014). A broad assessment of theory of mind in adolescence: The complexity of mindreading. *Consciousness and Cognition, 24, 84–97.* https://doi.org/10.1016/j.concog.2014.01.003

Caplan, S. E. (2010). Theory and measurement of generalized problematic internet use: A two-step approach. *Computers in Human Behavior, 26*(5), 1089–1097. https://doi.org/10.1016/j.chb.2010.03.012

Christiani, C. J., Jepsen, J., Thorup, A., Hemager, N., Ellersgaard, D., Spang, K. S., Burton, B. K., Gregersen, M., Søndergaard, A., Greve, A. N., Gantriis, D. L., Poulsen, G., Uddin, M. J., Seidman, L. J., Mors, O., Plessen, K. J., & Nordentoft, M. (2019). Social cognition, language, and social behavior in 7- year-old children at familial high-risk of developing schizophrenia or bipolar disorder: The Danish High Risk and Resilience Study VIA 7-A Population-Based Cohort Study. *Schizophrenia Bulletin, 45*(6), 1218–1230. https://doi.org/10.1093/schbul/sbz001

Coccaro, E. F., Sripada, C. S., Yanowitch, R. N., & Phan, K. L. (2011). Corticolimbic function in impulsive aggressive behavior. *Biological Psychiatry, 69,* 1153–1159.

Coccaro, E. F., Fanning, J. R., Keedy, S. K., & Lee, R. J. (2016). Social cognition in intermittent explosive disorder and aggression. *Journal of Psychiatric Research, 83,* 140–150. https://doi.org/10.1016/j.jpsychires.2016.07.010

Cohen Kadosh, K., Cohen Kadosh, R., Dick, F., & Johnson, M. H. (2011). Developmental changes in effective connectivity in the emerging core face network. *Cerebral Cortex, 21*(6), 1389–1394. https://doi.org/10.1093/cercor/bhq215

Cohen Kadosh, K., Johnson, M. H., Dick, F., Cohen Kadosh, R., & Blakemore, S. J. (2013). Effects of age, task performance, and structural brain development on face processing. *Cerebral Cortex, 23*(7), 1630–1642. https://doi.org/10.1093/cercor/bhs150

Didehbani, N., Kandalaft, M., Krawczyk, D., & Chapman, S. (2016). Virtual Reality Social Cognition Training for children with high functioning autism. *Computers in Human Behavior, 62,* 703–711. https://doi.org/10.1016/j.chb.2016.04.033

Dumontheil, I., Apperly, I. A., & Blakemore, S. J. (2010). Online usage of theory of mind continues to develop in late adolescence. *Developmental Science, 13*(2), 331–338. https://doi.org/10.1111/j.1467-7687.2009.00888.x

Durlak, J., Weissberg, R., Dymnicki, A., Taylor, R., & Schellinger, K. (2011). The impact of enhancing students' social and emotional learning: A meta- analysis of school-based universal interventions. *Child Development, 82,* 405–432. https://doi.org/10.1111/j.1467-8624.2010.01564.x

Fantuzzo, J., Bulotsky-Shearer, R., McDermott, P., McWayne, C., Frye, D., & Perlman, S. (2007). Investigation of dimensions of social-emotional classroom behavior and school readiness for low-income urban preschool children. *School Psychology Review, 36,* 44–62.

Frith, C. D., & Frith, U. (2007). Social cognition in humans. *Current Biology, 17*(16), R724–R732. https://doi.org/10.1016/j.cub.2007.05.068

Hamilton, A. F. C. (2013). Reflecting on the mirror neuron system in autism: A systematic review of current theories. *Developmental Cognitive Neuroscience, 3,* 91–105. https://doi.org/10.1016/j.dcn.2012.09.008

Happé, F. (1994). An advanced test of theory of mind: Understanding of story characters' thoughts and feelings by able autistic, mentally handicapped, and normal children and adults. *Journal of Autism and Developmental Disorders, 24*(2), 129–154.

Herzog, J. I., & Schmahl, C. (2018). Adverse childhood experiences and the consequences on neurobiological, psychosocial, and somatic conditions across the lifespan. *Frontiers in Psychiatry, 9,* 420. https://doi.org/10.3389/fpsyt.2018.00420

Imuta, K., Henry, J., Slaughter, V., Selcuk, B., & Ruffman, T. (2016). Theory of mind and prosocial behaviour in childhood: A meta-analytic review. *Developmental Psychology, 52*, 1192–1205. https://doi.org/10.1037/dev0000140

Macdonald, G., Livingstone, N., Hanratty, J., et al. (2016). *The effectiveness, acceptability and cost-effectiveness of psychosocial interventions for maltreated children and adolescents: An evidence synthesis* (Health Technology Assessment, No. 20.69). NIHR Journals Library. Appendix 5, Types of interventions. Available from: https://www.ncbi.nlm.nih.gov/books/NBK385382/

Maylor, E. A., Moulson, J. M., Muncer, A. M., & Taylor, L. A. (2002). Does performance on theory of mind tasks decline in old age? *British Journal of Psychology, 93*, 465–485. https://doi.org/10.1348/000712602761381358

Meinhardt-Injac, B., Daum, M. M., & Meinhardt, G. (2020). Theory of mind development from adolescence to adulthood: Testing the two-component model. *The British Journal of Developmental Psychology, 38*, 289–303. https://doi.org/10.1111/bjdp.12320

Michel, C., Wronski, C., Pauen, S., Daum, M. M., & Hoehl, S. (2017). Infants' object processing is guided specifically by social cues. *Neuropsychologia*. https://doi.org/10.1016/j.neuropsychologia.2017.05.022

Mitchell, R. L., & Phillips, L. H. (2015). The overlapping relationship between emotion perception and theory of mind. *Neuropsychologia, 70*, 1–10.

Morel, A., Peyroux, E., Leleu, A., Favre, E., Franck, N., & Demily, C. (2018). Overview of social cognitive dysfunctions in rare developmental syndromes with psychiatric phenotype. *Frontiers in Pediatrics, 6*, 102. https://doi.org/10.3389/fped.2018.00102

Mohammadzadeh, A., Tehrani-Doost, M., Khorrami, A., & Noorian, N. (2016). Understanding intentionality in children with attention-deficit/hyperactivity disorder. *Attention Deficit and Hyperactivity Disorders, 2*, 73–78. https://doi.org/10.1007/s12402-015-0187-9

Otte, R. A., Donkers, F. C., Braeken, M. A., & Van den Bergh, B. R. (2015). Multimodal processing of emotional information in 9-month-old infants I: emotional faces and voices. *Brain and Cognition, 95*, 99–106. https://doi.org/10.1016/j.bandc.2014.09.007

Peterson, C. C., & Slaughter, V. P. (2006). Telling the story of theory of mind: Deaf and hearing childrens narratives of mental state understanding. *British Journal of Developmental Psychology, 24*, 151–179. https://doi.org/10.1348/026151005X60022

Pons, F., Harris, P. L., & de Rosnay, M. (2004). Emotion comprehension between 3 and 11 years: Developmental periods and hierarchical organization. *European Journal of Developmental Psychology, 1*(2), 127–152. https://doi.org/10.1080/17405620344000022

Premack, D., & Woodruff, G. (1978). Does the Chimpanzee Have a Theory of Mind? Behavioural and Brain Sciences, 1, 515–526. https://doi.org/10.1017/S0140525X00076512

Reid, V. M., & Striano, T. (2005). Adult gaze influences infant attention and object processing: Implications for cognitive neuroscience. *European Journal of Neuroscience, 21*(6), 1763–1766. https://doi.org/10.1111/j.1460-9568.2005.03986.x

Reid, V. M., & Striano, T. (2007). The directed attention model of infant social cognition. *European Journal of Developmental Psychology, 4*(1), 100–110. https://doi.org/10.1080/17405620601005648

Rokita, K. I., Dauvermann, M. R., & Donohoe, G. (2018). Early life experiences and social cognition in major psychiatric disorders: A systematic review. *European Psychiatry, 53*, 123–133. https://doi.org/10.1016/j.eurpsy.2018.06.006

Ros, R., & Graziano, P. A. (2018). Social functioning in children with or at risk for attention deficit/hyperactivity disorder: A meta-analytic review. *Journal of Clinical Child and Adolescent Psychology, 47*(2), 213–235. https://doi.org/10.1080/15374416.2016.1266644

Shonkoff, J. P., & Garner, A. S. (2012). Committee on Psychosocial Aspects of Child and Family Health, Committee on Early Childhood, Adoption, and Dependent Care, & Section on Developmental and Behavioral Pediatrics. The lifelong effects of early childhood adversity and toxic stress. *Pediatrics, 129*(1), e232–e246. https://doi.org/10.1542/peds.2011-2663

Simmons, G. L., Hilton, D. C., Jarrett, M. A., Tomeny, T. S., & White, S. W. (2019). Considering equifinality in treatment planning for social impairment: Divergent paths in neurodevelopmental disorders. *Bulletin of the Menninger Clinic, 83*(3), 278–300. https://doi.org/10.1521/bumc.2019.83.3.278

Smith, A. (2009). The empathy imbalance hypothesis of autism: A theoretical approach to cognitive and emotional empathy in autistic development. *The Psychological Record, 59*. https://doi.org/10.1007/BF03395663

Striano, T., Henning, A., & Stahl, D. (2005). Sensitivity to social contingencies between 1 and 3 months of age. *Developmental Science, 8*(6), 509–518. https://doi.org/10.1111/j.1467-7687.2005.00442.x

Spada, M. M., & Marino, C. (2017). Metacognitions and emotion regulation as predictors of problematic internet use in adolescents. *Clinical Neuropsychiatry, 14*, 59–63.

Tarbox, S. I., & Pogue-Geile, M. F. (2008). Development of social functioning in preschizophrenia children and adolescents: A systematic review. *Psychological Bulletin, 134*, 561–583.

Tousignant, B., Sirois, K., Achim, A. M., Massicotte, E., & Jackson, P. L. (2017). A comprehensive assessment of social cognition from adolescence to adulthood. *Cognitive Development, 43*, 214–223. https://doi.org/10.1016/j.cogdev.2017.05.001

van der Meer, J. M., Oerlemans, A. M., van Steijn, D. J., Lappenschaar, M. G., de Sonneville, L. M., Buitelaar, J. K., & Rommelse, N. N. (2012). Are autism spectrum disorder and attention-deficit/hyperactivity disorder different manifestations of one overarching disorder? Cognitive and symptom evidence from a clinical and population-based sample. *Journal of the American Academy of Child and Adolescent Psychiatry, 51*(11), 1160–1172.e3. https://doi.org/10.1016/j.jaac.2012.08.024

Valle, A., Massaro, D., Castelli, I., & Marchetti, A. (2015). Theory of mind development in adolescence and early adulthood: The growing complexity of recursive thinking ability. *Europe's Journal of Psychology, 11*(1), 112–124. https://doi.org/10.5964/ejop.v11i1.829

Wellman, H. M. (1990). *The child's theory of mind*. Bradford Books/MIT.

Wellman, H. M., Fang, F., & Peterson, C. C. (2011). Sequential progressions in a theory-of-mind scale: longitudinal perspectives. *Child Development, 82*(3), 780–792. https://doi.org/10.1111/j.1467-8624.2011.01583.x

Widen, S. C., & Russell, J. A. (2003). A closer look at preschoolers' freely produced labels for facial expressions. *Developmental Psychology, 39*, 114–128. https://doi.org/10.1037/0012-1649.39.1.114

Widen, S. C., & Russell, J. A. (2010). Differentiation in preschooler's categories of emotion. *Emotion, 10*, 651–661. https://doi.org/10.1037/a0019005

Wimmer, H., & Perner, J. (1983). Beliefs about beliefs: Representation and constraining function of wrong beliefs in young children's understanding of deception. *Cognition, 13*, 103–128. https://doi.org/10.1016/0010-0277(83)90004-5

Wu, R., & Kirkham, N. Z. (2010). No two cues are alike: Depth of learning during infancy is dependent on what orients attention. *Journal of Experimental Child Psychology, 107*(2), 118–136. https://doi.org/10.1016/j.jecp.2010.04.014

Yavuz, M., Nurullayeva, N., Arslandogdu, S., Cimendag, A., Gunduz, M., & Yavuz, B. G. (2019). The relationships between the digital game addiction, alexithymia and metacognitive problems in adolescents. *Turkish Journal of Clinical Psychiatry, 22*, 254–259.

Chapter 11
Clinical Neuroscience Meets Second-Person Neuropsychiatry

Leonhard Schilbach and Juha M. Lahnakoski

Abstract Disturbances of social and affective processes are at the core of psychiatric disorders. Together with genetic predisposing factors, deprivation of social contact and dysfunctional relationships during development are some of the most important contributors to psychiatric disorders over the lifetime, while some developmental disorders manifest as aberrant social behavior early in life. That the cause of mental illness is rooted in the brain was long held as a truism, yet finding the causes for and neurobiological correlates of these conditions in the brain has proven and continues to be difficult (Venkatasubramanian G, Keshavan MS, Ann Neurosci 23:3–5. https://doi.org/10.1159/000443549, 2016). In clinical practice, psychiatric disorders are diagnosed based on categorical manuals, such as the DSM and ICD, which form a useful guide for clinical diagnosis and interventions. Yet, understanding the specific neural mechanisms leading to or characterizing distinct psychiatric conditions through this categorical approach has been slow (see, for example, Lynch CJ, Gunning FM, Liston C, Biol Psychiatry 88:83–94. https://doi.org/10.1016/j.biopsych.2020.01.012, 2020). Findings in the brain often do not seem to lend support to common mechanisms for the defined disorder categories. This is not particularly surprising because, in these diagnostic manuals, multiple combinations of symptoms can often lead to the same diagnosis, which is reflected in highly variable phenotypes of psychiatric disorders.

Keywords Psychiatric disorders · Second-person neuroscience · Neuropsychiatry

L. Schilbach (✉)
Independent Max Planck Research Group for Social Neuroscience, Max Planck Institute of Psychiatry, Munich-Schwabing, Germany

LVR Klinikum Düsseldorf/Kliniken der Heinrich-Heine-Universität Düsseldorf, Düsseldorf, Germany

Ludwig-Maximilians-Universität, Medical Faculty, Munich, Germany
e-mail: leonhard.schilbach@lvr.de

J. M. Lahnakoski
Forschungszentrum Jülich, Institute of Neurosciences and Medicine (INM), Jülich, Germany

© The Author(s) 2023
P. S. Boggio et al. (eds.), *Social and Affective Neuroscience of Everyday Human Interaction*, https://doi.org/10.1007/978-3-031-08651-9_11

Introduction

Disturbances of social and affective processes are at the core of psychiatric disorders. Together with genetic predisposing factors, deprivation of social contact and dysfunctional relationships during development are some of the most important contributors to psychiatric disorders over the lifetime, while some developmental disorders manifest as aberrant social behavior early in life. That the cause of mental illness is rooted in the brain was long held as a truism, yet finding the causes for and neurobiological correlates of these conditions in the brain has proven and continues to be difficult (Venkatasubramanian & Keshavan, 2016).

In clinical practice, psychiatric disorders are diagnosed based on categorical manuals, such as the DSM and ICD, which form a useful guide for clinical diagnosis and interventions. Yet, understanding the specific neural mechanisms leading to or characterizing distinct psychiatric conditions through this categorical approach has been slow (see, for example, Lynch et al. 2020). Findings in the brain often do not seem to lend support to common mechanisms for the defined disorder categories. This is not particularly surprising because, in these diagnostic manuals, multiple combinations of symptoms can often lead to the same diagnosis, which is reflected in highly variable phenotypes of psychiatric disorders. Coupled with the complexity of the brain and its capacity for compensating regional disturbances through plastic changes makes it harder still to find causes and neural mechanisms of heterogeneous disorder labels. Moreover, evaluating the low-level contributors to psychiatric disorders is complicated as animal models for psychiatric conditions often cannot capture the complexity of these disorders in humans.

Recently, calls have been made for transdiagnostic approaches, such as the Research Domain Criteria (RDoC) framework (Insel et al., 2010), where mental illness is approached through specific behavioral domains rather than lists of specific symptoms and diagnostic labels. However, the majority of clinical research still relies on the categorization of patients, and even studies explicitly applying the RDoC framework to analyze neuroimaging data have mainly focused on finding correlates for a limited subset of the domains (Carcone & Ruocco, 2017). To address the variability of disorder phenotypes and to delineate particularly relevant transdiagnostic domains, a focus on social and affective processes relevant for mental health and illness appears to be crucial for improving our understanding of the brain basis of psychiatric disorders in a way that maximally benefits the patients. This is motivated by the insight that social impairments are some of the most debilitating facets of psychiatric disorders and the conceptual consideration that ascriptions of psychopathology always make reference to intersubjective conventions, which has led to the construal of psychiatric disorders as "disorders of social interaction" (Schilbach, 2016).

In this chapter, we will outline major approaches of studying the brain basis of psychiatric disorders, mainly focusing on our own work and related functional magnetic resonance imaging (fMRI) studies while briefly considering evidence from structural and brain stimulation studies as well. Furthermore, we will discuss recent methodological developments inspired by the above-described focus on social interaction, which has been described as a possible convergence of clinical neuro-science and psychiatry that could be described as the development of a second-person neuropsychiatry. This development highlights the importance of quantitatively measuring behavioral characteristics of patients during real-life social interaction and moving toward studying active behavior in addition to passive processing of social and affective information (Lahnakoski et al., 2020; Schilbach, 2019). Finally, we will consider when and why it is important to measure also the brain function of two (or more) people during real-time interaction and how the quantification of behavior becomes even more important under such complex conditions.

Structural Abnormalities and Functional Connectivity Correlates of Psychiatric Disorders

If we follow the long-standing logic that psychiatric disorders are "disorders of the brain" (Insel & Cuthbert, 2015), it is reasonable to assume that these disorders are reflected as physical and functional abnormalities in the brains of the affected indi-viduals. In some cases, such direct links exist, as witnessed by specific deficiencies or behavioral alterations due to brain injury and lesions in specific brain areas. Brain lesions can also lead to psychiatric symptoms in some cases, and, for example, mood disorders are often reported after traumatic brain injury. Yet, findings on spe-cific, focal abnormalities appear to be inconclusive for most psychiatric conditions. One notable exception is a focal target which is a region in the subgenual cingulate cortex that, when stimulated intracranially, can lead to a reduction of symptoms at least in some patients suffering from treatment-resistant depression (Mayberg et al., 2005). This effect is likely not mediated only by changes in local activity, but in the way this region modulates activity in other brain regions through its connections.

In addition to studying focal differences, structural abnormalities of white matter bundles or functional hubs of the brain, i.e., brain regions with a high number of connections, can disturb the functional architecture of the brain. This is clearly vis-ible in some neurological conditions, such as multiple sclerosis that affects the myelin sheath of neurons, thereby disturbing the electrical conduction of signals between brain regions. Functional connectivity, usually measured through temporal correlations of hemodynamic activity with fMRI, is thought to reflect the organiza-tion of brain connections and their functional integration. Repeatable patterns of connectivity have been produced in a multitude of studies reflecting plausible func-tional networks. Often, this connectivity is studied in the absence of a task, with the (implicit or explicit) assumption that the connectivity reflects relatively stable prop-erties underlying anatomy and physiology. Indeed, the effects of task-induced activ-ity on the functional connectivity patterns are reasonably subtle compared with the large-scale network structure (Gratton et al., 2018; Simony et al., 2016). Some evi-dence exists that reliable group-level differences exist in psychiatric disorders, such as autism spectrum disorder (Holiga et al., 2019) across multiple studied popula-tions. Yet, the variability between individuals in local connectivity measures tends to be high both in patients and in control populations highlighting the difficulty of finding common neural underpinnings for these disorders. Moreover, the temporal fluctuations of connectivity have recently gained more interest leading to an ongoing debate on whether state transitions and meta-states of connectivity at shorter timescale are a reliable or a more sensitive predictor of psychopathology than time averaged connectivity or, alternatively, an artifactual property of the analyses methods on slow and noisy signals.

Recently, other approaches looking at more global network or subnetwork properties, rather than local differences, have been gaining more attention. For example, differences in the subnetwork structure of functional networks including limbic regions have been reported that seem reliable across samples both during resting state and movie viewing paradigms (Glerean et al., 2016). However, the implications of these findings at the level of an individual remain unclear. Most network analyses rely critically on thresholding of the connectivity matrices (Garrison et al., 2015), and network properties can change considerably by a small change in the selected threshold. This can be alleviated, for example, by using relative thresholds (Garrison et al., 2015) or considering different ranges of connectivity values separately rather than setting a single threshold (Bassett et al., 2012), which can help in detecting connectivity patterns that are predictive of psychopathology.

One recent development has combined lesion studies with connectivity measures, where functional connectivity in patient groups sharing similar symptoms yet having distinct focal brain lesions suggests the connectivity of the lesioned areas may be particularly important to determine the functional consequences for the patients. For example, two aspects of "free will," volition and agency, appear to be differentially affected depending on the connectivity of the lesion site (Darby et al., 2018), with the former being associated with lesions in regions that connect to the anterior cingulate cortex and the latter with regions connecting to the precuneus. These results suggest that aberrant structure or function of different sets of brain regions may potentially have common effects through their connections in a region that is not directly affected by the lesion. However, whether these findings prove helpful for patients suffering from psychiatric disorders remains unclear.

Importantly, it seems that differences in functional brain networks, compared to a healthy population, are highly overlapping between multiple psychiatric disorders. Indeed, rather than being disorder-specific, measures of general level of psychopathology, the so-called p factor (Caspi et al., 2014), can often explain much of the neuroimaging findings. It has been argued that there may be a common underlying contributor that predisposes individuals to developing a range of psychiatric disorders, which may also be reflected in the overlap of genetic findings across psychiatric disorders, which is supported by recent findings of shared neurobiological and cellular mechanisms of at least six different psychiatric disorders reported by the relevant working groups of the Enigma project (Patel et al., 2020). Controlling for these disorder-general correlates of psychiatric disorders may help in pinpointing the disorder-specific mechanisms. However, if the brain is studied through static anatomical and connectivity properties without any behavioral readouts beyond a categorical label, understanding the significance of these findings to the social life and general well-being of patients is not straight forward.

Stimulation-, Task-, and Model-Based Studies

Most of social cognitive neuroscience, particularly neuroimaging studies, have focused on simplified stimulus- and task-based designs. The goal here is to isolate and systematically manipulate particular constituent features or task components that together could enable more complex tasks to be performed. In a clinical context, one might then compare how strongly particular brain regions are activated by a given task across different diagnostic groups or if the activity level is correlated with certain symptom dimensions.

This approach has clear benefits for the interpretation of potential group differences because the observed brain activity can be linked to specific cognitive functions, in particular when mathematical modeling allows to predict brain activity change, which can be taken to suggest that the brain realizes similar computations to generate and control behavior. One example of this approach is a suite of recent studies by Henco and colleagues (2020b, b), in which they investigate the effect of implicit social cues (e.g., gaze shifts of a face) to bias decision-making in a probabilistic learning task, even though study participants were not asked to the social cues into account. Intriguingly, the way these social cues affect decision-making appears to be mechanistically different in individuals with borderline personality disorder and schizophrenia compared to both healthy controls and patients with major depressive disorder (Henco et al., In press), suggesting that the study of implicit social processes in combination with computational modeling might be particularly helpful in elucidating the neural mechanisms that differentiate these disorders.

Importantly, these kinds of experimental task use a fixed reward and learning schedule, which offers high levels of experimental control, and lend themselves to data analytic approaches that use mathematical models to describe cognitive and putatively neural mechanisms that underlie participants' behavior. Using hierarchical Gaussian filter models, Sevgi and colleagues (Sevgi et al., 2020) demonstrated how participants integrate social and nonsocial information to come up with their decisions and how this differs as a function of interindividual variance of autistic traits. The parameters derived from computational modeling can also be used to inform neuroimaging analysis, which has become known as model-based fMRI: here, it can, for instance, be assessed whether trial-to-trial changes of modeling parameters are related to brain activity changes. Using this approach, Henco et al. (2020a, b) demonstrated that interindividual differences in social belief computations, i.e., whether participants tend to use social cues during decision-making, even when not explicitly instructed to do so, were related to brain activity levels in the putamen and insula, areas that have previously been associated with habitual behaviors and interoception.

Naturalistic Passive Observation

While the conventional approaches described above have allowed us to gain completely new insights into relevant brain processes, many of them tend to rely on the assumption of "pure insertion," at least approximately (Friston et al., 1996). That is, it is assumed that effects of the manipulation of individual features or processes are essentially independent of each other and, in more complex or naturalistic conditions, these effects sum up to produce more complex processes or behaviors. In some cases, findings of simplified experiments generalize to more natural conditions, at least to some extent. For example, contrast edges of video images correlate with activity in the early visual cortex (Lahnakoski et al., 2012a), as might be expected based on the properties of edge-detecting cells in the region, but the amount of variance explained is relatively low. Careful consideration of a range of stimulus features can reveal insight into the organization of the brain networks of naturalistic social observation, for example, highlighting regions such as the posterior superior temporal sulcus and surrounding temporoparietal regions as potentially key regions for integrating multiple types of socially relevant information (Lahnakoski et al., 2012b) as well as building coherent temporal sequences of related events (Lahnakoski et al., 2017). The amplitudes of responses to emotionally arousing events in these regions appear to be also related to individual differences of the endogenous opioid system (Karjalainen et al., 2019), which may prove helpful for assessing potentially aberrant neurotransmitter function in psychopathology. Importantly, however, it is less clear how complex intuitive social processes can be deconstructed into more basic constituents. Arguably, more naturalistic social processes are only observable in complex situations, and the underlying processes may not be directly accessible through the stimulus properties, event descriptions, or even simple dimensional models of emotion alone. For example, recent findings have shown that when participants share a point of view toward movie events, either experimentally (Lahnakoski et al., 2014) or through friendship in everyday life (Parkinson et al., 2018), the similarity of the brain activity is increased compared with individuals who do not share a perspective or do not know each other. Such similarity between friends appears not to be reflected in functional connectivity during rest (McNabb et al., 2020), although more sensitive measures may yet reveal such associations. Moreover, naturalistic stimulation may provide benefits for detecting aberrant brain activity related to psychiatric disorders (Eickhoff et al., 2020), and prediction of behavioral traits may prove to be more successful using connectivity measures derived from, particularly social, natural viewing paradigms rather than resting state data (Finn & Bandettini, 2020). Thus, more ecologically valid dynamic stimulation may not only highlight brain-behavior associations but also highlight the types of content that best reveal these associations to guide us to further our understanding of naturalistic brain processes beyond simple models of stimulus features or general emotion dimensions (see, for example, Finn et al. 2020).

However, despite this potential benefit in highlighting individual differences, the use of naturalistic stimuli in the study of psychiatric disorders is still relatively rare. Some of the earliest studies have shown that, for example, individuals with ASD tend to show idiosyncratic patterns of both eye gaze and brain activity during natural viewing conditions (Hasson et al., 2009; Salmi et al., 2013). This highlights a potential difficulty in understanding the brain mechanisms underlying psychiatric disorders mentioned earlier; if patients with the same diagnosis are highly variable, then group contrasts, and predictions are likely to fail. It, thus, appears particularly important to further characterize the participants' behavior and experiences, as well as the contents of the stimuli that are particularly relevant for detecting the disorders. For example, during movie viewing, aberrant brain activity related to positive symptoms of first-episode psychosis patients appears to be particularly observable during surreal, fantasy scenes, which may share aspects of the patients' symptoms (Rikandi et al., 2017). Further work is required to discover the limits of passive observation studies and to what extent specific neural functions can be studied in complex conditions, with more limited experimental control. Likely, a fruitful approach is to iteratively alternate between more exploratory findings in naturalistic experiments, working backward toward more controlled conditions to design experiments to test specific hypotheses on the low-level mechanisms of psychiatric disorders, and testing the mechanistic predictions again in more naturalistic conditions, potentially in interactive tasks mimicking real-life situations where the presumed mechanism is particularly important (cf. Schilbach, 2019).

Interactive Experiments, Second-Person Neuroscience, and Neuropsychiatry

While investigating more naturalistic social situations is beneficial to understand complex social cognition, it has been pointed out that a fundamental difference may exist between situations of social observation, i.e., social cognition from an observer's point of view, as compared to situations of social interaction, i.e., social cognition from an interactor's point of view (Schilbach, 2014, 2016; Schilbach et al., 2013). Contrary to the conventional stimulus-response paradigms described above, social interactions are characterized by behavioral reciprocity. That is, social perception leads to actions that, in turn, will be responded to by the interaction partner (and so forth). In order to investigate how these social contingencies and the ensuing dynamics of social interaction modulate brain activity, we, therefore, need truly interactive tasks, which allow for the participant to engage in such reciprocal social interactions. Following the call for a truly social or second-person neuroscience, recent years have seen a growing number of studies that have focused on core social-interactive behaviors, such as studies in which participants perceive communicative cues to engage them in interaction (e.g., direct gaze) all the way to studies that include reciprocal, face-to-face interactions with a social partner (real or perceived; see Redcay and Schilbach (2019) for a recent review). In addition to increasing the ecological validity of the task used and making the social encounters more lifelike and dynamic, for example, using real video recordings in place of computer-generated avatars (Brandi et al., 2019), second-person neuroscience has also focused on scanning interacting brains, which has been described as hyperscanning (e.g., Bilek et al. 2015; Dumas et al. 2010). Findings from these studies have helped to gain striking new insights into the workings of "social brains," which, indeed, indicate that the neural mechanisms supporting social interaction do, in fact, differ from those during social observation. Findings converge on a set of brain regions and large-scale neural networks that appear to play key roles and interact in intricate ways in order to support social behavior during social interaction. In addition, the use of two-person experiments and hyperscanning techniques allows us to take a completely new look at how social behavior is realized across persons and brains and to investigate phenomena such as interpersonal synchrony, mimicry, and other forms of alignment in more ecologically valid contexts (Bolis et al., 2017; Schilbach, 2015). These developments constitute important steps in the advancement of social neuroscience and will continue to provide new insights into how activity in large-scale neural networks is modulated by social interactions and also open up new avenues for future research.

In addition to this, a second-person neuroscience may also be relevant for neuroimaging research in the field of psychiatry and could, therefore, contribute to what might be called a second-person neuropsychiatry (Schilbach, 2016): Here, it has been increasingly recognized that it is social interaction rather than passive observation that is often most difficult for patients suffering from psychiatric disorders. For example, an individual may well understand an emotion depicted in a movie as the conventions that have been developed by the artists working in the movie industry are highly efficient in conveying emotions, whereas in real life emotional cues may be much subtler. Moreover, in real-time interactions, there is little time for explicit interpretations of the socio-emotional states of the interaction partner but rather relies on a practical "know-how" of how to deal with them. In other words, people often automatically understand, empathize with, and predict the words or actions of their partner enabling them to act appropriately without explicit reasoning. This has been demonstrated by a study by von der Lühe and colleagues (von der Lühe et al., 2016), in which it was shown that patients with high-functioning autism are able to recognize and explicitly label actions even when they are depicted by impoverished point-light displays but fail to use this information to predict the subsequent action of a potential interaction partner. In other words, it was only the complexity of a dyadic social interaction situation that brought about autism-specific deficits in predicting subsequent actions rather than difficulties in action perception, which was found to be intact. Following this lead, it appears important to introduce new methods and techniques that help us to quantitatively assess behavior during real-life social interactions as this may help to understand how social interaction difficulties might be related to alterations of cross-brain rather than single-brain network activity (Bilek et al., 2017; Bolis & Schilbach, 2018).

Behavioral Characterization of Psychiatric Disorders in Individuals, Dyads, and Social Networks in Everyday Life

Studying constrained social interactions in the laboratory has clear benefits for interpretability compared with trying to measure interactions "in the wild," much like controlled task designs in neuroimaging studies often allow for more straightforward modeling and interpretation of results than more naturalistic experiments. Yet, constrained experiments can be rather poor approximations of real-life social behavior. Moreover, our initial systematic measures of behavior during dyadic interaction suggest that some behavioral characteristics of individuals may only manifest when they can interact freely, with minimal experimental constraints (Lahnakoski et al., 2020). Thus, enabling the systematic, quantifiable measurement of social behavior in natural interactions, i.e., interaction-based phenotyping, in the clinic as well as in the everyday life of patients may be crucial for understanding the individual as well as shared symptoms of psychiatric disorders (Schilbach, 2019). This type of extensive characterization of psychiatric disorders at the level of individual patients may be the key to disentangling general brain correlates of psychopathology from disorder- and symptom-specific brain mechanisms. Moreover, it may be the key to finally move toward individualized interventions in psychiatry, which to a large extent are still lacking.

Interestingly, behavioral measures, such as interindividual synchrony and mimicry, distance, gaze, and orienting of the face and the body, have been shown to be predictive of the subjective quality of interactions (Lahnakoski et al., 2020). Also, using measures of motion energy in videos between patients and their therapist, behavioral synchrony has shown promise in predicting short- and long-term therapeutic success for patients with schizophrenia (Ramseyer & Tschacher, 2014). It may also be possible to differentiate between patients with autism spectrum disorder (ASD) from control participants based on their behavioral synchrony with an interaction partner (Georgescu et al., 2019), although further work is needed to evaluate the practical applicability of these preliminary findings in larger cohorts. Importantly, however, using such simple measures of synchrony of motion lack specificity of what the people are doing during the interaction. Moreover, synchrony does not appear to always be useful for detecting differences in subjective interaction quality. In the study mentioned above (Lahnakoski et al., 2020), we showed that measures like distance and facial orienting behavior may be more indicative of the subjective enjoyment and effort invested into interactions, respectively. Moreover, these may be differently predictive in different conditions, so a single measure may not fit all questions.

While such systematic and quantitative characterizations of dyadic social interactions appear to be a fruitful avenue for evaluating, for example, dyadic behavior during interactions with a therapist in the clinic, the majority of social interaction problems manifest in everyday life. Anecdotally at least, patients may feel fine at the clinic and have severe relapses of symptoms after they are discharged and have to continue their daily lives. Thus, to get a picture of the causes of the daily difficulties patients face, beyond subjective evaluations, quantitative measurements should be extended to daily life of individuals. The recent widespread introduction of personal digital devices, such as smartphones, led to the development of digital phenotyping (Onnela & Rauch, 2016), where such devices, potentially complemented by, for example, wearable sensors, can be used to continuously measure the behavior of individuals in their everyday life. This approach can produce a wealth of data for detecting various social and behavioral characteristics of illnesses (Torous et al., 2016), which can be of great benefit for finding behavioral markers that may guide therapy and further scientific inquiry. Yet, much work is still required to detect consistent, meaningful patterns in this type of data. Moreover, pattern detections and behavioral prediction that are not informed by strong theoretical foundation cannot substitute a mechanistic understanding of the disorders.

On an optimistic note, the use of interaction-based phenotyping and other forms of digital phenotyping "in the wild" may help to investigate the social behavior and factors that are relevant and constitutive of psychiatric disorders. As the relevant classifications used in psychiatry today rely on intersubjective conventions of what should be considered as a nosological entity, the use of quantitative, data-driven approaches that integrate information about social, psychological, and biological factors may help to delineate disorder-general and disorder-specific profiles for what we take as separate disorders today. In addition, a major challenge for the future also lies in the definition of mechanistic models of psychiatric disorders that are grounded in the underlying neurophysiology and are able to make predictions of outcomes of specific disturbances of the system and interventions that alleviate such disturbances. So far, the existing models have yet to prove their usefulness in the larger scale. However, initial mechanistic insight into potential contributors to disordered social processing has started to shed light on underlying psychological mechanisms of psychiatric disorders. In the two studies mentioned earlier (Henco et al., 2020a, b, In press), we used a hierarchical learning models to demonstrate that not only do patients with schizophrenia and borderline personality disorder score lower in probabilistic learning task in the presence of implicit social cues but also expanded to the mechanisms of excessive weighting of social information during periods of uncertainty. Similar learning models can be used for various types of interactions. However, modeling unconstrained real-life interactions is a significant challenge for future research. Thus, a thorough exploration and systematic characterization of interactions seem crucial for guiding modeling efforts of social interaction disorders and eventually linking them to their underlying causes in the mind, brain, and body.

Eventually, to fully understand and empirically test the brain mechanisms of reciprocal social interactions, we will also need to not only correlate behavior with subsequent brain measures but also be able to measure the brains of two (or more) interacting individuals at the same time to directly link brain activity and behavior. Such hyperscanning studies have been slowly gaining momentum, as briefly described above. In this context, mobile electroencephalography (EEG) and functional near-infrared spectroscopy (fNIRS) offer the benefit of much reduced constraints on behavior compared to fMRI, although simultaneous fMRI experiments have been performed for some time, either by linking two separate MRI devices (Montague et al., 2002) or with specially designed head coils within one scanner

(Renvall et al., 2020). However, it is important to consider when it is necessary to measure multiple people at the same time and when it is sufficient to measure, for example, only one person during an interaction with another person outside of the scanner (cf. Redcay & Schilbach, 2019). Alternatively, brain imaging can be performed sequentially, by first measuring the brain activity and audio or video recording, e.g., a person telling a story (Smirnov et al., 2019) or performing hand actions (Smirnov et al., 2017), followed by a measurement of participants listening or viewing the recording. Hyperscanning studies are complex to run and analyze, and, thus, it may be counterproductive to design such studies when a simpler experimental design would suffice. Moreover, when people are measured while they participate in an interaction, it is particularly important to know how they are behaving (Hamilton, 2020) as no exact schedule for events during the interaction can be enforced. For example, neural synchrony, which to some extent appears to be associated with sharing "the social world," or state of mind with other people (Nummenmaa et al., 2018) may, during an interaction, also arise trivially when the interactants just look at the same stimulus at the same time. Thus, during an interaction, synchrony may arise in a similar manner as in the passive observation studies described above without any deeper sharing of mental states. For the latter, activity in the so-called mentalizing network of the brain has been implicated, in particular in situations of direct social interaction (Redcay & Schilbach, 2019). Moreover, because every interaction is different, direct comparisons based only on the brain activity of people are difficult to interpret without characterizing the interaction. Thus, a combination of detailed behavioral characterization and brain-based measures is crucial for a more complete understanding of the neural underpinnings of natural social interactions and disorders thereof in psychopathology.

Conclusions

In the past, reliance on heterogeneous disorder categories and an overemphasis on the brain have potentially limited the progress of our understanding of the behavioral and neural mechanisms of psychiatric disorders. Moreover, common predisposing mechanisms appear to be shared by multiple disorders, which can lead to nonspecific findings between disorders, and more specific measures of the disorders are required. While subjective mental suffering of patients is not directly accessible to researchers or therapists, disordered social interactions are some of the most severe symptoms of many psychiatric disorders that are, at least in part, detectable and measurable by an external observer. Differences in social behavior are often intuitively used by therapists while diagnosing and interacting with patients, yet rarely are these behavioral abnormalities systematically measured. By carefully characterizing individual behavioral manifestations of the disorders between patients, particularly in social interactions and everyday life, we may better understand the complex disorder phenotypes and their underlying mechanisms and, ultimately, move closer to individualized interventions.

References

Bassett, D. S., Nelson, B. G., Mueller, B. A., Camchong, J., & Lim, K. O. (2012). Altered resting state complexity in schizophrenia. *NeuroImage, 59*(3), 2196–2207. https://doi.org/10.1016/j.neuroimage.2011.10.002

Bilek, E., Ruf, M., Schäfer, A., Akdeniz, C., Calhoun, V. D., Schmahl, C., … Meyer- Lindenberg, A. (2015). Information flow between interacting human brains: Identification, validation, and relationship to social expertise. *Proceedings of the National Academy of Sciences of the United States of America, 112*(16), 5207–5212. https://doi.org/10.1073/pnas.1421831112

Bilek, E., Stößel, G., Schäfer, A., Clement, L., Ruf, M., Robnik, L., … Meyer- Lindenberg, A. (2017). State-dependent cross-brain information flow in borderline personality disorder. *JAMA Psychiatry, 74*(9), 949–957. https://doi.org/10.1001/jamapsychiatry.2017.1682

Bolis, D., & Schilbach, L. (2018, January 1). Observing and participating in social interactions: Action perception and action control across the autistic spectrum. *Developmental Cognitive Neuroscience, 29*, 168–175. https://doi.org/10.1016/j.dcn.2017.01.009

Bolis, D., Balsters, J., Wenderoth, N., Becchio, C., & Schilbach, L. (2017). Beyond autism: Introducing the dialectical misattunement hypothesis and a bayesian account of intersubjectivity. *Psychopathology, 50*, 355–372. https://doi.org/10.1159/000484353

Brandi, M.-L., Kaifel, D., Lahnakoski, J. M., & Schilbach, L. (2019). A naturalistic paradigm simulating gaze-based social interactions for the investigation of social agency. *Behavior Research Methods.* https://doi.org/10.3758/s13428-019-01299-x

Carcone, D., & Ruocco, A. C. (2017). Six Years of Research on the National Institute of Mental Health's Research Domain Criteria (RDoC) Initiative: A systematic review. *Frontiers in Cellular Neuroscience, 11*, 46. https://doi.org/10.3389/fncel.2017.00046

Caspi, A., Houts, R. M., Belsky, D. W., Goldman-Mellor, S. J., Harrington, H., Israel, S., … Moffitt, T. E. (2014). The p factor: One general psychopathology factor in the structure of psychiatric disorders? *Clinical Psychological Science, 2*(2), 119–137. https://doi.org/10.1177/2167702613497473

Darby, R. R., Joutsa, J., Burke, M. J., & Fox, M. D. (2018). Lesion network localization of free will. *Proceedings of the National Academy of Sciences of the United States of America, 115*(42), 10792–10797. https://doi.org/10.1073/pnas.1814117115

Dumas, G., Nadel, J., Soussignan, R., Martinerie, J., & Garnero, L. (2010). Inter-brain synchronization during social interaction. *PloS One, 5*(8), e12166. https://doi.org/10.1371/journal.pone.0012166

Eickhoff, S. B., Milham, M., & Vanderwal, T. (2020). Towards clinical applications of movie fMRI. *NeuroImage, 217*, 116860. https://doi.org/10.1016/j.neuroimage.2020.116860

Finn, E. S., & Bandettini, P. A. (2020). Movie-watching outperforms rest for functional connectivity-based prediction of behavior. *BioRxiv*, 2020.08.23.263723. https://doi.org/10.1101/2020.08.23.263723

Finn, E. S., Glerean, E., Khojandi, A. Y., Nielson, D., Molfese, P. J., Handwerker, D. A., & Bandettini, P. A. (2020). Idiosynchrony: From shared responses to individual differences during naturalistic neuroimaging. *NeuroImage, 215*, 116828. https://doi.org/10.1016/j.neuroimage.2020.116828

Friston, K. J., Price, C. J., Fletcher, P., Moore, C., Frackowiak, R. S. J., & Dolan, R. J. (1996). The trouble with cognitive subtraction. *NeuroImage, 4*(2), 97–104. https://doi.org/10.1006/nimg.1996.0033

Garrison, K. A., Scheinost, D., Finn, E. S., Shen, X., & Constable, R. T. (2015). The (in)stability of functional brain network measures across thresholds. *NeuroImage, 118*, 651–661. https://doi.org/10.1016/j.neuroimage.2015.05.046

Georgescu, A. L., Koehler, J. C., Weiske, J., Vogeley, K., Koutsouleris, N., & Falter-Wagner, C. (2019). Machine learning to study social interaction difficulties in ASD. *Frontiers in Robotics and AI, 6*, 132. https://doi.org/10.3389/frobt.2019.00132

Glerean, E., Pan, R. K., Salmi, J., Kujala, R., Lahnakoski, J. M., Roine, U., ... Jääskeläinen, I. P. (2016). Reorganization of functionally connected brain subnetworks in high-functioning autism. *Human Brain Mapping, 37*(3). https://doi.org/10.1002/hbm.23084

Gratton, C., Laumann, T. O., Nielsen, A. N., Greene, D. J., Gordon, E. M., Gilmore, A. W., ... Petersen, S. E. (2018). Functional brain networks are dominated by stable group and individual factors, not cognitive or daily variation. *Neuron, 98*(2), 439–452.e5. https://doi.org/10.1016/j.neuron.2018.03.035

Hamilton, A. (2020). Hype, hyperscanning and embodied social neuroscience. https://doi.org/10.31234/osf.io/rc9wp

Hasson, U., Avidan, G., Gelbard, H., Vallines, I., Harel, M., Minshew, N., & Behrmann, M. (2009). Shared and idiosyncratic cortical activation patterns in autism revealed under continuous real-life viewing conditions. *Autism Research, 2*(4), 220–231. https://doi.org/10.1002/aur.89

Henco, L., Diaconescu, A., Lahnakoski, J., Brandi, M.-L., Hörmann, S., Hennings, J., ... Mathys, C. (2020a). Aberrant computational mechanisms of social learning and decision-making in schizophrenia and borderline personality disorder. *PLoS Computational Biology.*

Henco, L., Brandi, M. L., Lahnakoski, J. M., Diaconescu, A. O., Mathys, C., & Schilbach, L. (2020b). Bayesian modelling captures inter-individual differences in social belief computations in the putamen and insula. *Cortex.* https://doi.org/10.1016/j.cortex.2020.02.024

Holiga, Š., Hipp, J. F., Chatham, C. H., Garces, P., Spooren, W., D'Ardhuy, X. L., ... Dukart, J. (2019). Patients with autism spectrum disorders display reproducible functional connectivity alterations. *Science Translational Medicine, 11*(481). https://doi.org/10.1126/scitranslmed.aat9223

Insel, T., Cuthbert, B., Garvey, M., Heinssen, R., Pine, D. S., Quinn, K., ... Wang, P. (2010, July). Research Domain Criteria (RDoC): Toward a new classification framework for research on mental disorders. *American Journal of Psychiatry, 167*, 748–751. https://doi.org/10.1176/appi.ajp.2010.09091379

Insel, T. R., & Cuthbert, B. N. (2015). Brain disorders? Precisely: Precision medicine comes to psychiatry. *Science, 348*(6234), 499–500.

Karjalainen, T., Seppälä, K., Glerean, E., Karlsson, H. K., Lahnakoski, J. M., Nuutila, P., ... Nummenmaa, L. (2019). *Opioidergic regulation of emotional arousal: A combined PET-fMRI study.* Cerebral Cortex. https://doi.org/10.1093/cercor/bhy281

Lahnakoski, J. M., Glerean, E., Salmi, J., Jääskeläinen, I. P., Sams, M., Hari, R., & Nummenmaa, L. (2012a, July). Naturalistic fMRI mapping reveals superior temporal sulcus as the hub for the distributed brain network for social perception. *Frontiers in Human Neuroscience.*

Lahnakoski, J. M., Salmi, J., Jääskeläinen, I. P., Lampinen, J., Glerean, E., Tikka, P., & Sams, M. (2012b). Stimulus-related independent component and voxel- wise analysis of human brain activity during free viewing of a feature film. *PLoS ONE, 7*(4), e35215. https://doi.org/10.1371/journal.pone.0035215

Lahnakoski, J. M., Glerean, E., Jääskeläinen, I. P., Hyönä, J., Hari, R., Sams, M., & Nummenmaa, L. (2014). Synchronous brain activity across individuals underlies shared psychological perspectives. *NeuroImage, 100C*, 316–324. https://doi.org/10.1016/j.neuroimage.2014.06.022

Lahnakoski, J. M., Jääskeläinen, I. P., Sams, M., & Nummenmaa, L. (2017). Neural mechanisms for integrating consecutive and interleaved natural events. *Human Brain Mapping.* https://doi.org/10.1002/hbm.23591

Lahnakoski, J. M., Forbes, P. A. G., McCall, C., & Schilbach, L. (2020). Unobtrusive tracking of interpersonal orienting and distance predicts the subjective quality of social interactions. *Royal Society Open Science, 7*(8), 191815. https://doi.org/10.1098/rsos.191815

Lynch, C. J., Gunning, F. M., & Liston, C. (2020, July 1). Causes and consequences of diagnostic heterogeneity in depression: Paths to discovering novel biological depression subtypes. *Biological Psychiatry, 88*, 83–94. https://doi.org/10.1016/j.biopsych.2020.01.012

Mayberg, H. S., Lozano, A. M., Voon, V., McNeely, H. E., Seminowicz, D., Hamani, C., ... Kennedy, S. H. (2005). Deep brain stimulation for treatment-resistant depression. *Neuron, 45*(5), 651–660. https://doi.org/10.1016/j.neuron.2005.02.014

McNabb, C. B., Burgess, L. G., Fancourt, A., Mulligan, N., FitzGibbon, L., Riddell, P., & Murayama, K. (2020). No evidence for a relationship between social closeness and similarity in resting-state functional brain connectivity in schoolchildren. *Scientific Reports, 10*(1). https://doi.org/10.1038/s41598-020-67718-8

Montague, P. R., Berns, G. S., Cohen, J. D., McClure, S. M., Pagnoni, G., Dhamala, M., ... Fisher, R. E. (2002, August 1). Hyperscanning: Simultaneous fMRI during linked social interactions. *NeuroImage, 16*, 1159–1164. https://doi.org/10.1006/nimg.2002.1150

Nummenmaa, L., Lahnakoski, J. M., & Glerean, E. (2018). Sharing the social world via intersubject neural synchronisation. *Current Opinion in Psychology, 24*, 7–14. https://doi.org/10.1016/j.copsyc.2018.02.021. Epub 2018 Mar 8. PMID: 29550395.

Onnela, J. P., & Rauch, S. L. (2016, June 1). Harnessing smartphone-based digital phenotyping to enhance behavioral and mental health. *Neuropsychopharmacology, 41*, 1691–1696. https://doi.org/10.1038/npp.2016.7

Parkinson, C., Kleinbaum, A. M., & Wheatley, T. (2018). Similar neural responses predict friendship. *Nature Communications, 9*(1), 1–14. https://doi.org/10.1038/s41467-017-02722-7

Patel, Y., Parker, N., Shin, J., Howard, D., French, L., Thomopoulos, S. I., ... Paus, T. (2020). Virtual histology of cortical thickness and shared neurobiology in 6 psychiatric disorders. *JAMA Psychiatry.* https://doi.org/10.1001/jamapsychiatry.2020.2694

Ramseyer, F., & Tschacher, W. (2014). Nonverbal synchrony of head- and body- movement in psychotherapy: Different signals have different associations with outcome. *Frontiers in Psychology, 5.* https://doi.org/10.3389/FPSYG.2014.00979

Redcay, E., & Schilbach, L. (2019). Using second-person neuroscience to elucidate the mechanisms of social interaction. *Nature Reviews Neuroscience, 20*(8), 495–505. https://doi.org/10.1038/s41583-019-0179-4

Renvall, V., Kauramäki, J., Malinen, S., Hari, R., & Nummenmaa, L. (2020). Imaging real-time tactile interaction with two-person dual-coil fMRI. *Frontiers in Psychiatry, 11*, 279. https://doi.org/10.3389/fpsyt.2020.00279

Rikandi, E., Pamilo, S., Mäntylä, T., Suvisaari, J., Kieseppä, T., Hari, R., ... Raij, T. T. (2017). Precuneus functioning differentiates first-episode psychosis patients during the fantasy movie Alice in Wonderland. *Psychological Medicine, 47*(3), 495–506. https://doi.org/10.1017/S0033291716002609

Salmi, J., Roine, U., Glerean, E., Lahnakoski, J., Nieminen-von Wendt, T., Tani, P., ... Sams, M. (2013). The brains of high functioning autistic individuals do not synchronize with those of others. *NeuroImage. Clinical, 3*, 489–497. https://doi.org/10.1016/j.nicl.2013.10.011

Schilbach, L. (2014). On the relationship of online and offline social cognition. *Frontiers in Human Neuroscience, 8*(MAY), 278. https://doi.org/10.3389/fnhum.2014.00278

Schilbach, L. (2015). Eye to eye, face to face and brain to brain: novel approaches to study the behavioral dynamics and neural mechanisms of social interactions. *Current Opinion in Behavioral Sciences, 3*, 130–135. https://doi.org/10.1016/j.cobeha.2015.03.006

Schilbach, L. (2016). Towards a second-person neuropsychiatry. *Philosophical Transactions of the Royal Society of London. Series B, Biological Sciences, 371*(1686), 20150081. https://doi.org/10.1098/rstb.2015.0081

Schilbach, L. (2019). Using interaction-based phenotyping to assess the behavioral and neural mechanisms of transdiagnostic social impairments in psychiatry. *European Archives of Psychiatry and Clinical Neuroscience, 269*(3), 273–274. https://doi.org/10.1007/s00406-019-00998-y

Schilbach, L., Timmermans, B., Reddy, V., Costall, A., Bente, G., Schlicht, T., & Vogeley, K. (2013). Toward a second-person neuroscience. *The Behavioral and Brain Sciences, 36*(4), 393–414. https://doi.org/10.1017/S0140525X12000660

Sevgi, M., Diaconescu, A. O., Henco, L., Tittgemeyer, M., & Schilbach, L. (2020). Social Bayes: Using Bayesian modeling to study autistic trait–related differences in social cognition. *Biological Psychiatry, 87*(2), 185–193. https://doi.org/10.1016/j.biopsych.2019.09.032

Simony, E., Honey, C. J., Chen, J., Lositsky, O., Yeshurun, Y., Wiesel, A., & Hasson, U. (2016). Dynamic reconfiguration of the default mode network during narrative comprehension. *Nature Communications, 7.* https://doi.org/10.1038/ncomms12141

Smirnov, D., Lachat, F., Peltola, T., Lahnakoski, J. M., Koistinen, O.-P., Glerean, E., ... Nummenmaa, L. (2017). Brain-to-brain hyperclassification reveals action- specific motor mapping of observed actions in humans. *PLoS ONE*. https://doi.org/10.1371/journal.pone.0189508

Smirnov, D., Saarimäki, H., Glerean, E., Hari, R., Sams, M., & Nummenmaa, L. (2019). Emotions amplify speaker–listener neural alignment. *Human Brain Mapping, 40*(16), 4777–4788. https://doi.org/10.1002/hbm.24736

Torous, J., Kiang, M. V., Lorme, J., & Onnela, J.-P. (2016). New tools for new research in psychiatry: A scalable and customizable platform to empower data driven smartphone research. *JMIR Mental Health, 3*(2), e16. https://doi.org/10.2196/mental.5165

Venkatasubramanian, G., & Keshavan, M. S. (2016, March 1). Biomarkers in psychiatry – A critique. *Annals of Neurosciences, 23*, 3–5. https://doi.org/10.1159/000443549

von der Lühe, T., Manera, V., Barisic, I., Becchio, C., Vogeley, K., & Schilbach, L. (2016). Interpersonal predictive coding, not action perception, is impaired in autism. *Philosophical Transactions of the Royal Society B: Biological Sciences, 371*(1693). https://doi.org/10.1098/rstb.2015.0373

Part IV
Methods Used in Social and Affective Neuroscience

Chapter 12
EEG and ERPs in the Study of Language and Social Knowledge

Alice Mado Proverbio

Abstract Event-related potentials (ERPs) represent the ideal methodological approach for investigating the time course of language reading and comprehension processes. In this chapter, various ERP components reflecting orthographic, phonological, semantic, and syntactic processing of written and auditory language are examined. Furthermore, data are shown of how ERPs can reflect stereotypes, prejudices and world knowledge, including people's social traits and attributes. In particular, several recent neuroimaging and electrophysiological studies are presented investigating the neural underpinnings of ethnic and sex biases (both in male and female individuals).

Keywords EEG and ERPs · Electrophysiology of language · Orthographic analysis · Stereotypes and prejudices

Introduction: EEG and ERP Signals of the Brain

The electromagnetic activity of the brain essentially translates into (i) electric fields/potentials and oscillatory magnetic fields, which constitute the electroencephalogram and the magnetoencephalogram, and (ii) variations of the electric and magnetic fields caused by nerve impulses induced by external or mental stimuli/events, which result in event-related potentials (ERPs) and event-related fields (ERF), respectively (for details, see the handbook by Zani and Proverbio (2003)).

The rhythmic EEG oscillations originate in the cortex, but their pacemaker is subcortical and is located in the thalamic nuclei. The electrical potentials of the brain can be detected on the scalp surface through the application of metallic sensors named electrodes, while the magnetic fields are measured by sensitive MEG gradiometers. The potential changes recorded at the scalp derive from the sum of both excitatory and inhibitory postsynaptic potentials of neurons whose apical

A. M. Proverbio (✉)
Department of Psychology, University of Milano-Bicocca, Milan, Italy
e-mail: mado.proverbio@unimib.it

© The Author(s) 2023
P. S. Boggio et al. (eds.), *Social and Affective Neuroscience of Everyday Human Interaction*, https://doi.org/10.1007/978-3-031-08651-9_12

dendrites are oriented perpendicular to the cortical surface (e.g., pyramidal cells or hyper-columns of the visual cortex).

In general, the typical EEG rhythms of waking state in the adult person have a fairly rapid oscillation frequency, which varies between 8 and 25 Hz (alpha and beta rhythms), while in the sleepiness and sleep states, the EEG rhythm progressively decreases reaching 1 Hz of frequency in the *slow-wave sleep* (SWS), known as delta rhythm.

ERPs consist in electric potential oscillations that occur in the brain of an individual in response to a stimulus administered in one of the different sensory modalities ("exogenous" potentials) or in relation to higher cognitive functions such as attention, motivation, emotions, and expectations ("endogenous" potentials; see Zani and Proverbio (2003)). ERPs, in whatever modality they are recorded (visual, acoustic, or somatosensory), appear as waveforms characterized by a series of positive and negative deflections whose polarity is marked by P and N letters of the alphabet and accompanied by increasing numbers indicating the temporal progression of appearance (latency in ms). Each of the ERP components can be considered as the manifestation of neural activity associated with specific stages of information transmission and processing within the brain.

Electrophysiology of Language

Figure 12.1 shows the time course of linguistic information processing based on data derived from the event-related potentials (ERPs) recording technique. ERPs represent a unique tool in the study and analysis of different stages of linguistic information processing, since they are characterized, on the one hand, by the lack of invasiveness typical of electroencephalographic (EEG) recording and, on the other hand, by an optimal temporal resolution (which may be <1 ms).

The temporal latency of a given deflection or peak (positive or negative voltage shift), visible in the waveform of the ERPs, therefore represents the occurrence of brain processing activity time-locked to a cognitive event (Zani & Proverbio, 2003). For example, the occurrence of a voltage deflection at about 70–80 ms at the scalp sites over the primary visual cortex reflects the arrival of incoming information to the visual cortex and the corresponding activations of neural populations involved in visual information sensory processing. In the same way, the occurrence of a large negative deflection at about 400 ms in response to semantically incomprehensible stimuli reflects semantic meaning analysis processes for a given word. ERPs recording in the study of language comprehension mechanisms were applied for the first time at the end of the 1970s by researchers working in the field of what has since become known as cognitive neuroscience. In 1968, Sutton discovered that the human brain elicited a large positive response to those stimuli that were selectively attended at a particular moment (identical in terms of physical characteristics to those disregarded). This implied that it was possible to study mental processes by observing their neurophysiological manifestations.

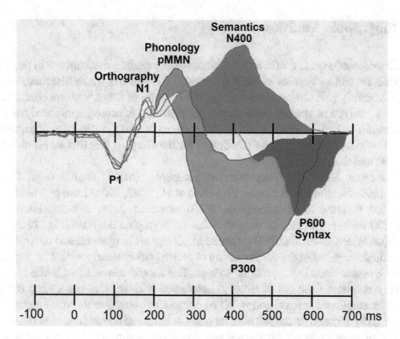

Fig. 12.1 Time course of cerebral activation during the processing of linguistic material as reflected by the latency of occurrence of various ERP components. Prelinguistic stimulus sensory processing occurs (P1 component) at about 100 ms poststimulus; orthographic analysis of written words (posterior N1 component) at 150–200 ms; phonologic/phonetic analysis at 200–300 ms, as revealed by phonological mismatch negativity (temporal and anterior pMMN) seen in response to phonologic incongruities (both visual and auditory); and a large centroparietal negativity at about 400 ms (N400), recorded in response to semantic incongruities and indexing lexical access mechanisms. The comprehension of meaningful sentences reaches consciousness between 300 and 500 ms (P300 component); finally, a second-order syntactic analysis is indexed by the appearance of a late positive deflection (P600) at about 600 ms poststimulus latency

To study language, Marta Kutas developed two different experimental paradigms. In the first, *rapid serial visual presentation* (RSVP), single words are consecutively presented in the center of a screen (Kutas, 1987) in order to simulate the process involved in the spontaneous reading of a sentence and to monitor the time course of semantic and syntactic comprehension processes while avoiding the horizontal ocular movements that normally go along with text reading. The second, quite popular, paradigm is called the *final word paradigm* (Kutas & Hillyard, 1980), and it is based on the presentation of a semantic or syntactic context of variable nature and complexity that is followed by a given terminal and critical word, to which brain potential is time-locked and which can be more or less congruent with the context or respectful of various word concatenation rules of a given language.

Orthographic Analysis

ERPs represent a quite useful tool for investigating reading mechanisms in that they provide several indices of what is occurring, millisecond by millisecond, in the brain, starting from stimulus onset: from the analysis of sensory visual characteristics (e.g., curved or straight lines, angles, circles, etc.) to orthographic analysis (letter recognition), to the analysis of complex patterns (words), and to their orthographic aspect (which, for example, greatly differs for the German, English, or Finnish languages) and their meaning.

Numerous ERPs and magnetoencephalography (MEG) studies (e.g., Bentin et al., 1999; Helenius et al., 1999; Proverbio et al., 2002, 2004) have provided clear evidence that the occipitotemporal N170 response (with a mean latency of 150–200 ms) specifically reflects stimulus orthographic analysis (Fig. 12.2). For example, Helenius et al. (1999) recorded MEG signals in dyslexic and control adult individuals engaged in the silent reading of words (either clearly visible or degraded with Gaussian noise) vs. symbolic strings. The results showed that while the first sensory response (100 ms of latency) associated with sensory processing did not differ in amplitude across groups, N170 component sensitive to orthographic factors, usually focused on the left inferior occipitotemporal cortex (i.e., over the *visual word form area*), was not lateralized and was considerably reduced in amplitude in dyslexic individuals.

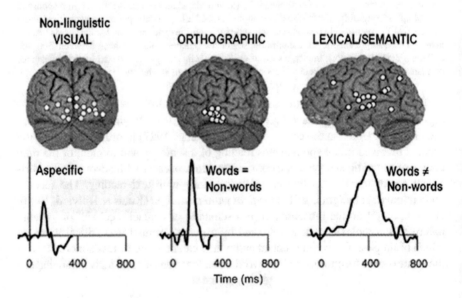

Fig. 12.2 Visual perception of words activates the left occipitotemporal cortex at about 170 ms poststimulus. This response is much larger to words than non-orthographic strings. Lexical processing reaches its peak at about 400 ms

In a recent study (Proverbio et al., 2007), we compared ERPs evoked by words and pseudo-words in their canonical orientation with those elicited by words and pseudo-words flipped horizontally. The aim was to assess whether the inversion of words deprived them of their linguistic properties, thus making them nonlinguistic stimuli. About 1300 Italian words and legal pseudo-words were presented to 18 right-handed Italian students engaged in a letter detection task. In order to identify the temporal latency of alphabetic letter processing and recognition, ERPs evoked by target and nontarget stimuli were compared. ERPs showed an early effect of word orientation at ~150 ms, with larger N1 amplitudes to rotated than to standard words. *Low-resolution brain electromagnetic tomography* (LORETA) localized this increase in N1 to flipped horizontally words primarily in the extrastriate cortex of the right occipital lobe (BA 18), which may indicate an effect of stimulus novelty. N1 was greater to target than to nontarget letters at left lateral occipital sites, thus reflecting the first stage of orthographic processing. LORETA revealed a strong focus of activation for this effect in the left fusiform gyrus (BA 37), which is consistent with the so-called visual word form area, corresponding to the left inferior occipitotemporal cortex.

Lexical Analysis

After accessing phonologic properties of words during reading, the brain is able to extract their semantic/lexical properties at about 300–400 ms of poststimulus latency (Federmeier et al., 2002), as indexed by N400 component. The amplitude of this component is generally greater over the right centroparietal areas at the scalp, but this does not correspond to the inner anatomical localization of semantic specialized areas. Intracranial recording studies have shown that N400 generators lie in the left temporal cortex, near the collateral sulcus and the anterior fusiform gyrus. In her original paper, Kutas (1980) used the *final word paradigm* to investigate the functional properties of N400 response and distinguished between the concepts of "semantic incongruence" and "subjective expectation" of sentence termination. Thus, Kutas postulated the existence of a *contextual constraint* generated by the overall semantic meaning of a sentence, which in itself would not be sufficient to explain the increased N400 effects. In order to illustrate this, let's take, for example, the following sentence: "She put sugar and lemon in her." The overall sentence meaning binds the terminal word to be a sort of drink and especially TEA. In the sentence "She put sugar and lemon in her BOOT," the final word elicits an N400 of noticeable magnitude because the contextual constraint has been macroscopically violated. However, in the sentence "She put sugar and lemon in her COFFEE," the final word still generates an N400 response but of lower amplitude as compared to the incongruent BOOT word. The negativity is generated because COFFE is semantically less related to sugar and lemon than the word TEA, but N400 to COFFEE is smaller than N400 to BOOT because the former belongs to the same semantic domain of TEA (drinks). This finding reflects the effect of the *contextual constraint*.

However, the contextual constraint alone cannot predict the whole process of semantic comprehension in reading.

Kutas also introduced the *cloze (closure) probability* factor, meant as the probability that a group of speakers might complete a certain sentence with a specific final word; this effect does not completely correspond to the contextual constraint, because it refers to subjective expectancy and not to semantic relatedness. For instance, in the sentence "He sent a letter without a STAMP," the final word is highly predictable, for many speakers, because it has a high cloze probability. Conversely, the final word of "There was anything wrong with the FREEZER" has a low cloze probability since, from a statistical point of view, not many speakers would complete this sentence in the same way. Therefore, while being not semantically incongruent with the context (therefore, not violating the semantic constraint), FREEZER would still elicit an N400 much larger than the word STAMP from the previous sentence, because it is completely unexpected and unpredictable for the speakers. In this vein, N400 paradigm might be advantageously used to investigate conceptual representation of social attributes in different groups of speakers, including stereotypes and prejudices (race-based, gender-based, etc.).

The N400 is a hallmark of the semantic integration mechanisms, and, as such, it is sensitive to the difficulty with which the reader/listener integrates the incoming sensory input with the previous context, based on the individual expectations. Although the maximum response peak to incompatible, unexpected, or low cloze probability words is reached around 400 ms, earlier ERP responses have also been reported to be sensitive to some lexical properties of words, such as their frequency of use. King and Kutas (1998) described an anterior negative component the *lexical processing negativity* (LPN), with a latency of about 280–340 ms, which seems very sensitive to the frequency of word occurrence.

In an ERP investigation (Proverbio et al., 2004) in which words, pseudo-words, and letter strings were presented during a phonetic decision task, the earliest effect of a lexical discrimination between words and pseudo-words was observed at about 200 ms poststimulus (Fig. 12.3). It would seem, then, that neural mechanisms of access to the lexical features of linguistic stimuli activate in parallel with the extraction of their orthographic and phonologic properties. Some studies also demonstrated a sensitivity to lexical properties of short, familiar words at latencies even earlier than 150 ms (e.g., Assadollahi & Pulvermüller, 2003; Pulvermüller et al., 2001). Finally, Proverbio et al. (2008) have shown that orthographic N170 response, generated within the left fusiform gyrus, manifested a sensitivity to sub-lexical properties (word frequency of use), being of greater amplitude in response to high- than low-frequency words.

Pragmatic Analysis

Just as the P300 represents an index at the scalp of neural mechanisms of *contextual updating*, that is, the updating of personal knowledge as a consequence of comparing the ongoing stimulus input with the information retained in long-term memory

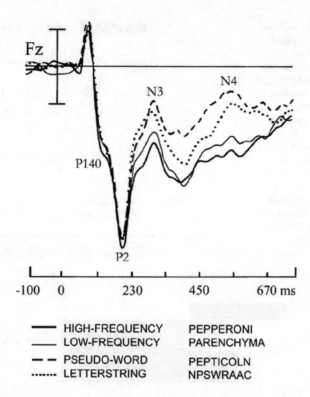

Fig. 12.3 ERP waveforms recorded in response to words, pseudo-words, and letter strings during a phonetic decision task (e.g., "Is phone/k/present in oranges?"). The first lexical effect was found at P2 level. (Adapted from Proverbio et al. (2004), with the permission of MIT Press)

(Donchin, 1987), conversely, an increased N400 represents a difficulty in integrating incoming inputs with previous knowledge, including world knowledge of pragmatic nature (scenarios, such as how to pay the bus ticket), and social knowledge (social conventions, cultural habits, rules about what is appropriate or not, etc.).

For example, let's first consider the classical case of a violation of the semantic constraint such as the one provided by the sentence "Jane told her brother that he was very...," followed by three possible terminal words:

A. FAST Congruent
B. SLOW Congruent
C. RAINY Incongruent

As extensively dealt with in the previous section, case C gives rise to N400 (depicted in Fig. 12.4, top) since "rainy-ness" is not a possible property of a person, which therefore makes it hard to integrate the meaning of the terminal word with the conceptual representation elicited by the aforementioned sentence. Since this incongruence is verifiable per se, independent of the context or the specific speakers, it is defined as a violation of the semantic constraint (which is to be distinguished by the cloze probability).

Fig. 12.4 ERPs recorded in response to terminal words completing a previous context (see the text for specific sentences) determining a violation of semantic constraint (case 1), contextual meaning (case 2), or pragmatic knowledge (case 3). Solid line, congruent word; dotted line, incongruent word. (Taken and adapted from studies of Hagoort and coauthors (Hagoorth et al., 2004; Van Berkum et al., 1999), with permission of authors)

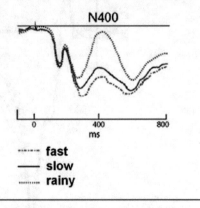

Violation of semantic constraint

N400

⋯⋯⋯ **fast**
— **slow**
⋯⋯⋯ **rainy**

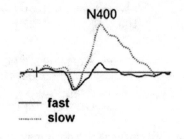

Violation of contextual meaning

N400

— **fast**
⋯⋯⋯ **slow**

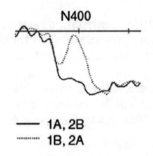

Violation of social knowledge

N400

— **1A, 2B**
⋯⋯⋯ **1B, 2A**

However, Hagoorth and coworkers (e. g., Van Berkum et al., 1999) discovered that the N400 is also sensitive to violations of the sentence meaning mediated by the context or by social knowledge. Let's consider, for instance, the sentence "At 7:00 a.m., Jane's brother had already taken a shower and had dressed too," followed by "Jane told her brother that he was incredibly...," completed by the terminal words:

A. FAST Congruent
B. SLOW Incongruent

Case B would give rise to a large N400 (depicted in Fig. 12.4, middle) since the conceptual representation of a fast and early-rising brother induced by the previous context is in striking contrast with the way his sister defines him. The semantic incongruence can be extended to implicit or pragmatic knowledge, such as social knowledge. Let's take, for instance, the sentence "On Sunday, I usually go to the park with..." pronounced by the voice of (1) a child and (2) an adult man and followed by two possible terminal words:

A. MY DADDY
B. MY WIFE

Final words 1B and 2A would elicit a wide N400 deflection (Fig. 12.4, bottom) in the absence of any violation of the semantic constraint or of the contextual constraint, thus indexing a pure violation of pragmatic and/or social knowledge. Indeed the adult male voice would not predict a "daddy" final, so as the childish voice would not predict a "wife" final. These predictions are not based on semantics but on our social knowledge, according to which a child is not usually married and an adult male does not typically use a "sugary" language.

Another study by Hagoort et al. (2004) provided a very interesting parallelism between violation of the semantic constraint and violation of the world knowledge. A typical example of *world knowledge* could be the direction in which that doors open (almost always inward but outward in case of anti-panic doors), a knowledge that is implicitly learned by means of repeated experience with the external world. Hagoort comparatively presented three types of sentences:

A. Dutch trains are yellow and very crowded
B. Dutch trains are white and very crowded
C. Dutch trains are sour and very crowded

In their study sentences, B (i.e., a violation of the world knowledge) and C (i.e., a semantic violation) elicited N400s of similar amplitudes and topographic distributions in Dutch participants, although these violations were extremely different in type. Everyone knows that a train cannot be acidic (semantic knowledge). Similarly, a Dutch person who has traveled by subway or railway would definitely be aware of the fact that the trains of their town and country are not white.

The difficulty in integrating the incoming information provided by sentences B and C with previous knowledge would stimulate cognitive processes observable at about 400 ms after critical word onset, in the form of an enhanced N400 response.

ERP Indices of Stereotypes and Prejudices

The N400 has also been found to be affected by personal semantics (Coronel & Federmeier, 2016), that is, by violations relative to subjective knowledge (i.e., personal preferences such as likes and dislikes) across a wide range of topics (including foods, sport teams, music, films, etc.).

A few studies have used the N400 response to investigate the neural representation of stereotypes (Bartholow et al., 2001, 2003). Osterhout and coauthors (1997) showed participant sentences referring to stereotypically male or female occupations and pronouns that did or did not match the gender stereotypically implied by the job (e.g., "The beautician put herself through school" vs. "The beautician put himself through school"). They found increased N400 responses in association with the prejudice violation.

Recently, Proverbio and coauthors (2017) used ERPs to investigate the detection of a discrepancy between gender-based occupational stereotypes and written material presented to 15 Italian viewers in a completely implicit task. No awareness or judgment about stereotypes was involved, no decision had to be made on sentence acceptability or congruence, and no prime words related to gender were presented (which might reveal the matter of the investigation). EEG was recorded while participants were engaged in a task that consisted in quickly pressing a response key to animal words while ignoring the overall study's purpose. Two hundred forty sentences that did or did not violate gender stereotypes were presented randomly mixed with 32 other sentences ending with an animal word. Final words violating gender stereotypes (such as "The notary is BREASTFEEDING" or "Here is the commissioner with HER HUSBAND") elicited a greater anterior N400 response and left anterior negativity (LAN) than words conforming to the gender stereotype (e.g., "The chemist put on a nice TIE") (see Fig. 12.5). LAN modulation suggests that gender stereotypes are processed automatically (as if they were morphosyntactic errors) and hints at how they are deeply rooted in our linguistic brain.

According to the inverse solution applied to incongruent minus congruent ERP difference waves recorded in the 350–450 time window, which corresponds to the N400 peak, the neural representation of gender-based stereotypes mostly involved the middle frontal gyrus (MFG), which is known to support the neural representation of stereotypes. The temporal/parietal junction (TPJ) supporting theory of mind (TOM) processes was also engaged, along with the superior and middle temporal gyri (STG and MTG) representing person information. The TPJ has been associated with the ability to attribute intentions and meanings to the behavior of others, which is part of TOM (Saxe, 2010; Young et al., 2010).

According to the neuroimaging literature, the medial frontal cortex (mdFC) represents social information that refers to others, particularly outgroup stereotyping and prejudice (Mitchell et al., 2006). In particular, sub-regions of the medial prefrontal cortex (mdPFC) would differentiate between thinking about the attributes and mental states of similar versus dissimilar others (Mahy et al., 2014). In a recent study on the neural bases of prejudice, it was found that the left cortical superior frontal gyrus (SFG, BA10) was particularly involved in representing negative prejudices related to others (Proverbio et al., 2016), which strongly fits with the current findings.

In that study, the neural bases and functional properties of social prejudices were investigated. During social interactions, we make inferences about people's personal characteristics based on their appearance. These inferences form a potential prejudice that can positively or negatively bias our interaction with them. This

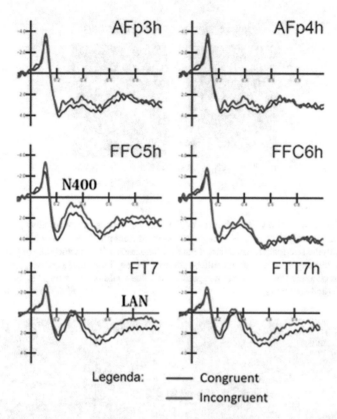

Fig. 12.5 N400 and LAN components elicited by incongruent (with respect to stereotypes) sentences over anterior scalp sites in Proverbio et al. (2017) study. (Courtesy of the authors)

ability was investigated by recording event-related potentials from 128 scalp sites in 16 volunteers. In the first session (encoding), they viewed 200 faces associated with a short fictional story that described anecdotal positive or negative characteristics about each person (see an example in Fig. 12.6).

In the second session (recognition), participants underwent an old/new memory test, in which they had to distinguish 100 new faces from the previously shown faces. ERP data relative to the encoding phase showed a larger anterior negativity in response to negatively (vs. positively) biased faces, indicating a deeper neural processing of faces with unpleasant social traits. In the recognition task, ERPs recorded in response to new faces elicited a larger FN400 than to old faces and to positive than negative faces. This piece of data indicates that negatively valenced faces were recognized as more familiar than positively valenced ones. Additionally, old faces elicited a larger old-new parietal response than new faces, in the form of an enlarged late positive component (LPC). An inverse solution swLORETA (applied to ERPs in the 450–550 ms poststimulus) indicated that remembering old faces was associated with the activation of right superior frontal gyrus (SFG), left middle temporal gyrus (mdTG), and right fusiform gyrus (FG). However, only negatively connoted

Negative bias   Positive bias

He savagely beats   He died during a
his wife.   robbery to protect a
                     child

Fig. 12.6 Examples of how Proverbio et al. (2016) induced a positive or negative prejudice about previously unknown persons. In the encoding task, faces were presented in association with a short story that provided fictional information about the character, such as an anecdote or personal information. The biographic information could be positive, thereby inducing a positive prejudice toward the depicted character, or vice versa, a negative prejudice could induce a negative bias. (Courtesy of Proverbio and coauthors)

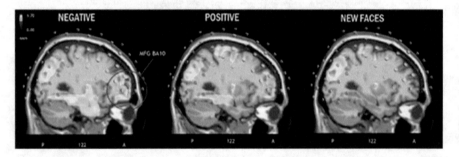

Fig. 12.7 Sagittal views of active sources during processing of negatively biased, positively biased, and new faces according to swLORETA analysis during the 450–550 ms time window. The images highlight the strong activation of the left middle frontal gyrus during memory recall of faces associated with a negative prejudice. (Taken from Proverbio et al. (2016) with permission from the authors. Creative Commons Public Domain picture)

faces strongly activated the limbic and parahippocampal areas and the left SFG (Fig. 12.7). Dissociation was found between familiarity (modulated by negative bias) and recollection (distinguishing old from new faces). Not only ERPs showed the existence of prejudices formed during the learning phase, but the latter were able to affect the recognition and memory recall of faces, with an advantage for negatively valenced social information.

Going back to gender-based prejudices, quite recently, Proverbio and coauthors (2018) showed that ERPs are so sensitive to social representations and constructs such as prejudices and stereotypes that it is possible to find differences within the population as a function of the different degree of prejudice possessed. In this study, the time course and the neural correlates involved in the representation of occupational gender bias were investigated by addressing two questions: first, if the bias varied as a function of participant's sex and, second, if there was a difference based on the gender of the character depicted in the phrases presented to participants. Sentences were created in a way that the gender of the character engaging in a given professional activity or behavior was made explicit only at the very end of the sentence (*final word paradigm*). An implicit paradigm was chosen to trigger the automatic activation of any mental function involved in the processing of gender stereotypes. This was carried out by recording electrophysiological responses in heterosexual Italian university students during the reading of hundreds of sentences depicting female and male characters and their professional attitudes (see Table 12.1 for some example of sentences carrying typical female or male stereotypes). The task consisted in responding as quickly and accurately as possible to animal words, that is, an implicit task designed in order to avoid social desirability processes. Brain responses of male and female participants totally unaware of study's purpose were compared as a function of whether the sentence was congruent or not with a gender stereotype.

EEG was recorded from 128 sites in 38 Italian participants. While looking for rare animal words, participants read 240 sentences, half of which expressed notions congruent with gender stereotypes, and the other half did not. ERPs were time-locked to critical words. Findings showed enhanced anterior N400 and occipitoparietal P600 responses to items that violated gender stereotypes, mostly in men (Fig. 12.8). The swLORETA analysis applied to N400 potentials in response to incongruent phrases showed that the most activated areas during stereotype processing were the right middle temporal (mdTG) and middle frontal gyri (mdFG), as well as the TPJ, as expected on the basis of previous literature (Fig. 12.9). The data hint at a gender difference in stereotyping, with men being more prejudicial especially when the depicted character was a male. One possible interpretation of these findings relies on the asymmetrical nature of occupational stereotypes, mostly rooted in the principle that females could not perform male professions because of a lack of strength or powerful attitude. Therefore, it is conceivable that women participants might disagree more easily with the stereotype being themselves women.

An asymmetry in gender bias, with a stronger prejudicial attitude in men, is not unknown in the literature. For example, an article summarizing data from more than 2.5 million completed IATs (Implicit Association Tests) and self-reports (Nosek & Smyth, 2007) showed that men are more prejudicial in terms of theories postulating that they have more social dominance (e.g., Sidanius & Pratto, 1999), attitudes toward gay vs. heterosexual people (e.g., Negy & Eisenman, 2005), and attitudes toward black vs. white people (e.g., Qualls et al., 1992). As for neuroscientific data, only in men it was shown that hostile sexism correlated with the activation of brain regions associated with mental state attributions (such as the medial prefrontal cortex (mdPFC), the posterior cingulate cortex (pCC), and the temporal poles) in Cikara et al.'s (2011) fMRI study.

Overall ERPs, and especially the amplitude of N400 component, proved to be extremely sensitive to violations of implicit stereotypes, thus allowing to tap at the representation of social attributes (such as stereotypes and prejudices) without the

Table 12.1 Example of sentence stimuli, relative to men or women, and violating or not current occupational gender stereotypes in which women engage more in care-related professions and men in strength–/power-related professions

Sentences incongruent with prejudices (men)
Prepared the tomato sauce and then shaved.
Gave up figure skating when he became father.
Hang the clothes out to dry and caught up with his wife.
We would be shopping all day if it was up to Johnny.
Lost his pipe leaving the ballet class.
That waiter wore a colorful skirt.
Sentences incongruent with prejudices (women)
After whitewashing, she was exhausted.
The notary is breastfeeding.
While changing the engine's oil, she stained herself.
The major's name is Josephine Nicolini.
The musical software engineer waxed her legs.
That boxer has just given birth.
Sentences congruent with prejudices (men)
The chemist put on a nice tie.
The motorist suffers from prostatitis.
The financial controller soiled his pants.
The lab technician ruined his scrubs.
Served with intelligence until he was cast out.
Once finished with the tile install, he was haggard.
Sentences congruent with prejudices (women)
Prepared a synchronized swimming choreography of which she is proud.
Works as a baby-sitter, and she is very maternal.
Fed the little girl and went to the lady hairdresser.
Laura enjoys working as a switchboard operator.
Loved kids so much that became an elementary school teacher.
She hurt herself doing needle work.

Because these stereotypes are a part of everybody's cultural heritage and learned early in life, people form implicit gender stereotypes, which automatically associate men and women with stereotypical traits, abilities, and roles, even when they disavow these traditional beliefs (e.g., Nosek et al. 2002). For instance, women are typically stereotyped as being nicer (Eagly & Mladinic, 1989) and are more likely to enact subordinate roles that require communal traits. The presence of gender stereotyping has been demonstrated for an extensive list of role nouns in Czech, English, French, German, Italian, Norwegian, and Slovak by Misersky et al. (2014). To determine whether the sentences actually represented (or violated) stereotypes for university students living in the Milan metropolitan area, the stimuli underwent validation, in which a group of Milan University students were asked to rate, by means of a 3-point Likert scale, how they reacted to reading the terminal word of the phrase. Scale units were as follows: 0 = Actually, I was a bit surprised. 1 = I do not know. 2 = I kind of saw that coming.

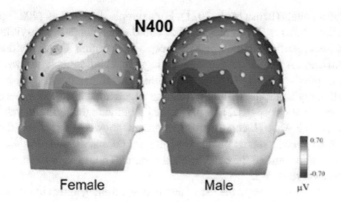

Fig. 12.8 Isocolor topographical maps (front view) of surface voltage measured in the 250–400 ms temporal window (N400 latency range) to incongruent stimuli as a function of participants' sex. It can be appreciated how N400 response to stereotypes violation was not found in female participants. This suggests that female participants were not surprised by final words that violated sex stereotypes. (Adapted from Proverbio et al., 2018)

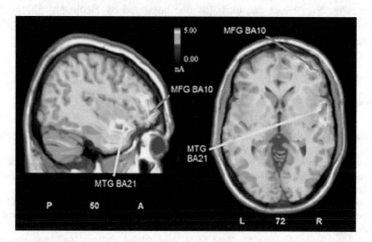

Fig. 12.9 Coronal and axial brain sections showing the location and strength of electromagnetic dipoles explaining the surface difference voltage obtained by subtracting ERPs to congruent from ERPs to incongruent stimuli in the 250–400 ms latency range, corresponding to the peak of N400. *L* left, *R*, right, *A* anterior, *P* posterior, *MTG* middle temporal gyrus, *MFG* middle frontal gyrus. (Taken for Proverbio et al., 2018)

problems related to social desirability processes and without participants being minimally aware of the study's purpose or experimental manipulation. For this reason, ERPs represent one core research technique for studying social cognition, including the representation of social attributes.

Indeed, N400 paradigm was also used for detecting implicit ethnic prejudices such as negative biases against rural migrant workers (Wang et al., 2011), unarmed Afro-Americans individuals (Correll et al., 2006), or non-Caucasian (other race

(OR)) professionals (Brusa et al., 2021). For example, Brusa et al. (2021) presented to Caucasian students 285 sentences that could either violate, non-violate, or be neutral with regard to stereotypical concepts concerning OR individuals (e.g., Asians, Africans, Arabs). No awareness or judgment about stereotypes was required. Participants passively read the sentences while engaged in a fictitious task, ignoring the overall study's purpose. Stimuli violating negative ethnic stereotypes elicited a large anterior N400 response, and participant's individual amplitude values of the N400-Difference Wave (Incongruent – Congruent) showed a direct correlation with the individual racism scores obtained at the *Subtle and Blatant Prejudice Scale*, administered at the end of the experimental session. The stronger the racial bias, the larger the N400 response.

These findings encourage the use of subjective, implicit, and explicit psychological scales to be correlated with physiological measures in the study of social stereotypes. Indeed, while the N400 paradigm allows to implicitly access the representation of racial or sexual stereotypes avoiding the activation of control processes guided by social desirability instances, the correlation between electrophysiological and behavioral measures can provide a wider and more complex view about psychological processes. Prejudices can exist (as demonstrated by electrophysiological signals) in the absence of conscious awareness and in contrast to a convinced voluntary progressive attitude of individuals.

References

Assadollahi, R., & Pulvermüller, F. (2003). Early influences of word length and frequency: A group study using MEG. *Neuroreport, 14*, 1183–1187.

Bartholow, B. D., Fabiana, M., Gratton, G., & Battencourt, B. A. (2001). A psychophysiological examination of cognitive processing of and affective responses to social expectancy violations. *Psychological Science, 12*(3), 197–204.

Bartholow, B. D., Pearson, M. A., Gratton, G., & Fabiani, M. (2003). Effects of alcohol on person perception: A social cognitive neuroscience approach. *Journal of Personality and Social Psychology, 85*(4), 627–638.

Bentin, S., Mouchetant-Rostaing, Y., Giard, M. H., et al. (1999). ERP manifestations of processing printed words at different psycholinguistic levels: Time course and scalp distribution. *Journal of Cognitive Neuroscience, 11*, 35–60.

Brusa, A., Bordone, G., & Proverbio, A. M. (2021, February, 23). Measuring implicit mental representations related to ethnic stereotypes with ERPs: An exploratory study. *Neuropsychologia, 155*, 107808. https://doi.org/10.1016/j.neuropsychologia.2021.107808

Cikara, M., Eberhardt, J. L., & Fiske, S. T. (2011). From agents to objects: Sexist attitudes and neural responses to sexualized targets. *Journal of Cognitive Neuroscience, 23*(3), 540–551.

Coronel, J. C., & Federmeier, K. D. (2016). The N400 reveals how personal semantics is processed: Insights into the nature and organization of self-knowledge. *Neuropsychologia, 84*, 36–43.

Correll, J., Urland, G. R., & Ito, T. A. (2006). Event-related potentials and the decision to shoot: The role of threat perception and cognitive control. *Journal of Experimental Social Psychology, 42*, 120–128.

Donchin, E. (1987). The P300 as a metric for mental workload. *Electroencephalography and Clinical Neurophysiology Supplement, 39*, 338–343.

Eagly, A. H., & Mladinic, A. (1989). Gender stereotypes and attitudes toward women and men. *Personality and Social Psychology Bulletin, 15*(4), 543–558.

Federmeier, K. D., Kluender, R., & Kutas, M. (2002). Aligning linguistic and brain views on language comprehension. In A. Zani & A. M. Proverbio (Eds.), *The cognitive electrophysiology of mind and brain*. Academic Press.

Hagoort, P., Hald, L., Bastiaansen, M., & Petersson, K. M. (2004). Integration of word meaning and world knowledge in language comprehension. *Science, 304*(5669), 438–441.

King, J. W., & Kutas, M. (1998). Neural plasticity in the dynamics of human visual word recognition. *Neuroscience Letters, 244*, 61–64.

Kutas, M. (1987). Event-related brain potentials (ERPs) elicited during rapid serial visual presentation of congruous and incongruous sentences. *Electroencephalography and Clinical Neurophysiology Supplement, 40*, 406–411.

Kutas, M., & Hillyard, S. A. (1980). Reading senseless sentences: Brain potentials reflect semantic incongruity. *Science, 207*(4427), 203–205.

Helenius, P., Tarkiainen, A., Cornelissen, P., Hansen, P. C., & Salmelin, R. (1999). Dissociation of normal feature analysis and deficient processing of letter- strings in dyslexic adults. *Cerebral Cortex, 9*(5), 476–483.

Mahy, C. E., Moses, L. J., & Pfeifer, J. H. (2014). How and where: Theory-of-mind in the brain. *Developmental Cognitive Neuroscience, 9*, 68–81.

Misersky, J., Gygax, P.M., Canal, P, Gabriel, U., Garnham, A Braun, F. et al. (2014*)*. Norms on the gender perception of role nouns in Czech, English, French, German, Italian, Norwegian, and Slovak. Behavior Research Methods, 46 (3), 841–871.

Mitchell, P., Macrae, C. N., & Banaji, M. R. (2006). Dissociable medial prefrontal contributions to judgments of similar and dissimilar others. *Neuron, 50*, 655–663.

Negy, C., & Eisenman, R. (2005). A comparison of African American and White college students' affective and attitudinal reactions to lesbian, gay, and bisexual individuals: An exploratory study. *Journal of Sex Research, 42*(4), 291–298.

Nosek, B. A., Banaji, M. R., & Greenwald, A. G. (2002). Harvesting implicit group attitudes and beliefs from a demonstration website. *Group Dynamics, 6*(1), 101–115.

Nosek, B. A., & Smyth, F. L. A. (2007). Multitrait-multimethod validation of the implicit association test: Implicit and explicit attitudes are related but distinct constructs. *Experimental Psychology, 54*(1), 14–29.

Osterhout, L., Bersick, M., & McLaughlin, J. (1997). Brain potentials reflect violations of gender stereotypes. *Memory and Cognition, 25*(3), 273–285.

Proverbio, A. M., Čok, B., & Zani, A. (2002). ERP measures of language processing in bilinguals. *Journal of Cognitive Neuroscience, 14*(7), 994–1017.

Proverbio, A. M., Vecchi, L., & Zani, A. (2004). From orthography to phonetics: ERP measures of grapheme-to-phoneme conversion mechanisms in reading. *Journal of Cognitive Neuroscience, 16*(2), 301–317.

Proverbio, A. M., Wiedemann, F., Adorni, R., Rossi, V., Del Zotto, M., & Zani, A. (2007). Dissociating object familiarity from linguistic properties in mirror word reading. *Behavioral and Brain Functions, 20*(3), 43.

Proverbio, A. M., Zani, A., & Adorni, R. (2008). The left fusiform area is affected by written frequency of words. *Neuropsychologia, 46*(9), 2292–2299.

Proverbio, A. M., La Mastra, F., & Zani, A. (2016). How negative social bias affects memory for faces: An electrical neuroimaging study. *PLoS One, 11*(9), e0162671.

Proverbio, A. M., Orlandi, A., & Bianchi, E. (2017). Electrophysiological markers of prejudice related to sexual gender. *Neuroscience, 1*(358), 1–12.

Proverbio, A. M., Alberio, A., & De Benedetto, F. (2018, November). Neural correlates of automatic beliefs about gender stereotypes: Males are more prejudicial. *Brain Lang, 186*, 8–16.

Pulvermüller, F., Assadollahi, R., & Elbert, T. (2001). Neuromagnetic evidence for early semantic access in word recognition. *The European Journal of Neuroscience, 13*(1), 201–205.

Qualls, R.C., Cox, M.B., Schehr, T.L. (1992). Racial attitudes on campus: Are there gender differ-ences? Journal of College Student Development, 33 (6), 524–530.

Saxe, R. (2010). The right temporo-parietal junction: A specific brain region for thinking about thoughts. In A. Leslie & T. German (Eds.), *The neuropsychology of proper names*. Handbook of Theory of Mind.

Sidanius, J., & Pratto, F. (1999). *Social dominance*. Cambridge University Press.

Young, L., Camprodon, J. A., Hauser, M., Pascual-Leone, A., & Saxe, R. (2010). Disruption of the right temporoparietal junction with transcranial magnetic stimulation reduces the role of beliefs in moral judgments. *Proceedings of the National Academy of Sciences, 107*, 6753–6758.

Van Berkum, J. J. A., Hagoort, P., & Brown, C. M. (1999). Semantic integration in sentences and discourse: Evidence from the N400. *Journal of Cognitive Neuroscience, 11*, 657–671.

Wang, L., Ma, Q., Song, Z., Shi, Y., Wang, Y., & Pfotenhauer, L. (2011). N400 and the activation of prejudice against rural migrant workers in China. *Brain Research, 1375*, 103–110.

Zani, A., & Proverbio, A. M. (Eds.). (2003). *The cognitive electrophysiology of mind and brain*. Academic Press/Elsevier.

Chapter 13
Brain Imaging Methods in Social and Affective Neuroscience: A Machine Learning Perspective

Lucas R. Trambaiolli, Claudinei E. Biazoli Jr, and João R. Sato

Abstract Machine learning (ML) is a subarea of artificial intelligence which uses the induction approach to learn based on previous experiences and make conclusions about new inputs (Mitchell, Machine learning. McGraw Hill, 1997). In the last decades, the use of ML approaches to analyze neuroimaging data has attracted widening attention (Pereira et al., Neuroimage 45(1):S199–S209, 2009; Lemm et al., Neuroimage 56(2):387–399, 2011). Particularly interesting recent applications to affective and social neuroscience include affective state decoding, exploring potential biomarkers of neurological and psychiatric disorders, predicting treatment response, and developing real-time neurofeedback and brain-computer interface protocols. In this chapter, we review the bases of the most common neuroimaging techniques, the basic concepts of ML, and how it can be applied to neuroimaging data. We also describe some recent examples of applications of ML-based analysis of neuroimaging data to social and affective neuroscience issues. Finally, we discuss the main ethical aspects and future perspectives for these emerging approaches.

Keywords Brain imaging methods · Machine learning · Neuroscience machine learning · Emotion/affective decoding · Neurofeedback

L. R. Trambaiolli (✉)
Basic Neuroscience Division, Mclean Hospital – Harvard Medical School,
Belmont, MA, USA
e-mail: ltrambaiolli@mclean.harvard.edu

C. E. Biazoli Jr · J. R. Sato
Center for Mathematics, Computing, and Cognition, Federal University of ABC,
São Bernardo do Campo, Brazil

Introduction

Machine learning (ML) is a subarea of artificial intelligence which uses the induction approach to learn based on previous experiences and make conclusions about new inputs (Mitchell, 1997). In the last decades, the use of ML approaches to analyze neuroimaging data has attracted widening attention (Pereira et al., 2009; Lemm et al., 2011). Particularly interesting recent applications to affective and social neuroscience include affective state decoding, exploring potential biomarkers of neurological and psychiatric disorders, predicting treatment response, and developing real-time neurofeedback and brain-computer interface protocols. In this chapter, we review the bases of the most common neuroimaging techniques, the basic concepts of ML, and how it can be applied to neuroimaging data. We also describe some recent examples of applications of ML-based analysis of neuroimaging data to social and affective neuroscience issues. Finally, we discuss the main ethical aspects and future perspectives for these emerging approaches.

Brain Imaging Methods

Most neuroimaging experiments in human social and affective neuroscience are based on two groups of techniques (Fig. 13.1) (Min et al., 2010). The first group comprises measurements of either electrical or magnetic features associated with the electrophysiological activity of neuronal assemblies. This group includes the electroencephalography (EEG) and the magnetoencephalography (MEG) data acquisitions. On the other hand, the second group comprises measurements of metabolic or hemodynamic features that are indirectly associated with neural activity. This second group of neuroimaging techniques includes functional magnetic resonance imaging (fMRI), functional near-infrared spectroscopy (fNIRS), and positron emission tomography (PET).

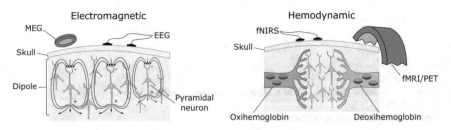

Fig. 13.1 Electromagnetic-based imaging approaches (left) use electric or magnetic sensors to capture the electromagnetic resultants from the neuronal and synaptic activity. Hemodynamic-based procedures (right) use light or magnetic sensors to measure the cerebral blood flow and oxygen consumption levels

Among the electromagnetic approaches, the EEG uses electrodes positioned over the scalp to record the sum of excitatory and inhibitory postsynaptic potentials in which the resulting dipoles are perpendicularly aligned to the scalp (Niedermeyer & da Silva, 2005). In consequence, its spatial resolution is limited and further compromised by volume conduction effects. However, its simplicity, low cost, and high temporal resolution (reaching the order of kilohertz in modern systems) make it one of the most common techniques in social and affective experiments. Similarly, MEG signals are resultant from the magnetic field generated by postsynaptic currents in apical dendrites (mainly those tangential to the skull) (Hansen et al., 2010). Despite presenting some mapping limitations similar to the EEG, MEG has a better spatial resolution, though restricted to superficial cortical sulci activity. Moreover, its higher cost and less availability when compared to EEG result in relatively fewer studies in human affective neuroscience using this technique (Min et al., 2010).

PET scanning is the pioneering metabolic and hemodynamic imaging approach. This technique uses an injected radioactive tracer to track brain tissue variations on blood flow and metabolic features associated with local neural activity (Maquet, 2000). However, with the emergence of noninvasive fMRI protocols, which did not depend on exogenous tracers, PET experiments became relatively less common in current research. The fMRI uses the paramagnetic properties of the deoxyhemoglobin molecules, which work as an endogenous tracer, to measure the blood-oxygen-level-dependent (BOLD) contrast effect (Ogawa et al., 1990). Both PET and fMRI acquisitions provide the highest spatial resolution among the brain imaging approaches, allowing the evaluation of both cortical and subcortical structures associated with social behavior and affective states (Liu et al., 2015). The worldwide availability of MRI scanners in clinical settings made it the most used neuroimaging technique in the last two decades. Among fMRI, main limitations in affective and social process research when compared with other approaches are its lower temporal resolution, scanner noise, and the setup that restrict movement (Doi et al., 2013). Hence, fMRI acquisition does not allow more naturalistic, out-of-the-laboratory protocols. As a complementary hemodynamics-based technique for more naturalistic settings, the fNIRS has the advances of portability, low cost, and a relatively good temporal-spatial ratio (Doi et al., 2013). This technique measures the absorption of near-infrared light by oxyhemoglobin and deoxyhemoglobin molecules in superficial layers of the brain tissue, during local neural activity (Ferrari & Quaresima, 2012). However, fNIRS acquisitions only cover brain layers close to the scalp, as is the case with MEG (Min et al., 2010), and with a sparse representation limited by the optodes arrangement.

In sum, each neuroimaging modality has advances and disadvantages, and the choice for a particular technique should be based on the specific research question. More recently, the use of multimodal setups emerged as a promising approach in the neuroimaging field. These approaches use two or more neuroimaging techniques aiming to combine its advantages and provide complementary and convergent information regarding the underlying neural phenomena (Liu et al., 2015). The most common combination involves at least one electromagnetic and one hemodynamic approach, such as EEG-fMRI, EEG-fNIRS, or EEG-fNIRS-fMRI. However, combinations into the same group of techniques are possible, such as EEG-MEG and fNIRS-fMRI.

Basic Concepts of Machine Learning

The primary aim of a machine learning algorithm is to *learn* (i.e., extract knowledge) from an original dataset (training set), validate its ability to make predictions in an independent dataset (validation set), and then make decisions or predictions in new samples (test set) (Mitchell, 1997). During the learning process, the decision model bases its conclusions on patterns observed on the features of the examples in the training set. Such features might include, for example, frequencies of neural activity during specific tasks, event-specific potentials for a particular set of stimuli, or the connectivity level between different brain areas (Rubinov & Sporns, 2010; Sakkalis, 2011).

Learning Process

Three main approaches might be used to guide the learning process, according to the presence or absence of labels for each example (i.e., instance or subject in the dataset) (Fig. 13.2). The first approach (which will be the focus of this chapter) is the supervised learning, where each instance has a corresponding label (e.g., patient

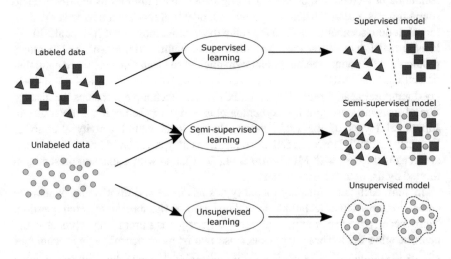

Fig. 13.2 Supervised learning methods use labeled examples to learn from data, while unsupervised learning methods extract patterns from data using unlabeled inputs. The recently proposed semi-supervised approach, otherwise, combines both labeled and unlabeled inputs during the learning process

or healthy subject). In this case, the objective is to develop models which can predict the desired labels with minimal error (Larranaga et al., 2006). Thus, during the learning process, the algorithm continually evaluates and adjusts the decision model until it reaches a near-to-optimal performance (Kuhn & Johnson, 2013). In unsupervised procedures, on the other hand, labels are not provided during the learning process. Here, the aim is to extract patterns exclusively based on similarities among groups of features (usually grouping examples according to these measures) (Larranaga et al., 2006). Finally, the third approach merges the characteristics from both previous methods. In this so-called semi-supervised approach, both labeled and unlabeled examples are used during the learning process. This approach takes advantage of the higher precision from the labeled training, as well as the lower computational cost from the non-labeled training (Cohen et al., 2004).

Validation Procedures

To converge in optimal decisions during the learning process, the decision model is continuously tested with a second dataset (i.e., the validation set) and, if necessary, remodeled using the training set (Kuhn & Johnson, 2013). The best approach for this procedure would be to train and validate the model with as much data as possible. However, due to experimental design constraints or to limited sample sizes, this task is commonly performed using somewhat suboptimal datasets (Lemm et al., 2011). Critically, to avoid variance and bias, the processes of training, validating, and testing the model *should not* be performed on the same data (Pereira et al., 2009). Different validation approaches have been proposed to overcome the issues raised by using limited datasets. One popular strategy for experiments using supervised learning is the cross-validation method (Lemm et al., 2011). In this approach, a small sample of the dataset is first split to be used as the test set, while the remaining part is further splitted into the train and the validation sets (Fig. 13.3a). This partitioning procedure is repeated several times to create different samples for each iteration (Lemm et al., 2011).

Different partitioning schemes might be used for this division. For example, in k-fold cross-validation (Fig. 13.3b), the dataset is divided into k disjoint subsets with equal size. Then, k-1 folds are used to train the model, and the remaining one is used for validation. This last step is repeated k times until all subsets are used as the validation set (Pereira et al., 2009). Another popular approach is the leave-one-out cross-validation (Fig. 13.3c), which is a particular case of k-fold cross-validation where k is equal to the number of examples.

Finally, in Monte Carlo cross-validation (Kuhn & Johnson, 2013) (Fig. 13.3d), the train and validation sets are composed by a fixed number of examples (e.g., X% for training and 100-X% for validating). Then, samples are randomly selected to form each set. This procedure might be repeated until all combinations are tested (high computational cost) or up to a predetermined number of permutations.

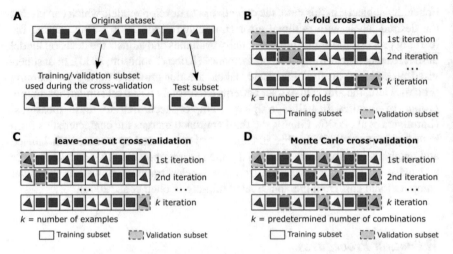

Fig. 13.3 Different steps and approaches for data splitting. (**a**) The first step of the validation process is to select a sample subset for testing purposes. Then, cross-validation approaches are used to split the remaining data into training and validation subsets. (**b**) During the k-fold cross-validation, data is split into k-folds of similar lengths. Then, the algorithm is validated k times, until all folds were used as the validation subset. (**c**) The leave-one-out cross-validation is a particular case of k-fold cross-validation, where each fold corresponds to a single example. (**d**) The Monte Carlo cross-validation performs a predetermined number of combinations, where the validation subset is composed of a fixed quantity of randomly selected samples

Dimensionality Reduction

In contrast to a limited number of examples, supervised models usually have a wide range of features associated with them. This growing abundance of assessed features relates to the improvement of brain imaging technologies and the development of new feature extraction methods. However, contrary to a common belief, the increasing high dimensionality of neuroimaging datasets does not necessarily lead to improved ML models. Indeed, much of these new features are redundant or irrelevant to the model design and might even cause a decrease in performance (Guyon & Elisseeff, 2003). With this in mind, dimensionality reduction strategies became a fundamental step for model building (Lemm et al., 2011).

As the learning approaches, feature selection (FS) methods can be grouped into unsupervised and supervised categories. The common spatial pattern (CSP), an example of supervised method, uses the class label to search for an optimal and reduced subset of features, where the maximum of relevant information is held (Lemm et al., 2011). On the other hand, unsupervised methods, such as the principal component analysis (PCA) and the independent component analysis (ICA), are mainly used for dimensionality and noise reduction based on projections to the more relevant factors or based on grouping of effects (Lemm et al., 2011). However, unlike the supervised category, unsupervised methods often require manual selection of relevant factors or groups.

Over the last decades, supervised FS methods have become popular in neuroscience (Huang, 2015). To select these optimal subsets of features, some topics should be established, such as the search strategy and the level of interaction with the ML algorithm.

Regarding the search strategy, two main approaches are possible, according to the subset composition. For the first strategy, all features are sorted according to some relevance criteria. Then, only those features with higher positions are selected to compose the subset (Huang, 2015). On the other strategy, subgroups are created with random features from the original feature set. Then, these subsets are evaluated according to its capacity to describe the whole dataset (Huang, 2015). The ideal FS algorithm would explore all combinations available to compose the feature subsets (i.e., to perform an exhaustive search) (Guyon & Elisseeff, 2003). However, due to the complexity of the problem and to computational limitations, it is common to establish a stop criterion that defines when the algorithm decides for one subset of features (e.g., when the model reaches a specific performance threshold or when the subset reaches a particular amount of features) (Guyon & Elisseeff, 2003).

According to the level of interaction with the ML model, feature selection algorithms might also be grouped into three approaches (Kohavi & John, 1997) (Fig. 13.4): filter, wrapper, and embedded. The filter approach is the most commonly used procedure. In this, the feature selection is performed before and independently to the model induction (Fig. 13.4a). For the wrapper approach, every feature set is submitted to the ML algorithm, and the model performance is used to evaluate the selected subset (Fig. 13.4b). Finally, embedded approaches merge the feature selection and the model induction steps, with the subsets being created internally by the ML model (Fig. 13.4c).

Types of Classifiers

Different types of classifiers are defined according to the specific assumptions made during the learning process (Pereira et al., 2009). For example, logic-based algorithms create successive layers in which instances are classified according to the values of a single feature. These algorithms might be described as a decision tree which is composed by nodes and branches (Fig. 13.5a). Each node has a particular rule that divides the instance into different branches according to the corresponding feature value (Murthy, 1998). The first node of the tree is the feature that best separates the training data, followed by nodes ordered by a decreasing predictive power until no more rules become necessary to classify the dataset correctly. This kind of algorithm tends to perform better when dealing with categorical features (Kotsiantis, 2007).

In perceptron-based algorithms, the perceptron calculates a linear combination of the input features and, further, sum all weighted inputs to make a decision. When the result is higher than a specified threshold, the instance is labeled as class A or marked as class B otherwise (Mitchell, 1997). These weights are randomly

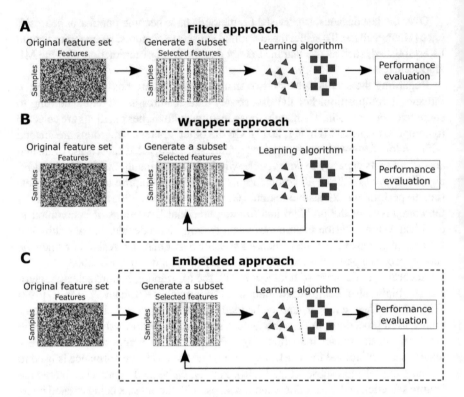

Fig. 13.4 Level of interaction between the feature selection algorithm and the classifier. (**a**) During the filter approach, the feature selection is performed before and apart from the classifier. (**b**) During the wrapper approach, every single feature subset is submitted to the classifier, and the classification performance is used to evaluate the sample. (**c**) During the embedded procedure, both the feature selection and the classifier algorithms are merged and happen simultaneously

established at first but optimized during the learning process until they reach near-to-optimal predictions (Mitchell, 1997). The perceptron approach, however, can only classify linearly separable inputs (Kotsiantis, 2007). To perform nonlinear discrimination, the use of artificial neural networks (ANN) was proposed. In this, multiple perceptrons are combined creating a complex network where the output from one single perceptron might be used as an input for several other perceptrons (Fig. 13.5b) (Zhang, 2000).

Unlike other classifiers, statistical-based algorithms provide the probability of the evaluated instance belonging to any given class (Kotsiantis, 2007). A classic example of this group of algorithms is the linear discriminant analysis (LDA) which explores linear combinations of features that best label instances into the desired classes (Fig. 13.5c) (Balakrishnama & Ganapathiraju, 1998).

Finally, support vector machines (SVM) compose a non-probabilistic method inspired by statistically based approaches. In this case, data is separated into two classes by a hyperplane (Vapnik, 1995). This hyperplane is defined trying to

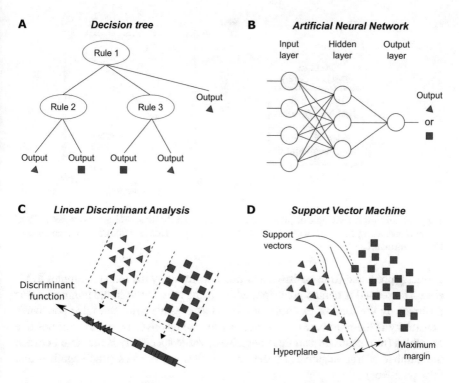

Fig. 13.5 Examples of classifiers commonly applied to neuroimaging studies. (**a**) A decision tree, (**b**) artificial neural networks, (**c**) linear discrimination analysis, (**d**) support vector machines

maximize its distance (margin) to the instances on either category (Fig. 13.5d) and, consequently, reducing the expected generalization error (Cristianini & Shawe-Taylor, 2000). For the classification of non-separable data, the dataset might be translated onto a higher-dimensional space using kernel methods, to apply the SVM-designed hyperplane (for more details about kernel methods, please refer to Cristianini & Shawe-Taylor, 2000).

Although multiclass classification approaches have been architected for the previously listed classifiers, binary classification (e.g., task vs. control group, task A vs task B, etc.) is most commonly applied in social and affective neuroscience studies.

Evaluating and Interpreting a Machine Learning Model

One easy way to evaluate the performance of a binary classifier is the use of a confusion matrix (or error matrix) (Sokolova & Lapalme, 2009). This matrix represents the relation between the actual and the predicted classes (Fig. 13.6a). Four main measures might be extracted from this matrix (Sokolova & Lapalme, 2009). The first measure, named accuracy, is the ratio between the number of examples

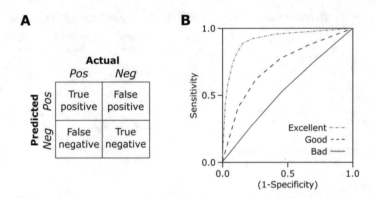

Fig. 13.6 Illustrative example of (**a**) a confusion matrix and (**b**) three different examples of ROC curves representing classifiers with excellent (dotted line), good (dashed line), and bad (continuous line) performances

correctly predicted (true positives and true negatives) by the total of samples available. The second is named precision, which is the ratio between the number of true positives by the total of examples predicted as positive (true and false positives). Sensitivity is the ratio between the number of true positives by the total of positive examples (true positives and false negatives), while specificity is the ratio between the number of true negatives by the total of negative samples (true negatives and false positives).

In general, an optimal model should present high sensitivity and specificity. However, real-world datasets tend to show an unbalance between these measures. To evaluate this aspect, the receiver operating characteristic curve (ROC curve) presents an illustrative plot of the discriminant ability of the binary classifier for different thresholds (Fawcett, 2006). This curve is plotted using the sensitivity of the classifier as the y-axis and the fall-out (i.e., 1-specificity) as the x-axis (Fig. 13.6b). Thus, the area under the ROC curve (AUC) describes the probability that the classifier will rank a random positive instance higher than a random negative example (Fawcett, 2006). In other words, when comparing the AUC of different classifiers, the higher the AUC, the better is the classifier average discriminative power.

Finally, linear classifiers such as the LDA and the linear SVM present weights relative to each variable. These weights describe how relevant each variable is to identify each class (Sato et al., 2009). In addition to performance measures, this information adds valuable clues regarding the neural basis of the studied mental process. For example, that specific frequencies in some brain areas are more related to one affective state than the other or that the volume of a subcortical structure might be a predictor of a given psychiatric disease.

Besides the evaluation methods listed in this chapter, other performance metrics might be used according to the characteristics of the ML algorithm and the experimental design. For a comparative review, please refer to Sokolova and Lapalme (2009).

ML Applications in Social and Affective Neuroscience

Computer-Aided Diagnosis

Psychiatric disorders are defined by the presence of specific set of symptoms. However, some symptoms are shared across disorders and a single patient might satisfy criteria for multiple disorders, or do not fit the requirements for any precise diagnosis (Huys et al., 2016). In this context, an increasingly popular application of ML in social and affective neuroscience is in the quest for imaging biomarkers of psychiatric disorders. This popularity is due to a recent focus on individualized medicine. Although classical statistical approaches provide biomarker descriptions at the group level, physicians should make clinical decisions about individuals (Orru et al., 2012). Thus, ML has been an active area of research to the development of potential computer-aided individualized diagnosis methods.

From this perspective, the use of structural MRI data combined with ML approaches is presenting promising results for the better comprehension of the obsessive-compulsive disorder (OCD). For example, Soriano-Mas et al. (2007) successfully classified patients with OCD from healthy control with more than 90% of accuracy based on brain structural features. Also, these data were used to predict the severity of obsessive-compulsive symptoms (Hoexter et al., 2013), as well as to list potential biomarkers using dimensionality reduction approaches (Trambaiolli et al., 2017).

In depressive spectrum disorders, structural MRI also achieved accuracies around the 90% threshold when classifying patients and controls (Mwangi et al., 2012), while functional MRI successfully discriminated between bipolar and unipolar depression with similar performances (Grotegerd et al., 2013). Also, structural and functional variations in affective-related brain regions, such as the amygdala, the insula, and the cingulate cortex, predicted symptom severity and treatment response (Siegle et al., 2006; Chen et al., 2007). Similarly, ML predictive approaches efficiently predicted the treatment response from patients with anxiety disorder for both pharmacological (Whalen et al., 2008) and cognitive behavioral (Doehrmann et al., 2013) therapies. However, it is important to emphasize that such findings had not yet reached clinical significance and are not currently incorporated in psychiatric practice.

Emotion/Affective Decoding

Brain decoding is the identification of someone's mental states based exclusively on measurements of their brain activity (Haynes & Rees, 2006). This stands on the idea that different neural activity patterns are associated with different mental states. Thus, decoding these patterns might be fundamental for our understanding of the neural basis of human cognition (Haynes & Rees, 2006). In this context, the ability from ML methods to identify and learn from patterns makes it a quite suitable approach for affective brain decoding.

A spectral power asymmetry over the frontal regions during emotion elicitation is a classical effect reported from EEG data analysis (Balconi et al., 2015). Applying an ML approach, Wang et al. (2014) reached more than 80% of predictive accuracy when distinguishing between positive and negative affective valences. Similar classification results were reported using fNIRS recordings over the prefrontal cortex when comparing positive or negative affective states with neutral states (Trambaiolli et al., 2018a). Also, the prefrontal activity even during resting state seems to be related with the emotional processing, since resting state frontal asymmetry predicts responsiveness to affective elicitation (Balconi et al., 2015).

However, human emotions involve complex networks comprising areas not accessed by the EEG or fNIRS spatial sampling and resolution. Using fMRI data, Baucom et al. (2012) achieved up to 90% accuracy in single participant classification between positive and negative valences using voxels from the medial and the ventrolateral prefrontal cortex, anterior cingulate, and amygdala, among other regions. Later, Lindquist et al. (2016) developed a meta-analytic study compiling data from 397 functional studies and different ML learning methods to investigate different hypotheses of network organization during the elicitation of affective valence. Their evidence suggests a single network composed by areas such as the dorsomedial prefrontal cortex, ventrolateral prefrontal cortex, supplementary motor area, anterior insula, amygdala, ventral striatum, and thalamus, which respond both for positive and negative valence, but with different patterns of activation depending on the affective state (Lindquist et al., 2016).

Neurofeedback

Due to the recent success of ML in decoding different mental states, this approach was also used to develop therapeutic applications, such as neurofeedback. Neurofeedback is a real-time procedure where a feedback of the neural activity in specific neural substrates is provided to the volunteer aiming to achieve the self-regulation of these areas or networks (Sitaram et al., 2017). Specifically, affective neurofeedback targets substrates related to emotional processing (Trambaiolli et al., 2018b) and might be useful as a nonpharmacological treatment for psychiatric symptoms or disorders, such as schizophrenia, major depressive disorder, attention-deficit/hyperactivity disorder, and obsessive-compulsive disorder (Fovet et al., 2015).

Different imaging methods allow different approaches to control affective networks. On the one hand, electrophysiological methods usually aim to control specific frequency bands in particular subsets of electrodes (Begemann et al., 2016; Enriquez-Geppert et al., 2017). For example, EEG alpha asymmetry in frontal electrodes was tested to reduce depressive symptoms, while central beta suppression and theta enhancement were applied to minimize inattention and impulsivity symptoms (Begemann et al., 2016).

On the other hand, hemodynamic methods use the upregulation or downregulation of the local blood flow in specific targets (Sulzer et al., 2013). For example, depressive patients who achieved self-control of the amygdala through fMRI-based neurofeedback showed reduced indices of anxiety and increased indices of happiness (Young et al., 2014), as well as a positive correlation between the symptom improvement and the reorganization of amygdala functional connectivity after the neurofeedback training (Young et al., 2018).

Social Neuroscience

Despite the indisputable importance of living in a structured society for human affective and cognitive processes, how the human brain works throughout simple to complex social contexts remains largely elusive (Babiloni & Astolfi, 2014).

In current social neuroscience, the possibility of simultaneously recording brain activity of two or more people interacting (i.e., hyperscanning) and of conceptualizing the connectivity emerging from such interactions (i.e., hyperconnectivity) has gained momentum (Montague et al., 2002). In this context, ML algorithms could be applied to modeling some level of a causal relation in social interactions mediated by interactions in brain activities (Konvalinka & Roepstorff, 2012). Anders et al. (2011) used fMRI records to predict the level of neural activity in romantic partners while experiencing the same emotional feelings. For this, the model was trained using data from one partner and used to successfully estimate the brain functional activation pattern of the other partner.

Another appealing field of research questions using ML approaches is the investigation of the neural correlates complexing social preferences and behaviors, such as friendship or engagement with political ideologies. For example, Kanai et al. (2011) applied a classifier to differentiate between participants with self-declared conservative or liberal political ideologies. Using the gray matter volume of the anterior cingulate and the right amygdala as inputs, the classifier reached near to 70% accuracy (Kanai et al., 2011). In another study, liberal and conservative participants were classified using functional MRI data with remarkable AUC values of more than 98% (Ahn et al., 2014).

Future Perspectives and Ethical Aspects

During the last decade, the neuroimaging community is making a continuous effort to create structured and standardized publicly available datasets, covering a wide range of samples and experiments (Poldrack & Gorgolewski, 2014). This action is fundamental to the development of optimized models for computer-aided diagnosis, for example. With larger samples, population heterogeneity, and standardized protocols, new ML models will be less susceptible to outliers and noise influence and will present higher generalization power (Schnack & Kahn, 2016). The extensive information resulting from these datasets will allow the use of ML approaches to confirm or to explore new aspects regarding the neural basis of affect and social interactions.

A promising instrumental evolution is the development of portable imaging devices, such as wearable EEG and fNIRS systems (Piper et al., 2014; von Lühmann et al., 2017). This technology allows studies outside the laboratory environment, leading to the observation of how the social brain acts in real-life situations (Balardin et al., 2017). Although ML algorithms should be adapted to deal with new levels of physiological (e.g., movement-related artifacts) and environmental (e.g., diverse magnetic fields) noises, a new range of naturalistic responses will be available for analysis. Neurofeedback applications would also be benefited by portable devices, with the possibility of location-independent training or the passive control of affect-driven software or equipment.

Another exciting prospect is the use of ML to develop new concepts of social interaction, such as the named collaborative brain-computer interfaces (BCI) (Wang & Jung, 2011). Following the idea of neurofeedback, in BCI, the user intends to control a computer exclusively based on their brain activity (Sitaram et al., 2017). Thus, collaborative BCI uses brain waves from multiple users to control one single machine, leading to increased task performances as high is the number of participants (Wang & Jung, 2011). Still, in the context of BCI, other social environments were created with the assistance of ML algorithms. For example, Rao et al. (2014) proposed the brain-to-brain interface in humans, where the EEG signals from one user were used to stimulate the brain of a second subject through transcranial magnetic stimulation (TMS). Later, this concept was expanded for the idea of a "brain-net," where the signals of some users (senders) were collaboratively merged to stimulate the brain of an independent participant (receiver) (Jiang et al., 2018).

The advance of ML applications in affective and social neuroscience also raises some ethical concerns. In clinical settings, for instance, the use of ML algorithms will only be possible after careful evaluation and when proper evidence for improvement in either diagnosis accuracy or treatment efficacy is in place. To date, no conclusion or clinical decision should be taken exclusively based on the ML output, and future applications surely will depend on the integration of ML procedures to expert knowledge (Fu & Costafreda, 2013). Decoding affective states is an essential tool for the understanding of the brain basis of the human mind, as well as for the development of therapeutic approaches such as neurofeedback. However, an essential ethical and legal aspect regarding brain decoding applications is ensuring privacy or non-consented commercial use of data or decoding results (Haynes, 2011).

Final Considerations

In this chapter, we introduced concepts of brain imaging and ML methods. Aside from describing learning and validation methods, dimensionality reduction and feature selection approaches, performance estimations, and currently popular classifiers, we purposefully focused on supervised methods. This choice was based on the facts that these are the best examples for an initial overview of the ML topic and the most popular approach in neuroimaging studies. We also described some uses of ML to social and affective neuroscience problems, from basic investigations to clinical and therapeutic applications. Promising prospects were also mentioned to contextualize the reader to cutting-edge advances in this area. Finally, we also highlighted some ethical aspects that might be carefully considered when developing applications of ML in social and affective neuroscience.

Acknowledgments JRS is grateful to Sao Paulo Research Foundation (FAPESP, Grants #2018/04654-9, #2018/21934-5 and #2021/05332-8).

References

Ahn, W. Y., Kishida, K. T., Gu, X., Lohrenz, T., Harvey, A., Alford, J. R., ... Montague, P. R. (2014). Nonpolitical images evoke neural predictors of political ideology. *Current Biology, 24*(22), 2693–2699.

Anders, S., Heinzle, J., Weiskopf, N., Ethofer, T., & Haynes, J. D. (2011). Flow of affective information between communicating brains. *NeuroImage, 54*(1), 439–446.

Babiloni, F., & Astolfi, L. (2014). Social neuroscience and hyperscanning techniques: Past, present and future. *Neuroscience and Biobehavioral Reviews, 44*, 76–93.

Balakrishnama, S., & Ganapathiraju, A. (1998). Linear discriminant analysis-a brief tutorial. *Institute for Signal and Information Processing, 18*, 1–8.

Balardin, J. B., Zimeo Morais, G. A., Furucho, R. A., Trambaiolli, L., Vanzella, P., Biazoli, C., Jr., & Sato, J. R. (2017). Imaging brain function with functional near-infrared spectroscopy in unconstrained environments. *Frontiers in Human Neuroscience, 11*, 258.

Balconi, M., Grippa, E., & Vanutelli, M. E. (2015). Resting lateralized activity predicts the cortical response and appraisal of emotions: An fNIRS study. *Social Cognitive and Affective Neuroscience, 10*(12), 1607–1614.

Baucom, L. B., Wedell, D. H., Wang, J., Blitzer, D. N., & Shinkareva, S. V. (2012). Decoding the neural representation of affective states. *NeuroImage, 59*(1), 718–727.

Begemann, M. J., Florisse, E. J., Van Lutterveld, R., Kooyman, M., & Sommer, I. E. (2016). Efficacy of EEG neurofeedback in psychiatry: A comprehensive overview and meta-analysis. *Translational Brain Rhythmicity, 1*(1), 19–29.

Chen, C. H., Ridler, K., Suckling, J., Williams, S., Fu, C. H., Merlo-Pich, E., & Bullmore, E. (2007). Brain imaging correlates of depressive symptom severity and predictors of symptom improvement after antidepressant treatment. *Biological Psychiatry, 62*(5), 407–414.

Cohen, I., Cozman, F. G., Sebe, N., Cirelo, M. C., & Huang, T. S. (2004). Semisupervised learning of classifiers: Theory, algorithms, and their application to human-computer interaction. *IEEE Transactions on Pattern Analysis and Machine Intelligence, 26*(12), 1553–1566.

Cristianini, N., & Shawe-Taylor, J. (2000). *An introduction to support vector machines and other kernel-based learning methods*. Cambridge University Press.

Doehrmann, O., Ghosh, S. S., Polli, F. E., Reynolds, G. O., Horn, F., Keshavan, A., ... Pollack, M. (2013). Predicting treatment response in social anxiety disorder from functional magnetic resonance imaging. *JAMA Psychiatry, 70*(1), 87–97.

Doi, H., Nishitani, S., & Shinohara, K. (2013). NIRS as a tool for assaying emotional function in the prefrontal cortex. *Frontiers in Human Neuroscience, 7*, 770.

Enriquez-Geppert, S., Huster, R. J., & Herrmann, C. S. (2017). EEG-neurofeedback as a tool to modulate cognition and behavior: A review tutorial. *Frontiers in Human Neuroscience, 11*, 51.

Fawcett, T. (2006). An introduction to ROC analysis. *Pattern Recognition Letters, 27*(8), 861–874.

Ferrari, M., & Quaresima, V. (2012). A brief review on the history of human functional near-infrared spectroscopy (fNIRS) development and fields of application. *NeuroImage, 63*(2), 921–935.

Fovet, T., Jardri, R., & Linden, D. (2015). Current issues in the use of fMRI-based neurofeedback to relieve psychiatric symptoms. *Current Pharmaceutical Design, 21*(23), 3384–3394.

Fu, C. H., & Costafreda, S. G. (2013). Neuroimaging-based biomarkers in psychiatry: Clinical opportunities of a paradigm shift. *Canadian Journal of Psychiatry, 58*(9), 499–508.

Grotegerd, D., Suslow, T., Bauer, J., Ohrmann, P., Arolt, V., Stuhrmann, A., … Dannlowski, U. (2013). Discriminating unipolar and bipolar depression by means of fMRI and pattern classification: A pilot study. *European Archives of Psychiatry and Clinical Neuroscience, 263*(2), 119–131.

Guyon, I., & Elisseeff, A. (2003). An introduction to variable and feature selection. *Journal of Machine Learning Research, 3*(Mar), 1157–1182.

Hansen, P., Kringelbach, M., & Salmelin, R. (Eds.). (2010). *MEG: An introduction to methods.* Oxford University Press.

Haynes, J. D. (2011). Brain reading: Decoding mental states from brain activity in humans. In Judy Illes and Barbara J. Sahakian (Eds.), *The Oxford handbook of neuroethics*, Oxford University Press, Oxford. pp. 3–13.

Haynes, J. D., & Rees, G. (2006). Neuroimaging: Decoding mental states from brain activity in humans. *Nature Reviews Neuroscience, 7*(7), 523.

Hoexter, M. Q., Miguel, E. C., Diniz, J. B., Shavitt, R. G., Busatto, G. F., & Sato, J. R. (2013). Predicting obsessive–compulsive disorder severity combining neuroimaging and machine learning methods. *Journal of Affective Disorders, 150*(3), 1213–1216.

Huang, S. H. (2015). Supervised feature selection: A tutorial. *Artificial Intelligence Research, 4*(2), 22.

Huys, Q. J., Maia, T. V., & Frank, M. J. (2016). Computational psychiatry as a bridge from neuroscience to clinical applications. *Nature Neuroscience, 19*(3), 404.

Jiang, L., Stocco, A., Losey, D. M., Abernethy, J. A., Prat, C. S., & Rao, R. P. (2018). BrainNet: A multi-person brain-to-brain interface for direct collaboration between brains. *arXiv, 1809.08632*.

Kanai, R., Feilden, T., Firth, C., & Rees, G. (2011). Political orientations are correlated with brain structure in young adults. *Current Biology, 21*(8), 677–680.

Kohavi, R., & John, G. H. (1997). Wrappers for feature subset selection. *Artificial Intelligence, 97*(1–2), 273–324.

Konvalinka, I., & Roepstorff, A. (2012). The two-brain approach: How can mutually interacting brains teach us something about social interaction? *Frontiers in Human Neuroscience, 6*, 215.

Kotsiantis, S. B. (2007). Supervised machine learning: A review of classification techniques. In I. G. Maglogiannis (Ed.), *Emerging artificial intelligence applications in computer engineering*, IOS Press, Amsterdam, Netherland. (Vol. 160, pp. 3–24).

Kuhn, M., & Johnson, K. (2013). *Applied predictive modeling.* Springer.

Larranaga, P., Calvo, B., Santana, R., Bielza, C., Galdiano, J., Inza, I., … Robles, V. (2006). Machine learning in bioinformatics. *Briefings in Bioinformatics, 7*(1), 86–112.

Lemm, S., Blankertz, B., Dickhaus, T., & Müller, K. R. (2011). Introduction to machine learning for brain imaging. *NeuroImage, 56*(2), 387–399.

Lindquist, K. A., Satpute, A. B., Wager, T. D., Weber, J., & Barrett, L. F. (2016). The brain basis of positive and negative affect: Evidence from a meta-analysis of the human neuroimaging literature. *Cerebral Cortex, 26*(5), 1910–1922.

Liu, S., Cai, W., Liu, S., Zhang, F., Fulham, M., Feng, D., … Kikinis, R. (2015). Multimodal neuroimaging computing: A review of the applications in neuropsychiatric disorders. *Brain Informatics, 2*(3), 167.

Maquet, P. (2000). Functional neuroimaging of normal human sleep by positron emission tomography. *Journal of Sleep Research, 9*(3), 207–232.

Min, B. K., Marzelli, M. J., & Yoo, S. S. (2010). Neuroimaging-based approaches in the brain–computer interface. *Trends in Biotechnology, 28*(11), 552–560.

Mitchell, T. M. (1997). *Machine learning*. McGraw Hill.

Montague, P. R., Berns, G. S., Cohen, J. D., McClure, S. M., Pagnoni, G., Dhamala, M., ... Fisher, R. E. (2002). Hyperscanning: Simultaneous fMRI during linked social interactions. *NeuroImage, 16*, 1159–1164.

Murthy, S. K. (1998). Automatic construction of decision trees from data: A multi-disciplinary survey. *Data Mining and Knowledge Discovery, 2*(4), 345–389.

Mwangi, B., Ebmeier, K. P., Matthews, K., & Douglas Steele, J. (2012). Multi-centre diagnostic classification of individual structural neuroimaging scans from patients with major depressive disorder. *Brain, 135*(5), 1508–1521.

Niedermeyer, E., & da Silva, F. L. (Eds.). (2005). *Electroencephalography: Basic principles, clinical applications, and related fields*. Lippincott Williams & Wilkins.

Ogawa, S., Lee, T. M., Kay, A. R., & Tank, D. W. (1990). Brain magnetic resonance imaging with contrast dependent on blood oxygenation. *Proceedings of the National Academy of Sciences of the United States of America, 87*(24), 9868–9872.

Orru, G., Pettersson-Yeo, W., Marquand, A. F., Sartori, G., & Mechelli, A. (2012). Using support vector machine to identify imaging biomarkers of neurological and psychiatric disease: A critical review. *Neuroscience and Biobehavioral Reviews, 36*(4), 1140–1152.

Pereira, F., Mitchell, T., & Botvinick, M. (2009). Machine learning classifiers and fMRI: A tutorial overview. *NeuroImage, 45*(1), S199–S209.

Piper, S. K., Krueger, A., Koch, S. P., Mehnert, J., Habermehl, C., Steinbrink, J., ... Schmitz, C. H. (2014). A wearable multi-channel fNIRS system for brain imaging in freely moving subjects. *NeuroImage, 85*, 64–71.

Poldrack, R. A., & Gorgolewski, K. J. (2014). Making big data open: Data sharing in neuroimaging. *Nature Neuroscience, 17*(11), 1510.

Rao, R. P., Stocco, A., Bryan, M., Sarma, D., Youngquist, T. M., Wu, J., & Prat, C. S. (2014). A direct brain-to-brain interface in humans. *PLoS One, 9*(11), e111332.

Rubinov, M., & Sporns, O. (2010). Complex network measures of brain connectivity: Uses and interpretations. *NeuroImage, 52*(3), 1059–1069.

Sakkalis, V. (2011). Review of advanced techniques for the estimation of brain connectivity measured with EEG/MEG. *Computers in Biology and Medicine, 41*(12), 1110–1117.

Sato, J. R., Fujita, A., Thomaz, C. E., Martin, M. D. G. M., Mourão-Miranda, J., Brammer, M. J., & Junior, E. A. (2009). Evaluating SVM and MLDA in the extraction of discriminant regions for mental state prediction. *NeuroImage, 46*(1), 105–114.

Schnack, H. G., & Kahn, R. S. (2016). Detecting neuroimaging biomarkers for psychiatric disorders: Sample size matters. *Frontiers in Psychiatry, 7*, 50.

Siegle, G. J., Carter, C. S., & Thase, M. E. (2006). Use of FMRI to predict recovery from unipolar depression with cognitive behavior therapy. *The American Journal of Psychiatry, 163*(4), 735–738.

Sitaram, R., Ros, T., Stoeckel, L., Haller, S., Scharnowski, F., Lewis-Peacock, J., ... Birbaumer, N. (2017). Closed-loop brain training: The science of neurofeedback. *Nature Reviews Neuroscience, 18*(2), 86.

Sokolova, M., & Lapalme, G. (2009). A systematic analysis of performance measures for classification tasks. *Information Processing and Management, 45*(4), 427–437.

Soriano-Mas, C., Pujol, J., Alonso, P., Cardoner, N., Menchón, J. M., Harrison, B. J., ... Gaser, C. (2007). Identifying patients with obsessive–compulsive disorder using whole-brain anatomy. *NeuroImage, 35*(3), 1028–1037.

Sulzer, J., Haller, S., Scharnowski, F., Weiskopf, N., Birbaumer, N., Blefari, M. L., ... Herwig, U. (2013). Real-time fMRI neurofeedback: Progress and challenges. *NeuroImage, 76*, 386–399.

Trambaiolli, L. R., Biazoli, C. E., Jr., Balardin, J. B., Hoexter, M. Q., & Sato, J. R. (2017). The relevance of feature selection methods to the classification of obsessive-compulsive disorder based on volumetric measures. *Journal of Affective Disorders, 222*, 49–56.

Trambaiolli, L. R., Biazoli, C. E., Cravo, A. M., & Sato, J. R. (2018a). Predicting affective valence using cortical hemodynamic signals. *Scientific Reports, 8*(1), 5406.

Trambaiolli, L. R., Biazoli, C. E., Cravo, A. M., Falk, T. H., & Sato, J. R. (2018b). Functional near-infrared spectroscopy-based affective neurofeedback: Feedback effect, illiteracy phenomena, and whole-connectivity profiles. *Neurophotonics, 5*(3), 035009.

Vapnik, V. (1995). *The nature of statistical learning theory*. Springer.

von Lühmann, A., Wabnitz, H., Sander, T., & Müller, K. R. (2017). M3BA: A mobile, modular, multimodal biosignal acquisition architecture for miniaturized EEG-NIRS-based hybrid BCI and monitoring. *IEEE Transactions on Biomedical Engineering, 64*(6), 1199–1210.

Wang, Y., & Jung, T. P. (2011). A collaborative brain-computer interface for improving human performance. *PLoS One, 6*(5), e20422.

Wang, X. W., Nie, D., & Lu, B. L. (2014). Emotional state classification from EEG data using machine learning approach. *Neurocomputing, 129*, 94–106.

Whalen, P. J., Johnstone, T., Somerville, L. H., Nitschke, J. B., Polis, S., Alexander, A. L., … Kalin, N. H. (2008). A functional magnetic resonance imaging predictor of treatment response to venlafaxine in generalized anxiety disorder. *Biological Psychiatry, 63*(9), 858–863.

Young, K. D., Zotev, V., Phillips, R., Misaki, M., Yuan, H., Drevets, W. C., & Bodurka, J. (2014). Real-time FMRI neurofeedback training of amygdala activity in patients with major depressive disorder. *PLoS One, 9*(2), e88785.

Young, K. D., Siegle, G. J., Misaki, M., Zotev, V., Phillips, R., Drevets, W. C., & Bodurka, J. (2018). Altered task-based and resting-state amygdala functional connectivity following real-time fMRI amygdala neurofeedback training in major depressive disorder. *Neuroimage: Clinical, 17*, 691–703.

Zhang, G. P. (2000). Neural networks for classification: A survey. *IEEE Transactions on Systems, Man, and Cybernetics, Part C (Applications and Reviews), 30*(4), 451–462.

Chapter 14
fMRI and fNIRS Methods for Social Brain Studies: Hyperscanning Possibilities

Paulo Rodrigo Bazán and Edson Amaro Jr

Abstract Recently, the "social brain" (i.e., how the brain works in social context and the mechanisms for our social behavior) has gained focus in neuroscience literature – largely due to the fact that recently developed techniques allow studying different aspects of human social cognition and its brain correlates. In this context, hyperscanning techniques (Montague et al., Neuroimage 16(4):1159–1164, 2002) open the horizon for human interaction studies, allowing for the evaluation of interbrain connectivity. These techniques represent methods for simultaneously recording signals from different brains when subjects are interacting. In this chapter, we will explore the potentials of functional magnetic resonance imaging (fMRI) and functional near-infrared spectroscopy (fNIRS), which are techniques based on blood-oxygen-level-dependent (BOLD) signal. We will start with a brief explanation of the BOLD response basic principles and the mechanisms involved in fMRI and fNIRS measurements related to brain function. We will then discuss the foundation of the social brain, based on the first studies, with one subject per data acquisition, to allow for understanding the new possibilities that hyperscanning techniques offer. Finally, we will focus on the scientific literature reporting fMRI and fNIRS hyperscanning contribution to understand the social brain.

Keywords fMRI · fNIRS · Social brain · Hyperscanning · Whole-brain coverage

P. R. Bazán (✉)
LIM-44, Departamento de Radiologia, Hospital das Clínicas da Faculdade de Medicina da Universidade de São Paulo, São Paulo, Brazil

Hospital Israelita Albert Einstein, São Paulo, Brazil
e-mail: paulo.bazan@usp.br; paulo.bazan@einstein.br

E. Amaro Jr
LIM-44, Departamento de Radiologia, Hospital das Clínicas da Faculdade de Medicina da Universidade de São Paulo, São Paulo, Brazil

© The Author(s) 2023
P. S. Boggio et al. (eds.), *Social and Affective Neuroscience of Everyday Human Interaction*, https://doi.org/10.1007/978-3-031-08651-9_14

231

Introduction

Recently, the "social brain" (i.e., how the brain works in social context and the mechanisms for our social behavior) has gained focus in neuroscience literature – largely due to the fact that recently developed techniques allow studying different aspects of human social cognition and its brain correlates. In this context, hyperscanning techniques (Montague et al., 2002) open the horizon for human interaction studies, allowing for the evaluation of interbrain connectivity. These techniques represent methods for simultaneously recording signals from different brains when subjects are interacting. In this chapter, we will explore the potentials of functional magnetic resonance imaging (fMRI) and functional near-infrared spectroscopy (fNIRS), which are techniques based on blood-oxygen-level-dependent (BOLD) signal. We will start with a brief explanation of the BOLD response basic principles and the mechanisms involved in fMRI and fNIRS measurements related to brain function. We will then discuss the foundation of the social brain, based on the first studies, with one subject per data acquisition, to allow for understanding the new possibilities that hyperscanning techniques offer. Finally, we will focus on the scientific literature reporting fMRI and fNIRS hyperscanning contribution to understand the social brain.

Hemodynamic Response and BOLD Signal

The relationship between neuronal activity and hemodynamic response is the basis underlying the blood-oxygen-level-dependent (BOLD) signal. One of the first models based on biomechanical properties related to blood volume, blood flow, and oxygen consumption was proposed by Buxton et al. (1998): the balloon model. This construct is a valid point to address how hemodynamics is related to neural function. Although local increase in metabolism related to neuronal activity increases the consumption of oxygen, the increase in supply of oxygen is higher than required by energy consumption needs. This is due to the fact that the transport from intravascular (hemoglobin linked) O2 to intraneuronal space depends on passive mechanisms related to differences in pressure gradients. We have also to add to that fact the increased blood volume and flow in the dynamics of the process. This equilibrium evolves over time and generates a small initial decrease in oxyhemoglobin/deoxyhemoglobin proportion, followed by a strong increase of this ratio. When the temporal dynamics are considered in this equation – and also based on the observations – the BOLD response associated with neuronal activity is a slow response that reaches its peak around 6 s after a stimulus. Moreover, this neurovascular coupling is associated with several pathways, including neuronal release of vasoactive mediators (e.g., nitric oxide), and pathways related to calcium activity in astrocytes (for reviews, read Longden et al. (2016) and Filosa et al. (2016)). Thus, the BOLD response evaluated with fMRI and fNIRS is an indirect measure of neural activity.

How Functional Magnetic Resonance Imaging Detects BOLD

The blood-oxygen-level-dependent (BOLD) signal was detected in MRI by Ogawa et al. (1990) and was first described in human brains in 1992 (Bandettini et al., 1992; Kwong et al., 1992; Ogawa et al., 1992). For a detailed history of the development of fMRI, we recommend the review article by Bandettini (2012), which celebrates 20 years of the technique. The detection of BOLD signal was possible because the hemoglobin has different magnetic properties when it is oxygenated, due to conformation changes related with the iron-oxygen binding site. While oxy-hemoglobin is diamagnetic (low interaction with magnetic induction), deoxyhemo-globin is paramagnetic and generates a distortion of the magnetic field around it. A distortion in the magnetic field causes a faster decrease in the hydrogen nuclear resonance signal. Higher concentrations of oxyhemoglobin relative to deoxyhemo-globin allow a more stable local magnetic field and therefore more signal. Hence, fMRI is a technique that measures relative signals based on the oxyhemoglobin and deoxyhemoglobin proportion. This highlights the importance of baseline control conditions for fMRI experiments as it is not an absolute measure; it is a relative measure. fMRI has a whole-brain coverage with good spatial resolution (in the order of millimeters, and can reach submillimetric resolution using ultrahigh field MR systems), however a relatively lower temporal resolution (in the order of seconds, mainly due to the slow temporal hemodynamic response, although MR systems are capable of acquiring data in the order of hundreds of milliseconds). The precision of MRI is related to gradients generated in the magnetic field, which alter the specific radio frequency absorbed by hydrogen nucleus. Therefore, it is highly sensitive to movement and requires participants to lay down inside the scanner. During fMRI, volunteers enter the scanner bore, a tunnel large enough to host a human body – and as such are not a natural environment, but rather may induce claustrophobia – and image acquisition depends on a head coil to detect the resonance signals. Also, there are several restrictions or exclusion criteria for participating in fMRI experiment, due to the intense magnetic field, such as pregnancy, pacemakers, magnetic prosthesis, tattoos (depending on the pigment used), and other situations that might induce risk for the participant.

How Functional Near-Infrared Spectroscopy Detects BOLD

The BOLD signal in the human brain was detected using fNIRS around the same period it was observed using fMRI (Hoshi & Tamura, 1993; Chance et al., 1993; Kato et al., 1993; Villringer et al., 1993). Therefore, similar celebrative reviews detail the history of fNIRS development (Ferrari & Quaresima, 2012; Scholkmann et al., 2014). As mentioned above, the oxygen bond to hemoglobin causes a conformational change which alters electromagnetic properties of the molecule. fNIRS depends on changes in light abortion in different near-infrared wavelengths related to oxygen bond to hemoglobin. Though different absorption rates of oxyhemoglobin and deoxyhemoglobin are observed in several parts of the electromagnetic light spectrum, near-infrared light is less absorbed by the skull and other tissues between the cortex and the scalp. In this way, fNIRS can be used to evaluate separately the cortical concentration of oxyhemoglobin and deoxyhemoglobin. For these measures, fNIRS uses optodes (similar to electrodes but with optical properties) as sources of light and as detectors of the light that is scattered through the brain tissue. A combination of source and detector forms an fNIRS channel (one source and four detectors could form four channels) located between the optodes. Duo to light scattering and absorption, fNIRS can only detect signal a few centimeters below the scalp, providing mainly cortical signal. It is also important to notice that tissue transparency to light depends on age, in a way that the skull in babies is more transparent than in adults. The recommended distance between optodes for adults is between 2.5 and 4 cm. Positioning optodes closer together (e.g., 0.8 cm) can be used to detect and later filter hemodynamic processes unrelated to local brain activity (Brigadoi & Cooper, 2015). Usually, these optodes are attached to a cap that follows the 10–20 coordinate system (and its variations) of electroencephalography (Jasper, 1958; Oostenveld & Praamstra, 2001). As an advantage, fNIRS can be portable and is less affected by movement, being more suitable for ecological and naturalistic studies. Also, since the volunteers can move to a certain degree, it is more suitable for studies with babies and young children. The temporal resolution of fNIRS systems depends on the number of sources used, since each source has to be turned on separately to avoid mixing signals from different regions in the detectors. Usually, the sampling rate can vary around 4 Hz to 60 Hz, providing fNIRS with higher temporal resolution than fMRI. This is useful for correction of cardiac artifacts, given the higher sampling rate diminishes aliasing artifacts, which are present in fMRI data. On the other hand, fNIRS has lower spatial resolution (in the order of centimeters) and does not have a whole-brain coverage (restricted to cortical signal), and the number of optodes available defines the cortical coverage level of the system.

Types of Experiment Design and Data Analysis for BOLD Studies

In task-based designs, fMRI and fNIRS can be used to identify regions with BOLD signal variation related to task variation (from a baseline control condition to the task of interest), based on BOLD response after a stimulus (event-related design), or due to a block of stimulus (block design). In these types of design, it is important to consider stimulus (or task) sequence, duration, number of repetitions, time between stimulus, and the hypothesis of which brain regions will be related to the task in order to define the design that will provide more statistical power for a general linear model analysis (the most common analysis in this context). For a review of study design, we suggest Amaro Jr and Barker (2006).

Alternatively, there are designs in which the subject sustains a brain state either by continuously performing a specific task or by remaining in resting state (with no specific task, only with the instruction to remain awake and not focus in anything in particular). These designs are used in connectivity studies, which explore signal relation between different brain regions and explore brain organization and communication between areas. For example, resting-state studies allowed the identification of intrinsic brain networks (Greicius et al., 2003; Fox et al., 2005; Damoiseaux et al., 2006). There are several connectivity measures, and they can be applied both to resting-state (Han et al., 2018) and to task-based (as event-related and block) studies as well (Friston, 2011). Some measures are data-driven, like independent component analysis, while others depend on previous hypothesis-driven models, as in the case of dynamic causal modeling. Even more simple calculations, as correlation index, can be used to study organization of brain networks. Graph theory can be applied using these connectivity measures to evaluate the characteristics of these networks.

Hyperscanning Design and Data Analysis

The term hyperscanning was first used by Montague et al. (2002) referring to measuring brain signal from interacting humans. In their study, two synchronized fMRI scanners were used to measure brain activity while pairs of volunteers interacted in a competitive game. Since then, hyperscanning has been performed with several techniques, such as electroencephalography (EEG), magnetoencephalography (MEG), fMRI, and fNIRS (Babiloni & Astolfi, 2014; Zhdanov et al., 2015; Wang et al., 2018). Different from acquiring data from subjects separately, hyperscanning offers the possibility of relating brain activities from different subjects preserving trial-specific characteristics. In other words, even though having a pair of volunteers perform the same task twice while measuring one subject at a time would provide a way of comparing brain activities, the data from each subject would have different specific trial characteristics as performance score or event-specific strategy or brain state during the trial; on the other hand, with hyperscanning, these are preserved, providing more information on brain signals during interaction. Also, hyperscanning can be used to expand the concepts of connectivity analysis from within to between brains, revealing more than what areas have more signal during interaction but also how these areas from different brains coordinate their activity.

There are some important technical aspects to consider when doing hyperscanning:

- Synchrony between equipment. In order to be able to take advantage of simultaneous recording, it is necessary to have precise synchrony of recordings from the different volunteers. As an example, if there is asynchrony or a lag between recordings, then correlation measures would be shifted; in case of causation methods, this could significantly alter the interpretation of leader and follower. It is also important to notice that sampling rate and the type of signal measured directly influence the required precision of synchrony. In fMRI, usually with whole-brain sampling rate of 2 s (and around 50 ms per slice), a lag of 50 ms (one slice) might be acceptable, also taking into account the slow hemodynamic response; in an EEG setup with 1000 Hz sampling rate measuring fast electric changes, 50 ms would represent a lag of 50 data points, and would not be acceptable. It is also important to notice that if the lag is constant, it can be corrected during analysis by shifting the time series appropriately. However, if the lag is variable and with no clear pattern, then it probably will not be correctable. Intranet- and Internet-based synchrony solutions were developed to allow for triggering MRI scanners in different buildings (Montague et al., 2002). Also, there are software solutions such as lab streaming layer (https://github.com/sccn/labstreaminglayer; Gramann et al., 2014; Ojeda et al., 2014), which allows for synchronizing fNIRS and other types of signals received by a computer. Another option would be to use only one system to get the data from two or more subjects (see Fig. 14.1 for hyperscanning options).

- Equivalent measures in each equipment. Hyperscanning requires having the same signal quality control routines for all systems used to assure equivalent measures in all participants. This is highlighted in MRI scans, since there is a great variability of sequences and parameters available in different scanners. Therefore, having the same version of scanner for each subject can help assuring comparable measures. A usual alternative in fNIRS is to split the optodes from one system between subjects, so equivalent quality control and synchrony are guaranteed. In the case of MRI, a similar solution would be to use one scanner with a dual-head coil (Lee et al., 2010, 2012; Lee, 2015a, b), although this would have implications for comfort and in the type of interaction between participants.

- Controlling prior interaction and types of interaction between participants. A few studies have evaluated that the level of relationship (strangers, friends, romantic relationship) could alter the level of interbrain connectivity, with lovers presenting more connectivity (Pan et al., 2017). Therefore, controlling how well participants know each other may be important. Also, during the hyperscanning protocols, participants can try to interact in different ways, for example, a pair might try to use verbal communication, while other might choose nonverbal if no specific instruction is provided. Moreover, eye contact can alter interbrain synchrony (Hirsch et al., 2017; Koike et al., 2019). Since different brain mechanisms could be participating in each type of interaction, it is important to prepare your experiment design and instruction to participants in order to select the type of interaction that is important for your study. Regarding interaction paradigms, they can be classified according to three axes: goal of interaction, divided in cooperative and competitive interactions; temporal structure and sequence of actions in time, which can be turn-based (as observed in chess or card games) or continuous interaction (e.g., coordinated singing); and dependency of actions, which can be in independent (usually in competitions) or interdependent (Liu & Pelowski, 2014). Decision-making in social context, as in the game theory para-

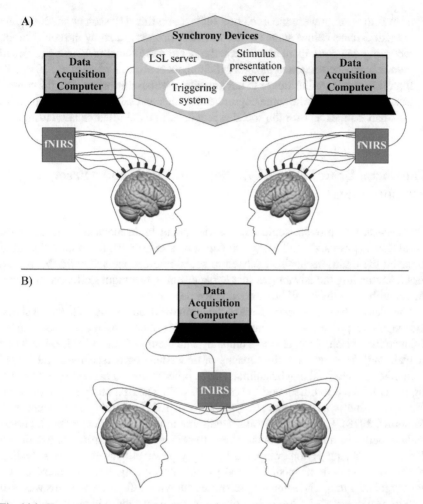

Fig. 14.1 Hyperscanning options: (**a**) using one data acquisition device for each subject and using synchrony devices to assure synchronized data acquisition; (**b**) sharing the same data acquisition device for all the subjects and providing synchrony between data from all subjects but limiting distance between subjects and reducing the number of channels (fNIRS). For schematic representation, fNIRS was presented in the image, but similar considerations are valid for fMRI hyperscanning. *LSL* lab streaming layer

digms (Wang et al., 2015), are good examples of interaction paradigms. We will further explore brain mechanism in some of these types of interaction in the section called "Hyperscanning Studies of the Social Brain."

- Having a good control condition. Since one of the main goals in hyperscanning is to measure interbrain connectivity and relating it to a specific task or condition, a proper control condition is required. For example, two volunteers receiving the same stimuli or performing the same task (e.g., watching a movie), with the same temporal structure, would have a level of synchrony between their brains, even if these data were recorded one subject at a time (Hasson et al., 2004). As we will discuss below, some studies used this property to study the social brain, recording one participant at a time, but repeating the social task, using recorded videos from one session in the other (Schippers et al., 2009; Lee et al., 2018). To filter this signal similarities or to have an adequate baseline condition, the control task should have all the same cognitive elements and with the same intensity as the main task, except for the social aspect on which the research is focusing.

The Social Brain: Empathy, Theory of Mind, and Mirror Neuron System

This section will provide some key theories about brain mechanism relevant for social tasks. For this, the following text will concentrate on the first approaches used to unveil the brain mechanisms related to social interactions and build the ground for understanding the advantages that hyperscanning techniques offer for studying the social brain, which will be covered in the next section.

The idea of the social brain refers to the brain mechanism related to social skills, and cognitive processes important for social interaction, as motor coordination (joint action or imitation), affective empathy, and theory of mind (ToM – also called mentalizing). ToM refers to the capacity of understanding perspectives and beliefs from other people (and maybe animals) or, in other words, to the ability of attributing a mental state to others and to oneself as well. The first study to use the term "theory of mind" was evaluating the possibility of ToM in chimpanzees (Premack & Woodruff, 1978). Later, using positron emission tomography, Fletcher et al. (1995) studied theory of mind in humans, related to stories that had a mind state explanation for a character action compared to simply physical causality stories, finding increased activity in left medial frontal gyrus, anterior cingulate gyrus, and posterior cingulate gyrus. The same idea of stories but with different modalities was also used by Gallagher et al. (2000), identifying increased response in medial prefrontal cortex and temporoparietal junction (TPJ). The TPJ is now one of the main regions associated with the social brain, as the medial prefrontal cortex is related to ToM (Frith & Frith, 2003). Based on these results, fNIRS studies usually focus on TPJ and prefrontal regions in the context of ToM, since most fNIRS studies do not have enough optodes for a whole-head coverage. In adults, it was proposed that even without specific instruction (in a video story task), TPJ is engaged and related to spontaneous detection of others' beliefs (Hyde et al., 2015). Also, the emotional state of the participant seems to affect lateral prefrontal cortex activity during ToM director task, in which the volunteer has to consider the perspective of another person (Himichi et al., 2015). However, evaluation of cognitive empathy and affective empathy in children is very challenging. In these cases, fNIRS inherent characteristics enable specific designs in children. For instance, a study using cartoon stories and verbal stories in 4–8-year-old children showed involvement of medial orbitofrontal regions, as well as dorsolateral prefrontal cortex in these tasks (Brink et al., 2011). In addition, other authors observed greater TPJ relationship with belief detection compared to desire intention in 6–10-year-old children (Bowman et al., 2015).

Another part of the social brain is identifying motor intentions and coordinating motor action (as in imitation). Evaluating mechanisms of imitation with fMRI, the inferior frontal gyrus and the superior parietal lobule were found to be related both to performing a finger-tapping and to observing the finger-tapping, and even more intense activity was found during imitation of movement (Iacoboni et al., 1999). These results were later associated with the mirror neuron system (MSN, Grèzes et al., 2003; Iacoboni et al., 2005; Jeon & Lee, 2018 for a review), which is related to internal representation of motor intentions from others and is formed by inferior frontal gyrus, inferior and superior parietal lobule, borders of superior temporal sulcus, and premotor cortex. The mirror neuron system was also identified with fNIRS in a more naturalistic table-setting task, which involved observation and execution (Sun et al., 2018). The mirror neural system is usually associated with mentalizing, as simulation-based system (Gallese & Goldman, 1998; Frith & Frith, 2006; Mahy et al., 2014), but there is still current debate to define the specific function and limit of each system in each social context and how these systems interact. Canessa et al. (2012) studied human subjects observing pictures with cooperative context (two persons caring an object) and pictures with affective context (two persons holding hands) using fMRI. These authors observed that both conditions seemed to engage the temporoparietal junction, but the ventromedial prefrontal cortex was more related to affective scenes, while inferior frontal gyrus and inferior parietal lobule were more associated with cooperative images. Moreover, the signal in these regions was related to empathy scores of participants. In a posterior study, the effective connectivity of these regions was evaluated indicating significant connectivity in these regions, but with different directionality according to the type of picture (Arioli et al., 2018).

Receiving and processing social feedback are also another important part of the social brain. In an fMRI study, the brain regions related to social influence on rating of emotional images were evaluated showing participation of the prefrontal cortex, borders of superior temporal sulcus, amygdala, and insula (Lin et al., 2018). In this study, participants rated images before and after seeing the average rating given by (simulated) group members, to check the social influence and induced rating adjustment. Another fMRI study evaluated positive and negative feedback of personal traits simulating a hyperscanning competitive group situation, but the feedbacks were predefined by the experimenter, and only one subject was actually being evaluated at a time (Dalgleish et al., 2017). In these experiments, an increased BOLD signal was observed in the ventromedial prefrontal cortex related to positive feedback, while both positive and negative feedback elicited responses in the anterior cingulate cortex and amygdala.

The aforementioned studies had a similar design in the sense that they used single subject tasks to study specific cognitive function relevant for interaction (Fig. 14.2a). However, this method has limitations, as it does not measure the brain signals during an actual interaction. As a first option to overcome this limitation, it is possible to evaluate the effect of an actual interaction on the brain activity of a subject (Fig. 14.2b). For example, Chauvigné et al. (2018) compared brain activity of professional dancers in different types of hand interactions (in a type of joint action), comparing leading and following brain activities and finding activity in the inferior frontal gyrus and premotor cortex during leading and in the ventromedial prefrontal cortex and borders of superior temporal sulcus in following. Rauchbauer et al. (2019) compared brain activity of subjects during conversation with another human to the activity during conversation with robots and observed higher temporal activity in human-human interactions. These results may indicate a starting point to understand human social cognition, as well as the social competence of robots interacting with humans.

Some studies also tried to compare the brain activity of participants that were scanned in different sessions, by recording a video of the first session and using it in the second session with the other volunteer. Lee et al. (2018) evaluated mother-child brain signal similarities during stress condition for the adolescent (the video of the child was recorded, and the mother was scanned observing the video of the adolescent during this stress condition), finding family relationship level impacting similarity in the insula and anterior cingulate cortex. Schippers et al. (2009) also used this video recording technique and found mirror neuron system and TPJ activity during decoding of gestures in a charades game.

Although the scientific evidence covered in the above paragraphs provided the basis for main theories proposed to explain social interaction, they were based on experimental designs unable to probe the relationship between neural systems during dynamic interpersonal interaction. This is necessary to understand how interacting brains regulate their function based on the other person's behavioral responses.

Hyperscanning Studies of the Social Brain

Key aspects of correlation between brain signals and also brain activity in real interactions can be better evaluated in hyperscanning setups (Fig. 14.2, panel c). A review of hyperscanning with fNIRS reported 20 fMRI and 7 fNIRS hyperscanning studies published up to spring of 2013 (Scholkmann et al., 2013). Based on PubMed and Web of Science search using keywords Hyperscanning and fMRI and Hyperscanning and fNIRS, we found 14 fMRI hyperscanning research articles from the beginning of 2014 to March of 2019 and 33 fNIRS hyperscanning articles, showing the increased applications of these techniques, specially fNIRS hyperscanning. It is important to mention that our literature search found a total of 23 fMRI (9 up to spring of 2013) studies and 40 fNIRS studies, indicating that the search parameters used by Scholkmann were different from ours – perhaps not only due to time differences – especially regarding fMRI hyperscanning. The following subsections will explore hyperscanning experiments with each technique.

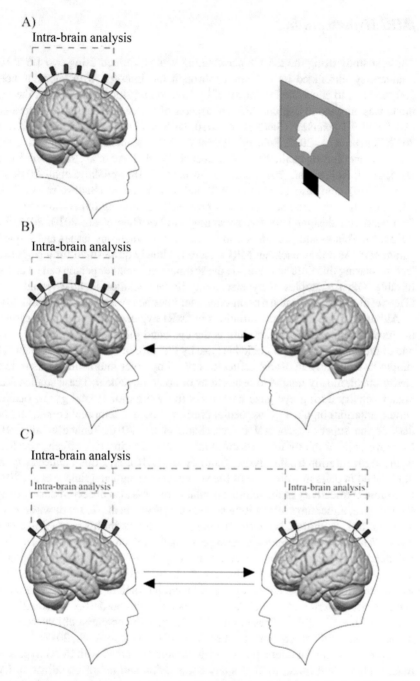

Fig. 14.2 Types of social brain experiments and analysis possibilities. (**a**) Single subject brain signal acquisition; single subject task. (**b**) Single subject brain signal acquisition; multi-subject task. (**c**) Multi-subject brain signal acquisition; multi-subject task

fMRI Hyperscanning

The first study using the term hyperscanning was performed using two 1.5 T MRI scanners synchronized by one server through the Internet, with latencies below 300–400 ms, to show the feasibility of hyperscanning studies. After that, the technique was adopted to explore different aspects of social interaction in combination of 1.5 and 3 T scanners (Saito et al., 2010; Krill & Platek, 2012; Fliessbach et al., 2012; Tanabe et al., 2012; Spiegelhalder et al., 2014; Stolk et al., 2014), two or more 3 T scanners (Tomlin et al., 2013; Morita et al., 2014; Trees et al., 2014; Bilek et al., 2015, 2017; Koike et al., 2016, 2019; Shaw et al., 2018; Špiláková et al., 2019; Abe et al., 2019), and also combining 3 T and 7 T scanners (Baecke et al., 2015). Alternatively, Ray Lee and colleagues proposed performing hyperscanning with a dual-head coil designed for hyperscanning studies (Lee et al., 2010, 2012; Lee, 2015a, b). This would provide a good solution for synchrony issues and sequence parameters. As a drawback, an MRI system is already quite small for a single subject, so sharing this little space inside the scanner with another person can be uncomfortable. Other examples of hyperscanning implementation include virtual reality (Trees et al., 2014) and using a brain-computer interface system (Baecke et al., 2015).

Although there are several variations in fMRI hyperscanning techniques, one of the most mentioned in the literature is the eye-cued joint action. In this approach, one of the volunteers has to guide his gaze by the gaze of the other participant. This simple task circumvents some difficulties faced by other interaction tasks in fMRI environment, mostly related to restrictions of body movement. These studies found higher activity during eye gaze cued tasks in the inferior frontal gyrus, occipital cortex, anterior cingulate gyrus/medial prefrontal cortex, temporal cortex, and borders of the superior temporal sulcus (Saito et al., 2010; Tanabe et al., 2012). Moreover, higher interbrain cross correlation, after filtering task effects, was found in the right inferior frontal gyrus (Saito et al., 2010), and this connectivity was diminished in pairs in which one of the volunteers had autism (Tanabe et al., 2012). Interbrain connectivity in the inferior frontal gyrus was also observed during simple mutual gaze, either days after a joint action task (Koike et al., 2016) or without joint attention task execution, but closer to the insular cortex in this case (Koike et al., 2019). With a different approach during eye-cued joint attention mutual gaze, using independent component analysis and evaluating the relation between components involved in the task from each volunteer of the pair, higher interbrain connectivity was detected in the right temporoparietal junction (Bilek et al., 2015). The normal connectivity pattern was also disrupted in patients with borderline personality disorder (Bilek et al., 2017). These examples illustrate the potential of hyperscanning in the context of disorders that affect social interaction (Ray et al., 2017).

Several other social interaction paradigms were evaluated with fMRI hyperscanning, such as joint force, used to study cooperation and motor coordination (Abe et al., 2019), finding higher interbrain connectivity in the right temporoparietal junction, and right temporoparietal junction signal increase during task was related to performance scores. Both Spiegelhalder et al. (2014) and Stolk et al. (2014) have evaluated verbal and signal communication (respectively) and found a similar result: brain signal synchrony in temporal lobes either to areas related to talking, in verbal communication, or to the pair temporal lobes, in case of signal communication.

Game theory tasks are also used in hyperscanning experiments. For instance, the ultimatum game is a turn-based game in which a proposer chooses a proportion of reward distribution between participants and then the responder has to decide either to accept the proposal, and the reward is distributed between participants as agreed, or to reject it, in which case none of the participants receive a reward. Using this type of paradigm, Fliessbach et al. (2012) found striatum and ventromedial prefrontal cortex activity in both participants in more generous proposals and in higher level of acceptance by responders. Another interesting finding is the anterior/middle cingulate cortex interbrain connectivity, which was correlated to reciprocity of the proposer, possibly related to judgment of proposals in situations with advantageous and disadvantageous inequities (Shaw et al., 2018). An important mechanism involved in the social brain decision is the reward mechanism, which seems to be associated with increased BOLD signal when the pair worked together to complete a task, as indicated in a study that used a maze task in which one of the participants saw the maze and gave instructors to the other, which had to drive through the maze without directly seeing it (Krill & Platek, 2012). In a group hyperscanning with groups of five participants at a time, the effect of social influence (provided by feedback about other participants' decision) on decision indicated that insular response was higher when individual decisions were different from other participants and also predicted the tendency of realignment to the group in the next decision. This insular activity could be related to embarrassment, as suggested by a study that evaluated self-face recognition while being observed by others, which also detected intra-brain connectivity between the anterior cingulate cortex and the dorsomedial prefrontal cortex when being observed (Morita et al., 2014).

As previously mentioned, social interaction tasks can have different structures related to temporal structure of the task, goal of task, and dependency of actions. These can evoke different neural systems, as evaluated by Špiláková et al. (2019). They studied goal and temporal structure effects with a pattern game in which a builder player had to form a pattern with disks in a virtual game table, while other participants could cooperate or try to avoid reaching the pattern (goal effect). In one condition, the players responded simultaneously, while in the other the builder started the turn-based interaction. Higher BOLD signal was found during cooperative tasks in the ventromedial prefrontal cortex, superior and middle temporal gyri, and orbitofrontal cortex, while competitive tasks were related to the dorsolateral prefrontal cortex, supplementary motor area, insula, and cerebellum. Also, simultaneous tasks had higher activity in the temporal lobe, insula, and motor areas. These highlight the importance of choosing the appropriate paradigm according to the desired interaction mechanism to be studied.

Together, these studies exemplify the possibilities of fMRI hyperscanning, using its high spatial resolution to try to identify specific roles for areas within a system. On the other hand, fMRI offers several restrictions for experiment design, which limit more naturalistic tasks.

fNIRS Hyperscanning

The first hyperscanning study using fNIRS technique involved joint action (Funane et al., 2011) in a task in which participants have to synchronize their motor response (usually a button press) after a stimulus. This task and small variations of it are probably the most used paradigms in hyperscanning fNIRS studies because they offer a simple model of interaction. The control condition in these studies usually is a competitive condition, in which the volunteers compete for the faster response after a signal to perform the movement. Using this kind of paradigm studies found increased prefrontal connectivity between brains associated with better performance (Funane et al., 2011) and similar results in the superior frontal cortex (Cui et al., 2012). The coherence between pairs in the right superior frontal cortex was affected but level of intimacy, when comparing lovers, friends, and stranger pairs, with lovers presenting higher coherence and better performance (Pan et al., 2017). Also, the gender of the pairs seems to affect interbrain connectivity and performance, with male-male pairs having better synchrony performance, although contrasting opposed interbrain connectivity results were found in different studies (Cheng et al., 2015; Baker et al., 2016). Interestingly, controlling the feedback after trials and runs could modulate performance and interbrain connectivity, with better performance associated with higher coherence in the superior prefrontal cortex and dorsolateral prefrontal cortex (Cui et al., 2012; Balconi et al., 2018). This effect was also evaluated from mother child dyads, also finding right dorsolateral prefrontal cortex interbrain connectivity related to joint action (Reindl et al., 2018), and different brain mechanisms might be associated with child gender (Miller et al., 2019). Simulated positive feedback could also alter connectivity and performance even in a competitive task (Balconi & Vanutelli, 2017).

A small variation of this task is the joint finger-tapping, in which participants have to synchronize finger-tapping movements, with the control conditions being metronome synchronization. Studies found increased interbrain connectivity in the right prefrontal cortex (Dai et al., 2018) and in the medial prefrontal cortex, when performing source-based analysis (Zhao et al., 2017). Adding an imitation component to the task with a definition of leader and follower to determine the finger-tapping rhythm, higher granger causality from leader to follower was detected in the left premotor cortex (Holper et al., 2012). Prefrontal cortex cortex involvement as in the studies mentioned above was observed by using the joint n-back task, a different type of joint task in which pairs performed dual n-back task (each participant in charge of a n-back) together (Dommer et al., 2012). These studies highlight the possibilities of joint action interaction and finger-tapping tasks, although they might be too associated with motor mechanisms, and therefore other types of tasks could be important to further explore the social brain.

A game theory paradigm with the ultimatum game was also evaluated with fNIRS hyperscanning indicating higher coherence between right temporoparietal junctions when performing the task face-to-face (Tang et al., 2016), probably associated with higher ToM. This is in agreement with studies with eye contact by itself, which can alter interbrain coherence in temporal regions (Hirsch et al., 2017). Face-to-face interaction also has an impact in communication mechanisms, presenting higher hyperconnectivity (interbrain connectivity) in the left inferior frontal cortex than during back-to-back dialogues and even than face-to-face monologues (Jiang et al., 2012). In a group communication, left temporoparietal junction hyperconnectivity distinguished a leader-follower pair, as opposed to two followers during group communication. In a four-group word game, synchrony was observed in frontopolar regions (Nozawa et al., 2016). There are several explanations for the difference in location of the synchrony since the studies used different fNIRS optode positioning and the communication tasks were different, as well as artifact handling during analysis.

In turn-based game interactions, poker game adaptation indicated the importance of temporoparietal junction for ToM when comparing human-human against human-computer competitions (Piva et al., 2017). TPJ was also related to higher-risk decision, which is assumed to engage more mentalizing due to more careful evaluation of the opponent (Zhang et al., 2017). This study further suggested there might be a gender difference in high-risk situations, with women presenting higher TPJ hyperconnectivity than men. Pattern game studies found higher mentalizing during competition and therefore increased hyperconnectivity in the inferior parietal lobule and further different inferior frontal gyrus participation, while borders of the superior temporal sulcus showed hyperconnectivity during both competitive and cooperative games (Liu et al., 2015, 2017). On the other hand, comparing obstructive and cooperative Jenga game indicated more hyperconnectivity in the right dorsolateral prefrontal cortex during cooperation (Liu et al., 2016). Therefore, competitive characteristics in some task-based games are completely different from competition control conditions in joint action or in the Jenga obstruction example, due to different strategies and different mentalizing requirement in each type of competition. These results suggest that rather than unified cooperation and competition mechanisms, different mirror neuron system, empathy, and ToM mechanisms might mediate cooperation and competition according to specific task design.

Working together to solve problems is a common situation in social interaction, and it is related to creativity. Studies with realistic resented problem task, in which participants have to provide as many solutions as possible to solve a realistic problem, showed higher hyperconnectivity in the dorsolateral prefrontal cortex and in the right temporoparietal junction associated with higher cooperation, either comparing to a competitive task or when evaluating creativity levels of dyads and also in the context of an experimenter acting as a participant and providing feedback for the ideas proposed by participants (Xue et al., 2018; Lu et al., 2018, 2019; Lu & Hao, 2019). Interestingly, when comparing creativity levels between pairs, low creativity individuals formed efficient dyads in cooperation with high interpersonal neural synchrony and good performance in the task, while other pairs with at least one participant classified as creative did not show the same cooperation and connectivity. Also, positive feedback enhanced interaction and cooperation, and negative feedback seemed to disrupt interaction.

Considering other close to real-life tasks, teacher and student interactions were evaluated showing higher interbrain connectivity in the left prefrontal cortex when information was transferred from teacher to student (Holper et al., 2013). Also, a feasibility study suggested that it is possible to perform the teacher-student experiment with a child student, finding student's prefrontal signal related to teacher's TPJ (Brockington et al., 2018). The same article explored the possibility of recording data from four students in a class and showed hyperconnectivity between students when they were paying attention to the teacher. Last, they showed that it is possible to perform these measures also combining with eye-tracking information. In another example of a real-life situation that can be studied with fNIRS hyperscanning, interbrain connectivity was evaluated between client and counselor during psychological counseling, indicating hyperconnectivity in the right TPJ compared to a chatting control situation, possibly related to required mentalizing for psychological counseling (Zhang et al., 2018).

Music is also highly related to social interactions and therefore could be a good task for fNIRS studies. A song-learning task found increased hyperconnectivity in the inferior frontal cortex, with directionality from teacher to learner (Pan et al., 2018). Similar results were observed related to cooperative singing and humming, regardless of face-to-face or non-face-to-face execution of the task (Osaka et al., 2014, 2015). Also, the right inferior frontal cortex seemed to be more associated with humming. Pairs of violinists in a leader-follower context presented higher temporoparietal junction and somatomotor signal when playing as follower (Vanzella et al., 2019), and a feasibility study suggested hyperconnectivity between violinists (Balardin et al., 2017). Another feasibility study suggested that multibrain hyperscanning could be represented as a multibrain network and evaluated as a graph and as feasibility proposed a data collection in nine participants simultaneously drumming (Duan et al., 2015).

A different approach to fNIRS hyperscanning was proposed by Duan et al. (2013). In their study, the authors have designed a neurofeedback platform. Neurofeedback has been studied as a tool for improvement of cognitive functions, especially in the context of brain disorders, though there is still debate about the efficacy and proper use of neurofeedback (Thibault et al., 2016; Kadosh & Staunton, 2019). Therefore, hyperscanning neurofeedback might present an opportunity for investigations on enhancement of social abilities or treatments of disorders which affect the social brain, but these possibilities should be addressed carefully.

These studies highlight the advantage of fNIRS for naturalistic and closer to real-life tasks, given its tolerance to movement, and portability (in some systems). However, it is also important to notice that some disagreement in the fNIRS hyperscanning literature can come from different optode position, given that studies with systems which allow whole cortical coverage are scarce, and possibly can be due to smaller spatial resolution, which implies studying larger cortical areas as a single region, compared to the specificity that fMRI studies can provide. Further, different preprocessing steps (mainly dealing with noise) could affect hyperconnectivity results, and there is still search for better analysis processes.

Conclusion and Future Perspectives

We have discussed the possibilities fMRI and fNIRS offer for studying the social brain. While fMRI provides high-resolution whole-brain coverage, fNIRS offers a great opportunity for real-life task studies. Studies recording one subject can help elucidate basic mechanisms that are engaged and combined during real interaction, as they offer easier possibilities for isolating and controlling cognitive aspects of the experiment. Meanwhile, hyperscanning provides an integrated look that can unveil new interactions between basic mechanisms and possibly new mechanisms (or better models) for social interaction related to different contexts. Moreover, hyperscanning can offer a new perspective in search for biomarkers and in understanding diseases and disorders that affect the social brain (Ray et al., 2017). Further studies should also focus on combining fMRI and fNRIS techniques such as EEG, eye-tracking, and possibly even MEG, to provide more information and higher temporal resolution, preserving spatial resolution, although newer analysis methods and hypothesis-driven models are also required for better use of the data (Koike et al., 2015). Another potentially interesting combination for fNIRS and fMRI hyperscanning are autonomic response measures, since there seems to be autonomic coupling during cooperation (Vanutelli et al., 2017). Controlling these autonomic responses in hemodynamic response-based systems could help interpret the data (Kadosh & Staunton, 2019).

References

Abe, M. O., Koike, T., Okazaki, S., Sugawara, S. K., Takahashi, K., Watanabe, K., & Sadato, N. (2019, May 1). Neural correlates of online cooperation during joint force production. *Neuroimage, 1*(191), 150–161. https://doi.org/10.1016/j.neuroimage.2019.02.003. Epub 2019 Feb 7.

Amaro, E., Jr., & Barker, G. J. (2006, April). Study design in fMRI: Basic principles. *Brain and Cognition, 60*(3), 220–232. Epub 2006 Jan 19.

Arioli, M., Perani, D., Cappa, S., Proverbio, A. M., Zani, A., Falini, A., & Canessa, N. (2018, March). Affective and cooperative social interactions modulate effective connectivity within and between the mirror and mentalizing systems. *Human Brain Mapping, 39*(3), 1412–1427. https://doi.org/10.1002/hbm.23930. Epub 2017 Dec 19.

Babiloni, F., & Astolfi, L. (2014). Social neuroscience and hyperscanning techniques: Past, present and future. *Neuroscience and Biobehavioral Reviews, 44*, 76–93. https://doi.org/10.1016/j.neubiorev.2012.07.006

Baecke, S., Lützkendorf, R., Mallow, J., et al. (2015). A proof-of-principle study of multi-site real-time functional imaging at 3T and 7T: Implementation and validation. *Scientific Reports, 5*, 8413. Published 2015 Feb 12. https://doi.org/10.1038/srep08413

Baker, J. M., Liu, N., Cui, X., et al. (2016). Sex differences in neural and behavioral signatures of cooperation revealed by fNIRS hyperscanning [published correction appears in Sci Rep. 2016 Aug 19;6:30512]. *Scientific Reports, 6*, 26492. Published 2016 Jun 8. https://doi.org/10.1038/srep26492

Balardin, J. B., Zimeo Morais, G. A., Furucho, R. A., et al. (2017). Imaging brain function with functional near-infrared spectroscopy in unconstrained environments. *Frontiers in Human Neuroscience, 11*, 258. Published 2017 May 17. https://doi.org/10.3389/fnhum.2017.00258

Balconi, M., & Vanutelli, M. E. (2017, August). Interbrains cooperation: Hyperscanning and self-perception in joint actions. *Journal of Clinical and Experimental Neuropsychology, 39*(6), 607–620. https://doi.org/10.1080/13803395.2016.1253666. Epub 2016 Nov 13.

Balconi, M., Vanutelli, M. E., & Gatti, L. (2018, June). Functional brain connectivity when cooperation fails. *Brain and Cognition, 123*, 65–73. https://doi.org/10.1016/j.bandc.2018.02.009. Epub 2018 Mar 8.

Bandettini, P. A. (2012, August 15). Twenty years of functional MRI: The science and the stories. *NeuroImage, 62*(2), 575–588. https://doi.org/10.1016/j.neuroimage.2012.04.026. Epub 2012 Apr 20.

Bandettini, P. A., Wong, E. C., Hinks, R. S., Tikofsky, R. S., & Hyde, J. S. (1992, June). Time course EPI of human brain function during task activation. *Magnetic Resonance in Medicine, 25*(2), 390–397.

Bilek, E., Ruf, M., Schäfer, A., et al. (2015). Information flow between interacting human brains: Identification, validation, and relationship to social expertise. *Proceedings of the National Academy of Sciences of the United States of America, 112*(16), 5207–5212. https://doi.org/10.1073/pnas.1421831112

Bilek, E., Stößel, G., Schäfer, A., et al. (2017). State-dependent cross-brain information flow in borderline personality disorder. *JAMA Psychiatry, 74*(9), 949–957. https://doi.org/10.1001/jamapsychiatry.2017.1682

Bowman, L. C., Kovelman, I., Hu, X., & Wellman, H. M. (2015). Children's belief- and desire-reasoning in the temporoparietal junction: Evidence for specialization from functional near-infrared spectroscopy. *Frontiers in Human Neuroscience, 9*, 560. Published 2015 Oct 7. https://doi.org/10.3389/fnhum.2015.00560

Brigadoi, S., & Cooper, R. J. (2015). How short is short? Optimum source-detector distance for short-separation channels in functional near-infrared spectroscopy. *Neurophotonics, 2*(2), 025005. https://doi.org/10.1117/1.NPh.2.2.025005

Brink, T. T., Urton, K., Held, D., et al. (2011). The role of orbitofrontal cortex in processing empathy stories in 4- to 8-year-old children. *Frontiers in Psychology, 2*, 80. Published 2011 Apr 28. https://doi.org/10.3389/fpsyg.2011.00080

Brockington, G., Balardin, J. B., Zimeo Morais, G. A., et al. (2018). From the laboratory to the classroom: The potential of functional near-infrared spectroscopy in educational neuroscience. *Frontiers in Psychology, 9*, 1840. Published 2018 Oct 11. https://doi.org/10.3389/fpsyg.2018.01840

Buxton, R. B., Wong, E. C., & Frank, L. R. (1998, June). Dynamics of blood flow and oxygenation changes during brain activation: The balloon model. *Magnetic Resonance in Medicine, 39*(6), 855–864.

Canessa, N., Alemanno, F., Riva, F., et al. (2012). The neural bases of social intention understanding: The role of interaction goals. *PLoS One, 7*(7), e42347. https://doi.org/10.1371/journal.pone.0042347

Chance, B., Zhuang, Z., UnAh, C., Alter, C., & Lipton, L. (1993). Cognition-activated low-frequency modulation of light absorption in human brain. *Proceedings of the National Academy of Sciences of the United States of America, 90*(8), 3770–3774.

Chauvigné, L. A. S., Belyk, M., & Brown, S. (2018). Taking two to tango: fMRI analysis of improvised joint action with physical contact. *PLoS One, 13*(1), e0191098. Published 2018 Jan 11. https://doi.org/10.1371/journal.pone.0191098

Cheng, X., Li, X., & Hu, Y. (2015, June). Synchronous brain activity during cooperative exchange depends on gender of partner: A fNIRS-based hyperscanning study. *Human Brain Mapping, 36*(6), 2039–2048. https://doi.org/10.1002/hbm.22754. Epub 2015 Feb 17.

Cui, X., Bryant, D. M., & Reiss, A. L. (2012). NIRS-based hyperscanning reveals increased interpersonal coherence in superior frontal cortex during cooperation. *NeuroImage, 59*(3), 2430–2437. https://doi.org/10.1016/j.neuroimage.2011.09.003

Dai, R., Liu, R., Liu, T., et al. (2018). Holistic cognitive and neural processes: A fNIRS-hyperscanning study on interpersonal sensorimotor synchronization. *Social Cognitive and Affective Neuroscience, 13*(11), 1141–1154. https://doi.org/10.1093/scan/nsy090

Dalgleish, T., Walsh, N. D., Mobbs, D., et al. (2017). Social pain and social gain in the adolescent brain: A common neural circuitry underlying both positive and negative social evaluation. *Scientific Reports, 7*, 42010. Published 2017 Feb 7. https://doi.org/10.1038/srep42010

Damoiseaux, J. S., Rombouts, S. A., Barkhof, F., et al. (2006). Consistent resting-state networks across healthy subjects. *Proceedings of the National Academy of Sciences of the United States of America, 103*(37), 13848–13853. https://doi.org/10.1073/pnas.0601417103

Dommer, L., Jäger, N., Scholkmann, F., Wolf, M., & Holper, L. (2012, October 1). Between-brain coherence during joint n-back task performance: A two-person functional near-infrared spectroscopy study. *Behavioural Brain Research, 234*(2), 212–222. https://doi.org/10.1016/j.bbr.2012.06.024. Epub 2012 Jun 29.

Duan, L., Liu, W. J., Dai, R. N., et al. (2013). Cross-brain neurofeedback: Scientific concept and experimental platform. *PLoS One, 8*(5), e64590. Published 2013 May 17. https://doi.org/10.1371/journal.pone.0064590

Duan, L., Dai, R. N., Xiao, X., Sun, P. P., Li, Z., & Zhu, C. Z. (2015). Cluster imaging of multi-brain networks (CIMBN): A general framework for hyperscanning and modeling a group of interacting brains. *Frontiers in Neuroscience, 9*, 267. Published 2015 Jul 28. https://doi.org/10.3389/fnins.2015.00267

Ferrari, M., & Quaresima, V. (2012, November 1). A brief review on the history of human functional near-infrared spectroscopy (fNIRS) development and fields of application. *NeuroImage, 63*(2), 921–935. https://doi.org/10.1016/j.neuroimage.2012.03.049. Epub 2012 Mar 28.

Filosa, J. A., Morrison, H. W., Iddings, J. A., Du, W., & Kim, K. J. (2016). Beyond neurovascular coupling, role of astrocytes in the regulation of vascular tone. *Neuroscience, 323*, 96–109. https://doi.org/10.1016/j.neuroscience.2015.03.064

Fletcher, P. C., Happé, F., Frith, U., Baker, S. C., Dolan, R. J., Frackowiak, R. S., & Frith, C. D. (1995, November). Other minds in the brain: A functional imaging study of "theory of mind" in story comprehension. *Cognition, 57*(2), 109–128.

Fliessbach, K., Phillipps, C. B., Trautner, P., et al. (2012). Neural responses to advantageous and disadvantageous inequity. *Frontiers in Human Neuroscience, 6*(165) Published 2012 Jun 8. https://doi.org/10.3389/fnhum.2012.00165

Fox, M. D., Snyder, A. Z., Vincent, J. L., Corbetta, M., Van Essen, D. C., & Raichle, M. E. (2005). The human brain is intrinsically organized into dynamic, anticorrelated functional networks. *Proceedings of the National Academy of Sciences of the United States of America, 102*(27), 9673–9678. https://doi.org/10.1073/pnas.0504136102

Friston, K. J. (2011). Functional and effective connectivity: A review. *Brain Connectivity, 1*(1), 13–36. https://doi.org/10.1089/brain.2011.0008

Frith, U., & Frith, C. D. (2003). Development and neurophysiology of mentalizing. *Philosophical Transactions of the Royal Society of London. Series B, Biological Sciences, 358*(1431), 459–473. https://doi.org/10.1098/rstb.2002.1218

Frith, C. D., & Frith, U. (2006, May 18). The neural basis of mentalizing. *Neuron, 50*(4), 531–534.

Funane, T., Kiguchi, M., Atsumori, H., Sato, H., Kubota, K., & Koizumi, H. (2011, July). Synchronous activity of two people's prefrontal cortices during a cooperative task measured by simultaneous near-infrared spectroscopy. *Journal of Biomedical Optics, 16*(7), 077011. https://doi.org/10.1117/1.3602853

Gallagher, H. L., Happé, F., Brunswick, N., Fletcher, P. C., Frith, U., & Frith, C. D. (2000). Reading the mind in cartoons and stories: An fMRI study of 'theory of mind' in verbal and nonverbal tasks. *Neuropsychologia, 38*(1), 11–21.

Gallese, V., & Goldman, A. (1998, December 1). Mirror neurons and the simulation theory of mind-reading. *Trends in Cognitive Sciences, 2*(12), 493–501.

Gramann, K., Ferris, D. P., Gwin, J., & Makeig, S. (2014). Imaging natural cognition in action. *International Journal of Psychophysiology, 91*(1), 22–29. https://doi.org/10.1016/j.ijpsycho.2013.09.003

Greicius, M. D., Krasnow, B., Reiss, A. L., & Menon, V. (2003). Functional connectivity in the resting brain: A network analysis of the default mode hypothesis. *Proceedings of the National Academy of Sciences of the United States of America, 100*(1), 253–258. https://doi.org/10.1073/pnas.0135058100

Grèzes, J., Armony, J. L., Rowe, J., & Passingham, R. E. (2003, April). Activations related to "mirror" and "canonical" neurones in the human brain: an fMRI study. *Neuroimage, 18*(4), 928–937.

Han, L. V., Wang, Z., Tong, E., et al. (2018). Resting-state functional MRI: Everything that non-experts have always wanted to know. *AJNR. American Journal of Neuroradiology, 39*(8), 1390–1399. https://doi.org/10.3174/ajnr.A5527

Hasson, U., Nir, Y., Levy, I., Fuhrmann, G., & Malach, R. (2004, March 12). Intersubject synchronization of cortical activity during natural vision. *Science, 303*(5664), 1634–1640.

Himichi, T., Fujita, H., & Nomura, M. (2015). Negative emotions impact lateral prefrontal cortex activation during theory of mind: An fNIRS study. *Social Neuroscience, 10*(6), 605–615. https://doi.org/10.1080/17470919.2015.1017112. Epub 2015 Mar 16.

Hirsch, J., Zhang, X., Noah, J. A., & Ono, Y. (2017, August 15). Frontal temporal and parietal systems synchronize within and across brains during live eye-to-eye contact. *Neuroimage, 157*, 314–330. https://doi.org/10.1016/j.neuroimage.2017.06.018. Epub 2017 Jun 12.

Holper, L., Scholkmann, F., & Wolf, M. (2012, October 15). Between-brain connectivity during imitation measured by fNIRS. *Neuroimage, 63*(1), 212–222. https://doi.org/10.1016/j.neuroimage.2012.06.028. Epub 2012 Jun 23.

Holper, L., Goldin, A. P., Shalóm, D. E., Battro, A. M., Wolfa, M., & Sigman, M. (2013). The teaching and the learning brain: A cortical hemodynamic marker of teacher–student interactions in the Socratic dialog. *International Journal of Educational Research, 59*, 1–10.

Hoshi, Y., & Tamura, M. (1993, February 5). Detection of dynamic changes in cerebral oxygenation coupled to neuronal function during mental work in man. *Neuroscience Letters, 150*(1), 5–8.

Hyde, D. C., Aparicio Betancourt, M., & Simon, C. E. (2015, December). Human temporal-parietal junction spontaneously tracks others' beliefs: A functional near-infrared spectroscopy study. *Human Brain Mapping, 36*(12), 4831–4846. https://doi.org/10.1002/hbm.22953. Epub 2015 Sep 14.

Iacoboni, M., Woods, R. P., Brass, M., Bekkering, H., Mazziotta, J. C., & Rizzolatti, G. (1999, December 24). Cortical mechanisms of human imitation. *Science, 286*(5449), 2526–2528.

Iacoboni, M., Molnar-Szakacs, I., Gallese, V., Buccino, G., Mazziotta, J. C., & Rizzolatti, G. (2005). Grasping the intentions of others with one's own mirror neuron system. *PLoS Biology, 3*(3), e79. https://doi.org/10.1371/journal.pbio.0030079

Jasper, H. H. (1958). The ten-twenty electrode system of the international federation. *Electroencephalography and Clinical Neurophysiology, 10*, 371–375.

Jeon, H., & Lee, S. H. (2018). From neurons to social beings: Short review of the mirror neuron system research and its socio-psychological and psychiatric implications. *Clinical Psychopharmacology and Neuroscience, 16*(1), 18–31. https://doi.org/10.9758/cpn.2018.16.1.18

Jiang, J., Dai, B., Peng, D., Zhu, C., Liu, L., & Lu, C. (2012, November 7). Neural synchronization during face-to-face communication. *The Journal of Neuroscience, 32*(45), 16064–16069. https://doi.org/10.1523/JNEUROSCI.2926-12.2012

Kadosh, K. C., & Staunton, G. (2019). A systematic review of the psychological factors that influence neurofeedback learning outcomes. *NeuroImage, 185*, 545–555. https://doi.org/10.1016/j.neuroimage.2018.10.021

Kato, T., Kamei, A., Takashima, S., & Ozaki, T. (1993, May). Human visual cortical function during photic stimulation monitoring by means of near-infrared spectroscopy. *Journal of Cerebral Blood Flow and Metabolism, 13*(3), 516–520.

Koike, T., Tanabe, H. C., & Sadato, N.. (2015, January). Hyperscanning neuroimaging technique to reveal the "two-in-one" system in social interactions. *Neuroscience Research, 90*, 25–32. https://doi.org/10.1016/j.neures.2014.11.006. Epub 2014 Dec 10.

Koike, T., Tanabe, H. C., Okazaki, S., Nakagawa, E., Sasaki, A. T., Shimada, K., Sugawara, S. K., Takahashi, H. K., Yoshihara, K., Bosch-Bayard, J., & Sadato, N. (2016, January 15). Neural substrates of shared attention as social memory: A hyperscanning functional magnetic resonance imaging study. *NeuroImage, 125*, 401–412. https://doi.org/10.1016/j.neuroimage.2015.09.076. Epub 2015 Oct 26.

Koike, T., Sumiya, M., Nakagawa, E., Okazaki, S., & Sadato, N. (2019). What makes eye contact special? Neural substrates of on-line mutual eye-gaze: A hyperscanning fMRI study. *eNeuro, 6*(1), ENEURO.0284–18.2019. Published 2019 Feb 28. https://doi.org/10.1523/ENEURO.0284-18.2019

Krill, A. L., & Platek, S. M. (2012). Working together may be better: Activation of reward centers during a cooperative maze task. *PLoS One, 7*(2), e30613. https://doi.org/10.1371/journal.pone.0030613

Kwong, K. K., Belliveau, J. W., Chesler, D. A., et al. (1992). Dynamic magnetic resonance imaging of human brain activity during primary sensory stimulation. *Proceedings of the National Academy of Sciences of the United States of America, 89*(12), 5675–5679.

Lee, R. F. (2015a). Emergence of the default-mode network from resting-state to activation-state in reciprocal social interaction via eye contact. *Conference Proceedings: Annual International Conference of the IEEE Engineering in Medicine and Biology Society, 2015*, 1821–1824. https://doi.org/10.1109/EMBC.2015.7318734

Lee, R. F. (2015b). Dual logic and cerebral coordinates for reciprocal interaction in eye contact [published correction appears in PLoS one. 2015;10(5):e0128480]. *PLoS One, 10*(4), e0121791. Published 2015 Apr 17. https://doi.org/10.1371/journal.pone.0121791

Lee, R. F., Dai, W., & Dix, W. (2010). A decoupled circular-polarized volume head coil pair for studying two interacting human brains with MRI. *Conference Proceedings: Annual International Conference of the IEEE Engineering in Medicine and Biology Society, 2010*, 6645–6648. https://doi.org/10.1109/IEMBS.2010.5627155

Lee, R. F., Dai, W., & Jones, J. (2012, October). Decoupled circular-polarized dual-head volume coil pair for studying two interacting human brains with dyadic fMRI. *Magnetic Resonance in Medicine, 68*(4), 1087–1096. https://doi.org/10.1002/mrm.23313. Epub 2011 Dec 28.

Lee, T. H., Qu, Y., & Telzer, E. H. (2018). Dyadic neural similarity during stress in mother-child dyads. *Journal of Research on Adolescence, 28*(1), 121–133. https://doi.org/10.1111/jora.12334

Lin, L. C., Qu, Y., & Telzer, E. H. (2018, October 16). Intergroup social influence on emotion processing in the brain. *Proceedings of the National Academy of Sciences of the United States of America, 115*(42), 10630–10635. https://doi.org/10.1073/pnas.1802111115. Epub 2018 Oct 3.

Liu, T., & Pelowski, M. (2014, September). A new research trend in social neuroscience: Towards an interactive-brain neuroscience. *PsyCh Journal, 3*(3), 177–188. https://doi.org/10.1002/pchj.56. Epub 2014 Apr 2.

Liu, T., Saito, H., & Oi, M. (2015, October). Role of the right inferior frontal gyrus in turn-based cooperation and competition: A near-infrared spectroscopy study. *Brain and Cognition, 99*, 17–23. https://doi.org/10.1016/j.bandc.2015.07.001. Epub 2015 Jul 17.

Liu, N., Mok, C., Witt, E. E., Pradhan, A. H., Chen, J. E., & Reiss, A. L. (2016). NIRS-based hyperscanning reveals inter-brain neural synchronization during cooperative Jenga game with face-to-face communication. *Frontiers in Human Neuroscience, 10*(82) Published 2016 Mar 8. https://doi.org/10.3389/fnhum.2016.00082

Liu, T., Saito, G., Lin, C., & Saito, H. (2017). Inter-brain network underlying turn-based cooperation and competition: A hyperscanning study using near-infrared spectroscopy. *Scientific Reports, 7*(1), 8684. Published 2017 Aug 17. https://doi.org/10.1038/s41598-017-09226-w

Longden, T. A., Hill-Eubanks, D. C., & Nelson, M. T. (2016). Ion channel networks in the control of cerebral blood flow. *Journal of Cerebral Blood Flow and Metabolism, 36*(3), 492–512. https://doi.org/10.1177/0271678X15616138

Lu, K., & Hao, N. (2019, March). When do we fall in neural synchrony with others? *Social Cognitive and Affective Neuroscience, 14*(3), pp. 253–261. https://doi.org/10.1093/scan/nsz012

Lu, K., Xue, K., Nozawa, T., & Hao, N. 2018 Cooperation makes a group be more creative. *Cerebral Cortex.* bhy215. https://doi.org/10.1093/cercor/bhy215

Lu, K., Qiao, X., & Hao, N. (2019, February). Praising or keeping silent on partner's ideas: Leading brainstorming in particular ways. *Neuropsychologia, 18;124*, 19–30. https://doi.org/10.1016/j. neuropsychologia.2019.01.004. Epub 2019 Jan 8.

Mahy, C. E., Moses, L. J., & Pfeifer, J. H. (2014, July). How and where: Theory-of-mind in the brain. *Developmental Cognitive Neuroscience, 9*, 68–81. https://doi.org/10.1016/j.dcn.2014.01.002. Epub 2014 Jan 25.

Miller, J. G., Vrtička, P., Cui, X., Shrestha, S., Hosseini, S. M. H., Baker, J. M., & Reiss, A. L. (2019, February). Inter-brain synchrony in mother-child dyads during cooperation: An fNIRS hyperscanning study. *Neuropsychologia, 18*(124), 117–124. https://doi.org/10.1016/j.neuropsychologia.2018.12.021. Epub 2018 Dec 27.

Montague, P. R., Berns, G. S., Cohen, J. D., McClure, S. M., Pagnoni, G., Dhamala, M., Wiest, M. C., Karpov, I., King, R. D., Apple, N., & Fisher, R. E. (2002, August). Hyperscanning: Simultaneous fMRI during linked social interactions. *NeuroImage, 16*(4), 1159–1164.

Morita, T., Tanabe, H. C., Sasaki, A. T., Shimada, K., Kakigi, R., & Sadato, N. (2014). The anterior insular and anterior cingulate cortices in emotional processing for self-face recognition. *Social Cognitive and Affective Neuroscience, 9*(5), 570–579. https://doi.org/10.1093/scan/nst011

Nozawa, T., Sasaki, Y., Sakaki, K., Yokoyama, R., & Kawashima, R. (2016, June). Interpersonal frontopolar neural synchronization in group communication: An exploration toward fNIRS hyperscanning of natural interactions. *NeuroImage, 133*, 484–497. https://doi.org/10.1016/j. neuroimage.2016.03.059. Epub 2016 Apr 1.

Ogawa, S., Lee, T. M., Kay, A. R., & Tank, D. W. (1990). Brain magnetic resonance imaging with contrast dependent on blood oxygenation. *Proceedings of the National Academy of Sciences of the United States of America, 87*(24), 9868–9872.

Ogawa, S., Tank, D. W., Menon, R., et al. (1992). Intrinsic signal changes accompanying sensory stimulation: Functional brain mapping with magnetic resonance imaging. *Proceedings of the National Academy of Sciences of the United States of America, 89*(13), 5951–5955.

Ojeda, A., Bigdely-Shamlo, N., & Makeig, S. (2014). MoBILAB: An open source toolbox for analysis and visualization of mobile brain/body imaging data. *Frontiers in Human Neuroscience, 8*, 121. Published 2014 Mar 5. https://doi.org/10.3389/fnhum.2014.00121

Oostenveld, R., & Praamstra, P. (2001). The five percent electrode system for high-resolution EEG and ERP measurements. *Clinical Neurophysiology, 112*(4), 713–719.

Osaka, N., Minamoto, T., Yaoi, K., Azuma, M., & Osaka, M. (2014, March). Neural synchronization during cooperated humming: A hyperscanning study using fNIRS. *Procedia – Social and Behavioral Sciences, 126*(21), pp. 241–243.

Osaka, N., Minamoto, T., Yaoi, K., Azuma, M., Shimada, Y. M., & Osaka, M. (2015). How two brains make one synchronized mind in the inferior frontal cortex: fNIRS-based hyperscanning during cooperative singing. *Frontiers in Psychology, 6*, 1811. Published 2015 Nov 26. https://doi.org/10.3389/fpsyg.2015.01811

Pan, Y., Cheng, X., Zhang, Z., Li, X., & Hu, Y. (2017, February). Cooperation in lovers: An fNIRS-based hyperscanning study. *Human Brain Mapping, 38*(2):831–841. https://doi.org/10.1002/hbm.23421. Epub 2016 Oct 4.

Pan, Y., Novembre, G., Song, B., Li, X., & Hu, Y. (2018, December). Interpersonal synchronization of inferior frontal cortices tracks social interactive learning of a song. *NeuroImage, 183*, 280–290. https://doi.org/10.1016/j.neuroimage.2018.08.005. Epub 2018 Aug 4.

Piva, M., Zhang, X., Noah, J. A., Chang, S. W. C., & Hirsch, J. (2017). Distributed neural activity patterns during human-to-human competition. *Frontiers in Human Neuroscience, 11*, 571. Published 2017 Nov 23. https://doi.org/10.3389/fnhum.2017.00571

Premack, D., & Woodruff, G. (1978). Does the chimpanzee have a theory of mind? *Behavioral and Brain Sciences, 1*(4), 515–526. https://doi.org/10.1017/S0140525X00076512

Rauchbauer, B., Nazarian, B., Bourhis, M., Ochs, M., Prévot, L., & Chaminade, T. (2019, April 29). Brain activity during reciprocal social interaction investigated using conversational robots as control condition. *Philosophical Transactions of the Royal Society of London. Series B, Biological Sciences, 374* (1771), 20180033.

Ray, D., Roy, D., Sindhu, B., Sharan, P., & Banerjee, A. (2017). Neural substrate of group mental health: Insights from multi-brain reference frame in functional neuroimaging. *Frontiers in Psychology, 8*, 1627. Published 2017 Sep 28. https://doi.org/10.3389/fpsyg.2017.01627

Reindl, V., Gerloff, C., Scharke, W., & Konrad, K. (2018, September). Brain-to-brain synchrony in parent-child dyads and the relationship with emotion regulation revealed by fNIRS-based hyperscanning. *NeuroImage, 178*, 493–502. https://doi.org/10.1016/j.neuroimage.2018.05.060. Epub 2018 May 26.

Saito, D. N., Tanabe, H. C., Izuma, K., et al. (2010). "Stay tuned": Inter-individual neural synchronization during mutual gaze and joint attention. *Frontiers in Integrative Neuroscience, 4*, 127. Published 2010 Nov 5. https://doi.org/10.3389/fnint.2010.00127

Schippers, M. B., Gazzola, V., Goebel, R., & Keysers, C. (2009). Playing charades in the fMRI: Are mirror and/or mentalizing areas involved in gestural communication? *PLoS One, 4*(8), e6801. Published 2009 Aug 27. https://doi.org/10.1371/journal.pone.0006801

Scholkmann, F., Holper, L., Wolf, U., & Wolf, M. (2013). A new methodical approach in neuroscience: Assessing inter-personal brain coupling using functional near-infrared imaging (fNIRI) hyperscanning. *Frontiers in Human Neuroscience, 7*, 813. https://doi.org/10.3389/fnhum.2013.00813

Scholkmann, F., Kleiser, S., Metz, A. J., Zimmermann, R., Mata Pavia, J., Wolf, U., & Wolf, M. (2014, January). A review on continuous wave functional near-infrared spectroscopy and imaging instrumentation and methodology. *NeuroImage, 15*(85 Pt 1), 6–27. https://doi.org/10.1016/j.neuroimage.2013.05.004. Epub 2013 May 16.

Shaw, D. J., Czekóová, K., Staněk, R., et al. (2018). A dual-fMRI investigation of the iterated Ultimatum Game reveals that reciprocal behaviour is associated with neural alignment. *Scientific Reports, 8*(1), 10896. Published 2018 Jul 18. https://doi.org/10.1038/s41598-018-29233-9

Spiegelhalder, K., Ohlendorf, S., Regen, W., Feige, B., Tebartz van Elst, L., Weiller, C., Hennig, J., Berger, M., & Tüscher, O. (2014, January 1). Interindividual synchronization of brain activity during live verbal communication. *Behavioural Brain Research, 258*, 75–79. https://doi.org/10.1016/j.bbr.2013.10.015. Epub 2013 Oct 18.

Špiláková, B., Shaw, D. J., Czekóová, K., & Brázdil, M. (2019). Dissecting social interaction: Dual-fMRI reveals patterns of interpersonal brain-behaviour relationships that dissociate among dimensions of social exchange [published online ahead of print, 2019 Jan 15]. *Social Cognitive and Affective Neuroscience, 14*(2), 225–235. https://doi.org/10.1093/scan/nsz004

Stolk, A., Noordzij, M. L., Verhagen, L., et al. (2014). Cerebral coherence between communicators marks the emergence of meaning. *Proceedings of the National Academy of Sciences of the United States of America, 111*(51), 18183–18188. https://doi.org/10.1073/pnas.1414886111

Sun, P. P., Tan, F. L., Zhang, Z., Jiang, Y. H., Zhao, Y., & Zhu, C. Z. (2018). Feasibility of functional near-infrared spectroscopy (fNIRS) to investigate the Mirror neuron system: An experimental study in a real-life situation. *Frontiers in Human Neuroscience, 12*, 86. Published 2018 Mar 5. https://doi.org/10.3389/fnhum.2018.00086

Tanabe, H. C., Kosaka, H., Saito, D. N., et al. (2012). Hard to "tune in": Neural mechanisms of live face-to-face interaction with high-functioning autistic spectrum disorder. *Frontiers in Human Neuroscience, 6*, 268. Published 2012 Sep 27. https://doi.org/10.3389/fnhum.2012.00268

Tang, H., Mai, X., Wang, S., Zhu, C., Krueger, F., & Liu, C. (2016). Interpersonal brain synchronization in the right temporo-parietal junction during face-to-face economic exchange. *Social Cognitive and Affective Neuroscience, 11*(1), 23–32. https://doi.org/10.1093/scan/nsv092

Thibault, R. T., Lifshitz, M., & Raz, A. (2016, January 1). The self-regulating brain and neurofeedback: Experimental science and clinical promise. *Cortex, 74*, 247–261. https://doi.org/10.1016/j.cortex.2015.10.024

Tomlin, D., Nedic, A., Prentice, D. A., Holmes, P., & Cohen, J. D. (2013). The neural substrates of social influence on decision making. *PLoS One, 8*(1), e52630. https://doi.org/10.1371/journal.pone.0052630

Trees, J., Snider, J., Falahpour, M., et al. (2014). Game controller modification for fMRI hyper-scanning experiments in a cooperative virtual reality environment. *MethodsX, 1*, 292–299. Published 2014 Nov 4. https://doi.org/10.1016/j.mex.2014.10.009

Vanutelli, M. E., Gatti, L., Angioletti, L., & Balconi, M. (2017). Affective synchrony and auto-nomic coupling during cooperation: A hyperscanning study. *BioMed Research International*. Article ID 3104564. https://doi.org/10.1155/2017/3104564

Vanzella, P., Balardin, J. B., Furucho, R. A., et al. (2019). fNIRS responses in professional violinists while playing duets: Evidence for distinct leader and follower roles at the brain level. *Frontiers in Psychology, 10*, 164. Published 2019 Feb 5. https://doi.org/10.3389/fpsyg.2019.00164

Villringer, A., Planck, J., Hock, C., Schleinkofer, L., & Dirnagl, U. (1993, May 14). Near infrared spectroscopy (NIRS): A new tool to study hemodynamic changes during activation of brain function in human adults. *Neuroscience Letters, 154*(1–2), 101–104.

Wang, Y., Yang, L. Q., Li, S., & Zhou, Y. (2015). Game theory paradigm: A new tool for investigat-ing social dysfunction in major depressive disorders. *Frontiers in Psychiatry, 6*, 128. Published 2015 Sep 15. https://doi.org/10.3389/fpsyt.2015.00128

Wang, M. Y., Luan, P., Zhang, J., Xiang, Y. T., Niu, H., & Yuan, Z. (2018). Concurrent map-ping of brain activation from multiple subjects during social interaction by hyperscanning: A mini-review. *Quantitative Imaging in Medicine and Surgery, 8*(8), 819–837. https://doi.org/10.21037/qims.2018.09.07

Xue, H., Lu, K., & Hao, N. (2018, May). Cooperation makes two less-creative individuals turn into a highly-creative pair. *NeuroImage, 15*(172), 527–537. https://doi.org/10.1016/j.neuroim-age.2018.02.007. Epub 2018 Feb 8.

Zhang, M., Liu, T., Pelowski, M., Jia, H., & Yu, D. (2017, December). Social risky decision-making reveals gender differences in the TPJ: A hyperscanning study using functional near-infrared spectroscopy. *Brain and Cognition, 119*, 54–63. https://doi.org/10.1016/j.bandc.2017.08.008. Epub 2017 Sep 8.

Zhang, Y., Meng, T., Hou, Y., Pan, Y., & Hu, Y. (2018, December). Interpersonal brain synchroniza-tion associated with working alliance during psychological counseling. *Psychiatry Research: Neuroimaging, 30*(282), 103–109. https://doi.org/10.1016/j.pscychresns.2018.09.007. Epub 2018 Sep 28.

Zhao, Y., Dai, R. N., Xiao, X., Zhang, Z., Duan, L., Li, Z., & Zhu, C. Z. (2017, February 1). Independent component analysis-based source-level hyperlink analysis for two-person neu-roscience studies. *Journal of Biomedical Optics, 22*(2), 27004. https://doi.org/10.1117/1.JBO.22.2.027004

Zhdanov, A., Nurminen, J., Baess, P., et al. (2015). An internet-based real-time audiovisual link for dual MEG recordings. *PLoS One, 10*(6), e0128485. Published 2015 Jun 22. https://doi.org/10.1371/journal.pone.0128485

Chapter 15
Modulating the Social and Affective Brain with Transcranial Stimulation Techniques

Gabriel Rego, Lucas Murrins Marques ⓘ, Marília Lira da Silveira Coêlho ⓘ, and Paulo Sérgio Boggio ⓘ

Abstract Transcranial brain stimulation (TBS) is a term that denotes different non-invasive techniques which aim to modulate brain cortical activity through an external source, usually an electric or magnetic one. Currently, there are several techniques categorized as TBS. However, two are more used for scientific research, the transcranial magnetic stimulation (TMS) and the transcranial direct current stimulation (tDCS), which stimulate brain areas with a high-intensity magnetic field or a weak electric current on the scalp, respectively. They represent an enormous contribution to behavioral, cognitive, and social neuroscience since they reveal how delimited brain cortical areas contribute to some behavior or cognition. They have also been proposed as a feasible tool in the clinical setting since they can modulate abnormal cognition or behavior due to brain activity modulation. This chapter will present the standard methods of transcranial stimulation, their contributions to social and affective neuroscience through a few main topics, and the studies that adopted those techniques, also summing their findings.

Keywords Transcranial magnetic stimulation · Transcranial direct current stimulation · TMS · tDCS

G. Rego · M. L. da Silveira Coêlho · P. S. Boggio (✉)
Social and Cognitive Neuroscience Laboratory, Developmental Disorders Program,
Center for Health and Biological Sciences, Mackenzie Presbyterian University,
São Paulo, Brazil
e-mail: boggio@mackenzie.br

L. M. Marques
Instituto de Medicina Fisica e Reabilitacao, Hospital das Clinicas HCFMUSP,
Faculdade de Medicina, Universidade de Sao Paulo, Sao Paulo, Brazil

© The Author(s) 2023
P. S. Boggio et al. (eds.), *Social and Affective Neuroscience of Everyday Human Interaction*, https://doi.org/10.1007/978-3-031-08651-9_15

Introduction

Transcranial brain stimulation (TBS) is a term that denotes different noninvasive techniques which aim to modulate brain cortical activity through an external source, usually an electric or magnetic one. Currently, there are several techniques categorized as TBS. However, two are more used for scientific research, the transcranial magnetic stimulation (TMS) and the transcranial direct current stimulation (tDCS), which stimulate brain areas with a high-intensity magnetic field or a weak electric current on the scalp, respectively. They represent an enormous contribution to behavioral, cognitive, and social neuroscience since they reveal how delimited brain cortical areas contribute to some behavior or cognition. They have also been proposed as a feasible tool in the clinical setting since they can modulate abnormal cognition or behavior due to brain activity modulation. This chapter will present the standard methods of transcranial stimulation, their contributions to social and affective neuroscience through a few main topics, and the studies that adopted those techniques, also summing their findings.

Essentials of Transcranial Electrical and Magnetic Stimulation

Transcranial magnetic stimulation (TMS) first appeared in 1985, at the beginning, adopted to investigate nervous propagation along the corticospinal tract and peripheral nerves (Rossini & Rossi, 2007) and investigate the brain function excitability of different brain areas (Hallet, 2007). It consists of a coil and one or two generators (also called stimulators), which generate current pulses converted on the coil in a magnetic field. When positioned on the scalp, the coil delivers a magnetic pulse that creates a transient electric field in cortical areas underneath, activating neural networks through axonal depolarization or impairing neural activity through post-excitatory inhibition, i.e., "silent period" (Chen et al., 1999; Lefaucheur et al., 2014). It is possible to apply single or paired (e.g., double or triple pulses) TMS to investigate intracortical circuits and their relation to behavior and cognition (Ni et al., 2011). In addition to these procedures, there is also possible to use repetitive TMS (rTMS) to excite or inhibit a cortical area depending on the parameters adopted, mainly the frequency of pulses delivered. Studies with the motor cortex established that low-frequency stimulation (≤ 1 Hz) is usually inhibitory while high-frequency stimulation (≥ 5 Hz) is excitatory, but a variation in these effects can occur due to differences in intensity and duration of rTMS. It is also important to highlight those differences in effect depend on other parameters such as type of the coil, distance and orientation to the head and the waveform, intensity, and frequency of magnetic pulse (Lefaucheur et al., 2014).

Another common neuromodulator adopted in neuroscience similar to the TMS is the transcranial direct current stimulation (tDCS). An initial version of the tDCS (named "medical battery") appeared during the nineteenth century to treat several ailments. Nevertheless, only at the beginning of the twenty-first century were its mechanisms deeply investigated and have been broadly adopted as a research tool (Wexler, 2017). tDCS consists of a low-intensity direct current (about 1 to 3 mA) applied on the brain by positioning two or more electrodes onto the scalp, forming an electrical circuit. While the minimum is two electrodes to close the circuit, it is possible to find assembles with more electrodes, just as in high-definition tDCS. The electrodes vary in format and size, usually ranging between 10 and 40 cm^2 in a round or square format. As observed in studies investigating motor cortex excitability, the stimulation's typical effect is enhanced excitability in cortical areas below anodic and inhibition below the cathodic electrode. However, differences in these effects can occur in brain areas other than the motor cortex, like those related to higher cognitive processing, where a linear effect between intensity and cortical excitability or inhibition seems not to be the rule. Also, the effect can vary accordingly to (i) the montage adopted, (ii) the size and orientation of the electrodes, (iii) the intensity and duration of stimulation, (iv) individual characteristics (e.g., gender, age, anatomical differences), and (v) if tDCS is applied during an active state (performing an activity of interest) or in a resting state (Giordano et al., 2017; Sellaro et al., 2017).

Despite the similarities here presented, there are also apparent differences between TMS and tDCS. For instance, TMS stimulation is more focal since the cortical target is circumscribed to an area about 2 or 3 cm^2 when using a coil (Lozano & Hallett, 2013). Conversely, the usual tDCS montage's cortical target is broader, but it is possible to target a narrow area employing high-definition tDCS. Regarding the TMS, its stimulation is more intense than tDCS, so it is possible to interrupt neural activity or stimulate an action potential with TMS, while tDCS can only modulate ongoing activity. Nevertheless, since TMS has a higher intensity, there is also a risk of seizures not present in tDCS stimulation, although reports in the scientific literature indicate it is rare (less than 1 seizure per 60,000 sessions) if safety guidelines are adopted (Lerner et al., 2019). Another critical question to research and clinics is that tDCS is easier to apply than TMS because it has fewer parameters. Moreover, tDCS is considerably inexpensive compared to TMS.

It is also essential to present some relevant limitations to both techniques. First, such techniques are more focused when modulating cortical areas of the brain but are not widely used to stimulate subcortical areas. The stimulation of subcortical areas is usually indirect, employing a tDCS current passing those areas (yet with limited focus) or in response to some cortical region's stimulation by tDCS or TMS, such as stimulation of frontal areas to modulate the activity of subcortical areas as in the case of emotion regulation. TMS also has some coil models (e.g., H-coil, halo coil, or double-cone coil) that allow deep stimulation but also with less focus when compared to cortical targets.

Finally, concerning some practical aspects of using TBS, both techniques are usually applied prior ("offline") or concomitant ("online") to some cognitive or behavioral task. It is essential to consider safety aspects when using such techniques, such as avoiding applying such techniques in participants with epilepsy, metallic implants on the head, or pacemakers. Concerning tDCS, it is relevant to ascertain the skin's integrity where the electrodes will be applied; besides, some participants report reactions of severe discomfort and skin irritation. Here, we present only a few more superficial aspects of both techniques. Bearing in mind that such techniques require different preparations and care, we recommend reading specific articles on practical aspects in applying and preparing experiments for tDCS (Woods et al., 2016) and TMS (Hannula & Ilmoniemi, 2017). In the following topics, we will address the use of both techniques in social and affective neuroscience, as well as their main findings.

Social Neuroscience

Social neuroscience is an interdisciplinary field that aims to understand the neurobiology of social cognition and behavior in humans and animals – first created from the merge of social psychology, neuroscience, and social sciences. It aims to investigate brain structures and their functioning on various social processes, such as communication, cooperation, empathy, moral judgment, prejudice, social learning, social decision-making, social perception, and so on (Cacioppo & Cacioppo, 2013; Lieberman, 2007). This section will present a few social neuroscience topics that adopted tDCS or TMS and demonstrate how these approaches clarified the brain processes related to prejudice, social decision-making, and moral judgment.

Prejudice

Prejudice is the attitude toward others based on their group membership, and it is intrinsically related to affective and cognitive processes, such as social categorization and stereotyping (Amodio, 2014). Negative beliefs about the outgroup influence choices, judgments, and behaviors (Sellaro et al., 2015) and can give rise to discrimination and prejudice to outgroup members (Amodio, 2014). In contrast, individuals judge more positively members of the same group when compared to another racial group, a phenomenon called ingroup favoritism (Taylor & Doria, 1981).

Different cerebral cortical regions are involved in prejudice, mainly associated with social perception and evaluation (Gamond et al., 2017). One of the primary brain areas associated with prejudice is the medial prefrontal cortex (MPFC), an area involved in several cognitive activities, such as social perception, categorization, stereotyping, and regulation/control of behavioral responses in social contexts (Amodio & Frith, 2006; Amodio, 2014; Sellaro et al., 2015). Sellaro et al. (2015) investigated the causal role of MPFC in stereotype neutralization using tDCS. In this study, participants performed the implicit racial attitude task (racial IAT) while submitted to a tDCS protocol (anodal, cathodal, or sham) of 1 mA intensity targeting the MPFC. Anodal stimulation decreased implicit bias when compared to cathodal stimulation or sham. Sellaro's study was the first to demonstrate MPFC's causal role in cognitive control in overcoming negative judgment concerning another social group.

Another area associated with prejudice is the cerebellum. Recent studies have demonstrated the cerebellum's functional connectivity to the MPFC and other cortical regions like the temporoparietal junction during social judgments related to body reading, action sequencing, and mentalizing behavior (see Van Overwalle et al. 2015). One study by Gamond et al. (2017) evaluated the cerebellum and dorsomedial prefrontal cortex (dMPFC) roles in participants' implicit attitudes, where Caucasian participants had to categorize valence of positive/negative primed by ingroup or outgroup faces while receiving TMS. The behavioral experiment (without neuromodulation) showed ingroup bias with faster categorization for positive adjectives primed by the ingroup faces. However, both the dMPFC and the right cerebellum modulation interfered with this effect, preventing the ingroup bias. The results suggest that both brain areas play a causal role in social cognition processes, such as implicit social attitudes for ingroup members.

Finally, another study demonstrated TMS over dMPFC interfered with one's ability to discriminate emotions expressed by ingroup members. These findings suggest a causal role of dMPFC in recognizing ingroup emotions (Gamond & Cattaneo, 2016). In summary, those studies have demonstrated, using tDCS and TMS, the crucial role of MPFC and related cortical areas (e.g., cerebellum) in social cognitive processes such as social group categorization and recognition.

Social Decision-Making

Social decision-making is a social neuroscience topic that aims to comprehend the neural mechanisms of choosing between alternatives in a social context (Sanfey, 2007). tDCS and TMS have been adopted in social decision-making to investigate brain areas' causal role (mainly prefrontal) during cooperation or competition situations simulated through simple games derived from behavioral economics. The selection of brain targets to modulate through tDCS or TMS is usually based on correlational studies previously conducted with neuroimage techniques, pointing to the probable involvement of a cortical area in some aspect of social decision-making.

Two main areas investigated through tDCS and TMS in social decision-making are the dorsolateral prefrontal cortex (DLPFC) and medial prefrontal cortex (MPFC). Several tDCS and TMS studies targeted DLPFC in social decision-making through ultimatum game (Knoch et al., 2006, 2008; Ruff et al., 2013), Trust Game (Knoch et al., 2009; Wang et al., 2016) and public goods game (Li et al., 2018; Liu et al., 2017). Overall, the findings point to the right DLPFC role in implementing controlled cognition to identify contextual social norms or expectations and orient adaptive behavior to comply with those norms (Sanfey et al., 2014). One study investigated left DLPFC role in supporting people, showing that this area's enhanced excitability led to increased prosocial behavior. The authors hypothesized that this area could be related to the management of emotional information by controlled cognition (Balconi & Canavesio, 2014). Although these studies have been clearly showing the involvement of DLPFC in social decision-making, it is still not clear the specific role of right and left areas in social decision-making.

Another area investigated in social decision-making is the MPFC, usually detected in neuroimage studies. One study with MPFC investigated its role on unfairness acceptance when unfair proposals were committed to oneself compared to a third party, showing that inhibition of this area led to a higher acceptance rate of unfair proposals and implying a causal role of MPFC in process fairness in situations involving self (Civai et al., 2014). In another study, Klucharev et al. (2011) evaluated the role of MPFC in social conforming on an attractiveness decision task, where participants should rate the attractiveness of models presented in photos. In this task, the downregulation of MFPC diminished social conformation, indicating that this area is related to social learning related to others' expectations in decision-making, as indicated by recent studies (Apps & Sallet, 2017; Sanfey et al., 2015). In summary, it appears that MFPC recruits controlled cognition to implement decisions related to oneself and calculate others' expectations in the context.

Moral Judgment

Moral judgment is the topic studying the judgment of right and wrong mainly respective to situations involving harm. Thus, most of the studies investigate moral judgment considering dilemmas such as the trolley problem (Thomson, 1984), in which the participant should decide between preserving individual rights from a single person and saving many others. This kind of task typically evaluates moral judgment in a utilitarian-deontological axis, considering cognitive reasoning relative to harm aversion. Furthermore, considering cortical brain regions, the majority of the studies investigated the modulation of two cortical structures, DLPFC and ventral MPFC (or just VMPFC), considering their role in other social phenomena (Boggio et al., 2016a; Darby & Pascual-Leone, 2017; Di Nuzzo et al., 2018).

Considering DLPFC modulation, one study by Tassy et al. (2011) investigated brain neuromodulation during moral judgment. The authors performed low-frequency rTMS (known to generate cortical inhibition) over the right DLPFC during a moral dilemma judgment task, where the participant should judge whether he or she considered an immoral attitude acceptable. The authors observed a significant increase in utilitarian judgments (i.e., "the most good for most people") during active TMS compared to shame. In this way, this finding points out the significant role of this structure in moral judgment, specifically, in controlling emotional processes usually related to decreased utilitarian decisions. Similarly, Jeurissen et al. (2014) also demonstrated that low-frequency rTMS over right DLPFC was associated with moral judgment modulation in personal dilemmas (leading to less utilitarian responses), but not in impersonal or nonmoral dilemmas. The authors explained that the personal moral dilemmas are more emotionally salient; thus, the study suggested the right DPLFC role in cognitive control, probably dampening emotion processing and consequently enhancing utilitarian responses.

Regarding the use of tDCS, Kuehne et al. (2015) performed a task very similar to Jeurissen et al. (2014) concerning dilemmas with personal involvement nevertheless sought to modulate contralateral homologous region, that is, the left DLPFC. For this purpose, they performed three experimental conditions: two active conditions with target electrode (anodal or cathodal) over the left DPLFC and reference over the right parietal cortex and one sham stimulation. The authors found that only anodal condition presented significant moral judgment modulation compared to sham condition, showing a decrease in utilitarian judgments (greater frequency of deontological judgments), thus highlighting this structure's role in the left hemisphere in the process of moral reasoning. However, in a recent study conducted by Zheng et al. (2018), an opposite effect was found, where the authors performed balanced bilateral tDCS over the DLPFC, with left anodal, left cathodal, and sham conditions. A significant decrease in the utilitarian judgment was observed for dilemmas with personal involvement during the anodal on the right hemisphere (cathodal at left hemisphere), which is compatible with the results found by Jeurissen et al. (2014). However, these findings revealed the need for experimental standardization since the positioning of the reference electrode can significantly impact tDCS effects.

Considering VMPFC role, many social neuroscience studies had shown this area involvement in empathy processes (Shamay-Tsoory et al., 2003), theory of mind (Shamay-Tsoory et al., 2005), and moral judgment (Greene, 2007; Moll & de Oliveira-Souza, 2007).

The first work with neuromodulation of ventral medial prefrontal cortex (VMPFC) and moral judgment conducted by Fumagalli et al. (2010) investigated VMPFC modulation employing tDCS, with anodal and cathodal over this area or over the occipital cortex (control condition) and reference electrode over right deltoid. They performed a judgment task of moral dilemmas with personal involvement, without personal involvement, and nonmoral dilemmas. The authors observed that brain modulation was only effective in female participants (which already presented low levels of utilitarian judgments in comparison to the male participants at the baseline), who presented a greater frequency of utilitarian responses after anodal tDCS on VPFC and lower frequency after cathodal tDCS, compared to baseline trials (Fumagalli et al., 2010). It is worth noting that the authors did not find any significant effect for tDCS in the occipital cortex or sham condition. In this way, these findings indicate that the neuromodulation of the VPFC may impact moral judgment and also highlights probably differences in brain circuitry for emotion processing between men and women (Fumagalli et al., 2010), as previously presented in the literature (Boggio et al., 2008). More recently, in a complementary way, Yuan et al. (2017) used a picture judgment task to assess moral judgment and arousal rating. Participants who received anodal tDCS on VMPFC (with reference electrode in the right deltoid) significantly increased moral judgment and arousal rating compared to sham condition. The authors did not evaluate differences in sex that could complement (Fumagalli et al., 2010) findings. Finally, recent work by Riva et al. (2018) investigated VMPFC modulation during a moral dilemma task, with the active electrode (anodal or cathodal) over VMPFC and the reference electrode over the occipital area. The findings revealed similar effects to Fumagalli et al. (2010) and Yuan et al. (2017), where participants receiving anodal tDCS over VMPFC had a higher frequency of utility judgments.

Overall, the findings regarding DLPFC and VMPFC's neuromodulation highlight these structures' essential causal role in moral judgment processes. However, all these findings represent tasks of moral dilemmas, such as the trolley/train problem (Thomson, 1984), which only measures the participant's judgment concerning a deontological-utilitarian axis, without taking into account the different moral foundations (Graham et al., 2013).

Affective Neuroscience

Besides social phenomena, some studies have sought to understand several cortical brain structures' specific role on affective phenomena, such as facial expression recognition and emotion regulation. It is a consensus that social and affective phenomena are closely intertwined (Boggio et al., 2016a, b), thus hindering the exclusive study of one of them. The following topics present the main findings regarding neuromodulation to understand two of the main topics from affective neuroscience:

Emotional Face Recognition

One crucial use of neuromodulation was to investigate brain networks involved in the recognition of emotional facial expressions. One of the main areas investigated is the medial prefrontal cortex (MPFC). Some of the studies assessed low-frequency TMS on dorsal MPFC (Balconi et al., 2011; Balconi & Bortolotti, 2012; Harmer et al., 2001), where inhibition of this area by neuromodulation specifically impaired recognition of facial expressions of anger and fear. TMS may have interfered with the dorsal anterior cingulate cortex's activity, usually responsive to negative valence emotions. Another possibility is related to the role of MFPC on other brain areas via top-down regulation, as indicated by one study coupling TMS and EEG where magnetic pulses delivered over right MPFC led to altered electroencephalographic early evoked potentials detected at temporal and occipital regions (Mattavelli et al., 2013).

In addition to the MPFC, other studies also assessed orbital and dorsolateral prefrontal areas to investigate their role in processing facial expressions. For example, Nitsche et al. (2012) applied anodal and cathodal tDCS over the left DLPFC (reference electrode positioned on the contralateral supraorbital region). They found enhanced performance in healthy subjects answering a facial expression identification task markedly for positive valence emotions and anodal tDCS. In another study, Willis et al. (2015) applied anodal tDCS over the right orbitofrontal cortex with reference over P3 (left parietal cortex). Compared to sham, active tDCS enhanced performance on facial expression recognition. It is essential to notice that implying those specific prefrontal regions in emotion recognition is not so straightforward since tDCS is not so focal and the reference electrode could also interfere in the results. For example, Heberlein et al. (2008) investigated patients with prefrontal lesions in diverse regions, where they found that only patients with ventromedial lesions had impaired facial expression recognition and emotional expression. Besides, tDCS over prefrontal regions could indirectly act over ventromedial regions, leading to confounding results about what region is related to emotion recognition. One way to solve this problem is to use TMS, which is more focal. In a study by Ferrari et al. (2017), TMS was applied over the right or left DLPFC, and they found that both stimulations interfered in recognition of facial expressions, irrespective of emotion, similar to what Nitsche et al. (2012) found with tDCS. Thus, it is possible to implicate DLPFC in emotion recognition.

Other regions in the frontal lobe also investigated through neuromodulation methods are the supplementary motor area's anterior region (pre-SMA) and primary motor area (M1). Regarding the pre-SMA, Rochas et al. (2013) inhibited its activity through low-frequency TMS and investigated recognition of faces expressing happiness, anger, or fear. In this case, left pre-SMA disruption impaired recognition of happy faces but did not affect fear or angry faces. In addition to Nitsche et al. (2012), it is possible to hypothesize the left hemisphere's implication in processing positive valence, in line with previous neuroimaging and behavioral studies (Root et al., 2006). However, Ferrari et al. (2017) did not detect this, and there is still controversy in the literature supporting this lateralized valence theory (Root et al., 2006). The study by Rochas et al. (2013) hypothesized that disrupting pre-SMA led to impaired emotion recognition due to the mirror neuron system, i.e., disrupting the motor simulation of an expression in motor areas could also impair emotion recognition, similar to presented in simulation theories of emotion recognition (Gallese & Sinigaglia, 2011; Goldman & Sripada, 2005). Another study indicated the role of MNS in emotional face recognition, which found a positive correlation between cortex excitability of M1 (assessed by TMS) in response to movement observation and performance in facial expression recognition (Enticott et al., 2008).

Another critical region investigated in facial expression recognition is the temporal lobe, given the vital role of the superior temporal sulcus in processing dynamic facial features, such as eye gaze and facial expressions (Furl et al., 2014). Three studies by the same group investigated the contribution of the right occipital face area (rOFCA) compared to the right somatosensory cortex (rSC) (Pitcher et al., 2008) or the right posterior superior temporal sulcus (rpSTS) (Pitcher, 2014; Pitcher et al., 2014) in dynamic face processing. They found that all those areas contribute to recognizing facial expressions, with rOFA responsible for early processing of facial features (less than 100 ms), while rSC and rpSTS were responsible for posterior processing, despite still in the automatic domain (between 100 and 170 ms).

Furthermore, although rOFA stimulation disrupted facial expression perception, this area appeared to be more related to the processing of static facial features, whereas rpSTS stimulation disrupted precisely dynamic face recognition. Summing, these results indicate a network of the occipital and temporal area responsible for processing dynamic features of facial expressions (Pitcher et al., 2008, 2014; Pitcher, 2014).

Another relevant study that neuromodulated the temporal lobe is from Boggio et al. (2008), using tDCS and finding opposite effects between women and men. In this study, they applied anodal tDCS over the left temporal and the reference over the contralateral region, which led to women's enhanced performance in detecting sad faces, while men performed worse due to stimulation. This study indicates differences among men and women in how the brain processes recognize basic emotions, which is specifically problematic given other studies have not evaluated gender as a factor in their analysis.

Finally, two other studies investigated emotional face recognition. Ferrucci et al. (2012) stimulated the cerebellum through anodal and cathodal tDCS, where both polarities led to better performance in recognition of faces expressing emotions of negative valence. In another experiment, Cecere et al. (2013) inhibited the left occipital region through cathodal tDCS, while participants responded to a go/no-go task with images of fearful and happy faces. This experiment investigated the occipital cortex's role in integrating explicit and implicit stimuli (i.e., subliminal visual stimuli) showed to the left and right visual fields, respectively; it also investigated how unconscious emotional stimuli could facilitate behavior in a go/no-go task (correctly react to targets pressing a button). This study demonstrated a facilitation effect in the go/no-go task when explicit and implicit were congruent (showing the same expression of happiness or fear). However, after occipital cortex disruption by tDCS, this congruent facilitation disappeared, and implicit detection of fearful faces facilitated behavior, but only when the target was happy faces (similar to hemianopsia patients). The study demonstrated cortical (occipital cortex) role and subcortical routes in processing implicit visual information, showing occipital role in processing high-order level information regarding congruence, while subcortical routes' role was relevant for processing implicit fear stimuli.

In sum, neuromodulation studies indicate the existence of different systems between basic emotions, as suggested by neuroimage studies (Tettamanti et al., 2012; Diano et al., 2017), and it can vary between men and women. Modulation techniques also helped to elucidate the role of several brain areas (e.g., cerebellum, temporal, occipital, and frontal lobes) and of the MNS system in emotion recognition.

Emotion Regulation

Emotion regulation is the capacity to modify oneself or someone else emotional responses in order to intensify (upregulation) or diminish (downregulation) current emotion (Gross, 2014). Many studies on this topic focused on the emotional reappraisal strategy, i.e., a technique to change the cognitive label of specific emotional content. This preference is because this strategy is more effective in modulating the long-term emotional response (Gross, 2014), besides presenting a direct relation with cognitive control and the brain structures involved in this control (Ochsner et al., 2012), mainly the dorsolateral prefrontal cortex (DLPFC) and ventrolateral prefrontal cortex (VLPFC), due to the critical role of these structures on cognitive control, attentional orientation, response inhibition (Ochsner et al., 2012), and mediating amygdala's activity (Wager et al., 2008).

One relevant study on this topic is by Feeser et al. (2014). They investigated the role of right DLPFC anodal tDCS in using emotion reappraisal strategy (cathodal electrode positioned at contralateral supraorbital region). They found a significant increase in cognitive control measured by arousal ratings and skin conductance response (SCR). The typical variation according to reappraisal, i.e., higher for upregulation and lower for downregulation compared to observation only, was potentialized with anodal stimulation of DLPFC. These findings clarify the significant role of the right DLPFC in cognitive control and emotion regulation through a reappraisal of negative valence content. In the same line, Pripfl and Lamm (2015) and Rêgo et al. (2015) also found a significant impact of right DLPFC anodal stimulation on cognitive control. However, contrary to Pripfl and Lamm (2015), Rêgo et al. (2015) also found that left anodal DLPFC condition significantly modulates emotion regulation, possibly due to increased attentional control, following Plewnia et al. (2015).

Thus, it seems that there is a misunderstanding between studies and relative to the neuromodulation of hemispheric sides. With this in mind, Marques et al. (2018) performed a study in order to investigate bilateral balanced DLPFC in two conditions compared to sham: (i) anodal left and cathodal right and (ii) anodal right and cathodal left. They did not find any significant impact of DLPFC tDCS on the emotional reappraisal of negative pictures. Notwithstanding, in a second study, they performed the same experimental procedures; however, over VLPFC, they found that left anodal VLPFC tDCS significantly impacted emotion reappraisal of negative pictures, increasing valence (more positive) regardless of emotion regulation strategy. Furthermore, they found a significant impact of left anodal VLPFC tDCS on the cardiac inter-beat interval, increasing cardiac recruitment on the first seconds of emotional processing, indicating that this neuromodulation condition significantly increased participants' cognitive engagement, and also leading to an increased valence estimation.

Thus, following the discussion of Paulo S Boggio et al. (2016b), these findings indicate several particularities of each mentioned brain structure on emotion regulation, as the role of DLPFC on cognitive control (Ochsner et al., 2012) and VLPFC on attentional control (Wager et al., 2008). Future studies should standardize the experimental protocol between studies due to significant discrepancies in the literature related to electrode size, current intensity, cathode positioning, and emotion regulation tasks. Moreover, as highlighted by Kim et al. (2019), future studies should also use TMS as an exciting technique to address both DLPFC and VLPFC's role in emotion regulation.

Conclusions

To conclude, the transcranial stimulation methods, tDCS, and TMS have been an important tool to investigate cortical circuits' role in several social, like prejudice, social decision-making, and moral judgment, and affective processes, like emotion recognition and regulation. Those techniques were essential to demonstrate several brain areas' role in a plethora of previously described processes in neuroimaging studies. TMS studies could also demonstrate the role of different areas in a brain network across time, which is very relevant to indicate how the brain integrates complex information among several cortical areas.

The observed cognitive and behavioral effects in response to brain modulation are of great relevance since they can indicate the future use of these neuromodulation techniques to modulate brain activity noninvasively in clinical patients with social or affective disorders to ameliorate their clinical condition.

References

Amodio, D. M., & Frith, C. D. (2006). Meeting of minds: The medial frontal cortex and social cognition. *Nature Reviews Neuroscience, 7*(4), 268.

Amodio, D. M. (2014). The neuroscience of prejudice and stereotyping. *Nature Reviews Neuroscience, 15*(10), 670.

Apps, M. A., & Sallet, J. (2017). Social learning in the medial prefrontal cortex. *Trends in Cognitive Sciences, 21*(3), 151–152.

Balconi, M., Bortolotti, A., & Gonzaga, L. (2011). Emotional face recognition, EMG response, and medial prefrontal activity in empathic behaviour. *Neuroscience Research, 71*(3), 251–259.

Balconi, M., & Bortolotti, A. (2012). Detection of the facial expression of emotion and self-report measures in empathic situations are influenced by sensorimotor circuit inhibition by low-frequency rTMS. *Brain Stimulation, 5*(3), 330–336.

Balconi, M., & Canavesio, Y. (2014). High-frequency rTMS on DLPFC increases prosocial attitude in case of decision to support people. *Social Neuroscience, 9*(1), 82–93.

Boggio, P. S., Rêgo, G. G., Marques, L. M., & Costa, T. L. (2016a). Social psychology and noninvasive electrical stimulation. *European Psychologist.*

Boggio, P. S., Rêgo, G. G., Marques, L. M., & Costa, T. L. (2016b). Transcranial direct current stimulation in social and emotion research. In *Transcranial direct current stimulation in neuropsychiatric disorders* (pp. 143–152). Springer.

Boggio, P. S., Rocha, R. R., da Silva, M. T., & Fregni, F. (2008). Differential modulatory effects of transcranial direct current stimulation on a facial expression go-no-go task in males and females. *Neuroscience Letters, 447*(2–3), 101–105.

Cacioppo, J. T., & Cacioppo, S. (2013). Social neuroscience. *Perspectives on Psychological Science, 8*(6), 667–669.

Cecere, R., Bertini, C., & Làdavas, E. (2013). Differential contribution of cortical and subcortical visual pathways to the implicit processing of emotional faces: A tDCS study. *Journal of Neuroscience, 33*(15), 6469–6475.

Chen, R., Lozano, A. M., & Ashby, P. (1999). Mechanism of the silent period following transcranial magnetic stimulation evidence from epidural recordings. *Experimental Brain Research, 128*(4), 539–542.

Civai, C., Miniussi, C., & Rumiati, R. I. (2014). Medial prefrontal cortex reacts to unfairness if this damages the self: A tDCS study. *Social Cognitive and Affective Neuroscience, 10*(8), 1054–1060.

Darby, R. R., & Pascual-Leone, A. (2017). Moral enhancement using noninvasive brain stimulation. *Frontiers in Human Neuroscience, 11*, 77.

Diano, M., Tamietto, M., Celeghin, A., Weiskrantz, L., Tatu, M. K., Bagnis, A., ... & Costa, T. (2017). Dynamic changes in amygdala psychophysiological connectivity reveal distinct neural networks for facial expressions of basic emotions. *Scientific Reports, 7*(1), 1–13.

Di Nuzzo, C., Ferrucci, R., Gianoli, E., Reitano, M., Tedino, D., Ruggiero, F., & Priori, A. (2018). How brain stimulation techniques can affect moral and social behaviour. *Journal of Cognitive Enhancement, 2*(4), 335–347.

Enticott, P. G., Johnston, P. J., Herring, S. E., Hoy, K. E., & Fitzgerald, P. B. (2008). Mirror neuron activation is associated with facial emotion processing. *Neuropsychologia, 46*(11), 2851–2854.

Feeser, M., Prehn, K., Kazzer, P., Mungee, A., & Bajbouj, M. (2014). Transcranial direct current stimulation enhances cognitive control during emotion regulation. *Brain Stimulation, 7*(1), 105–112.

Ferrari, C., Gamond, L., Gallucci, M., Vecchi, T., & Cattaneo, Z. (2017). An exploratory TMS study on prefrontal lateralization in valence categorization of facial expressions. *Experimental Psychology*.

Ferrucci, R., Giannicola, G., Rosa, M., Fumagalli, M., Boggio, P. S., Hallett, M., ... Priori, A. (2012). Cerebellum and processing of negative facial emotions: Cerebellar transcranial DC stimulation specifically enhances the emotional recognition of facial anger and sadness. *Cognition & Emotion, 26*(5), 786–799.

Fumagalli, M., Vergari, M., Pasqualetti, P., Marceglia, S., Mameli, F., Ferrucci, R., ... Pravettoni, G. (2010). Brain switches utilitarian behavior: Does gender make the difference? *PloS One, 5*(1), e8865.

Furl, N., Henson, R. N., Friston, K. J., & Calder, A. J. (2014). Network interactions explain sensitivity to dynamic faces in the superior temporal sulcus. *Cerebral Cortex, 25*(9), 2876–2882.

Gallese, V., & Sinigaglia, C. (2011). What is so special about embodied simulation? *Trends in Cognitive Sciences, 15*(11), 512–519.

Gamond, L., & Cattaneo, Z. (2016). The dorsomedial prefrontal cortex plays a causal role in mediating ingroup advantage in emotion recognition: A TMS study. *Neuropsychologia, 93*, 312–317.

Gamond, L., Ferrari, C., La Rocca, S., & Cattaneo, Z. (2017). Dorsomedial prefrontal cortex and cerebellar contribution to ingroup attitudes: A transcranial magnetic stimulation study. *European Journal of Neuroscience, 45*(7), 932–939.

Giordano, J., Bikson, M., Kappenman, E. S., Clark, V. P., Coslett, H. B., Hamblin, M. R., ... Nitsche, M. A. (2017). Mechanisms and effects of transcranial direct current stimulation. *Dose-Response, 15*(1), 1559325816685467.

Goldman, A. I., & Sripada, C. S. (2005). Simulationist models of face-based emotion recognition. *Cognition, 94*(3), 193–213.

Graham, J., Haidt, J., Koleva, S., Motyl, M., Iyer, R., Wojcik, S. P., & Ditto, P. H. (2013). Moral foundations theory: The pragmatic validity of moral pluralism. In *Advances in experimental social psychology* (Vol. 47, pp. 55–130). Elsevier.

Greene, J. D. (2007). Why are VMPFC patients more utilitarian? A dual-process theory of moral judgment explains. *Trends in Cognitive Sciences, 11*(8), 322–323.

Gross, J. J. (2014). Emotion regulation: Conceptual and empirical foundations.

Hallett, M. (2007). Transcranial magnetic stimulation: A primer. *Neuron, 55*(2), 187–199.

Hannula, H., & Ilmoniemi, R. J. (2017). Basic principles of navigated TMS. In *Navigated transcranial magnetic stimulation in neurosurgery* (pp. 3–29). Springer.

Harmer, C. J., Thilo, K. V., Rothwell, J. C., & Goodwin, G. M. (2001). Transcranial magnetic stimulation of medial–frontal cortex impairs the processing of angry facial expressions. *Nature Neuroscience, 4*(1), 17.

Heberlein, A. S., Padon, A. A., Gillihan, S. J., Farah, M. J., & Fellows, L. K. (2008). Ventromedial frontal lobe plays a critical role in facial emotion recognition. *Journal of Cognitive Neuroscience, 20*(4), 721–733.

Jeurissen, D., Sack, A. T., Roebroeck, A., Russ, B. E., & Pascual-Leone, A. (2014). TMS affects moral judgment, showing the role of DLPFC and TPJ in cognitive and emotional processing. *Frontiers in Neuroscience, 8*, 18.

Kim, J. U., Weisenbach, S. L., & Zald, D. H. (2019). Ventral prefrontal cortex and emotion regulation in aging: A case for utilizing transcranial magnetic stimulation. *International Journal of Geriatric Psychiatry, 34*(2), 215–222.

Klucharev, V., Munneke, M. A., Smidts, A., & Fernández, G. (2011). Downregulation of the posterior medial frontal cortex prevents social conformity. *Journal of Neuroscience, 31*(33), 11934–11940.

Knoch, D., Pascual-Leone, A., Meyer, K., Treyer, V., & Fehr, E. (2006). Diminishing reciprocal fairness by disrupting the right prefrontal cortex. *Science, 314*(5800), 829–832.

Knoch, D., Schneider, F., Schunk, D., Hohmann, M., & Fehr, E. (2009). Disrupting the prefrontal cortex diminishes the human ability to build a good reputation. *Proceedings of the National Academy of Sciences, 106*(49), 20895–20899.

Knoch, D., Nitsche, M. A., Fischbacher, U., Eisenegger, C., Pascual-Leone, A., & Fehr, E. (2008). Studying the neurobiology of social interaction with transcranial direct current stimulation—the example of punishing unfairness. *Cerebral Cortex, 18*(9), 1987–1990.

Kuehne, M., Heimrath, K., Heinze, H.-J., & Zaehle, T. (2015). Transcranial direct current stimulation of the left dorsolateral prefrontal cortex shifts preference of moral judgments. *PloS One, 10*(5), e0127061.

Lefaucheur, J. P., André-Obadia, N., Antal, A., Ayache, S. S., Baeken, C., Benninger, D. H., … Devanne, H. (2014). Evidence-based guidelines on the therapeutic use of repetitive transcranial magnetic stimulation (rTMS). *Clinical Neurophysiology, 125*(11), 2150–2206.

Lerner, A. J., Wassermann, E. M., & Tamir, D. I. (2019). Seizures from Transcranial Magnetic Stimulation 2012–2016: Results of a survey of active laboratories and clinics. *Clinical Neurophysiology*.

Li, J., Liu, X., Yin, X., Li, S., Wang, P., Niu, X., & Zhu, C. (2018). Transcranial direct current stimulation of the right lateral prefrontal cortex changes a priori normative beliefs in voluntary cooperation. *Frontiers in Neuroscience, 12*.

Lieberman, M. D. (2007). Social cognitive neuroscience: A review of core processes. *Annual Review of Psychology, 58*, 259–289.

Liu, X., Li, J., Wang, G., Yin, X., Li, S., & Fu, X. (2017). Transcranial direct current stimulation of the rLPFC shifts normative judgments in voluntary cooperation. *Neuroscience Letters*.

Lozano, A. M., & Hallett, M. (2013). *Brain stimulation* (Vol. 116). Newnes.

Marques, L. M., Morello, L. Y., & Boggio, P. S. (2018). Ventrolateral but not Dorsolateral Prefrontal Cortex tDCS effectively impact emotion reappraisal–effects on Emotional Experience and Interbeat Interval. *Scientific Reports, 8*(1), 15295.

Mattavelli, G., Rosanova, M., Casali, A. G., Papagno, C., & Lauro, L. J. R. (2013). Top-down interference and cortical responsiveness in face processing: A TMS-EEG study. *Neuroimage, 76*, 24–32.

Moll, J., & de Oliveira-Souza, R. (2007). Moral judgments, emotions and the utilitarian brain. *Trends in Cognitive Sciences, 11*(8), 319–321.

Ni, Z., Müller-Dahlhaus, F., Chen, R., & Ziemann, U. (2011). Triple-pulse TMS to study interactions between neural circuits in human cortex. *Brain Stimulation, 4*(4), 281–293.

Nitsche, M. A., Koschack, J., Pohlers, H., Hullemann, S., Paulus, W., & Happe, S. (2012). Effects of frontal transcranial direct current stimulation on emotional state and processing in healthy humans. *Frontiers in Psychiatry, 3*, 58.

Ochsner, K. N., Silvers, J. A., & Buhle, J. T. (2012). Functional imaging studies of emotion regulation: A synthetic review and evolving model of the cognitive control of emotion. *Annals of the New York Academy of Sciences, 1251*(1).

Plewnia, C., Schroeder, P. A., Kunze, R., Faehling, F., & Wolkenstein, L. (2015). Keep calm and carry on: Improved frustration tolerance and processing speed by transcranial direct current stimulation (tDCS). *PloS One, 10*(4), e0122578.

Pripfl, J., & Lamm, C. (2015). Focused transcranial direct current stimulation (tDCS) over the dorsolateral prefrontal cortex modulates specific domains of self-regulation. *Neuroscience Research, 91*, 41–47.

Rêgo, G. G., Lapenta, O. M., Marques, L. M., Costa, T. L., Leite, J., Carvalho, S., ... Boggio, P. S. (2015). Hemispheric dorsolateral prefrontal cortex lateralization in the regulation of empathy for pain. *Neuroscience Letters, 594*, 12–16.

Pitcher, D., Garrido, L., Walsh, V., & Duchaine, B. C. (2008). Transcranial magnetic stimulation disrupts the perception and embodiment of facial expressions. *Journal of Neuroscience, 28*(36), 8929–8933.

Pitcher, D. (2014). Facial expression recognition takes longer in the posterior superior temporal sulcus than in the occipital face area. *Journal of Neuroscience, 34*(27), 9173–9177.

Pitcher, D., Duchaine, B., & Walsh, V. (2014). Combined TMS and fMRI reveal dissociable cortical pathways for dynamic and static face perception. *Current Biology, 24*(17), 2066–2070.

Riva, P., Manfrinati, A., Sacchi, S., Pisoni, A., & Lauro, L. J. R. (2018). Selective changes in moral judgment by noninvasive brain stimulation of the medial prefrontal cortex. *Cognitive, Affective, & Behavioral Neuroscience*, 1–14.

Rochas, V., Gelmini, L., Krolak-Salmon, P., Poulet, E., Saoud, M., Brunelin, J., & Bediou, B. (2013). Disrupting pre-SMA activity impairs facial happiness recognition: An event-related TMS study. *Cerebral Cortex, 23*(7), 1517–1525.

Root, J. C., Wong, P. S., & Kinsbourne, M. (2006). Left hemisphere specialization for response to positive emotional expressions: A divided output methodology. *Emotion, 6*(3), 473.

Rossini, P. M., & Rossi, S. (2007). Transcranial magnetic stimulation: Diagnostic, therapeutic, and research potential. *Neurology, 68*(7), 484–488.

Ruff, C. C., Ugazio, G., & Fehr, E. (2013). Changing social norm compliance with noninvasive brain stimulation. *Science, 342*(6157), 482–484.

Sanfey, A. G., Civai, C., & Vavra, P. (2015). Predicting the other in cooperative interactions. *Trends in Cognitive Sciences, 19*(7), 364–365.

Sanfey, A. G. (2007). Social decision-making: insights from game theory and neuroscience. Science, 318(5850), 598–602. https://doi.org/10.1126/science.11429960.

Sanfey, A. G., Stallen, M., & Chang, L. J. (2014). Norms and expectations in social decision-making. *Trends in cognitive sciences, 18*(4), 172–174.

Sellaro, R., Derks, B., Nitsche, M. A., Hommel, B., van den Wildenberg, W. P., van Dam, K., & Colzato, L. S. (2015). Reducing prejudice through brain stimulation. *Brain Stimulation, 8*(5), 891–897.

Sellaro, R., Nitsche, M. A., & Colzato, L. S. (2017). Transcranial direct current stimulation. In *Theory-driven approaches to cognitive enhancement* (pp. 99–112). Springer.

Shamay-Tsoory, S. G., Tomer, R., Berger, B. D., & Aharon-Peretz, J. (2003). Characterization of empathy deficits following prefrontal brain damage: The role of the right ventromedial prefrontal cortex. *Journal of Cognitive Neuroscience, 15*(3), 324–337.

Shamay-Tsoory, S. G., Tomer, R., Berger, B. D., Goldsher, D., & Aharon-Peretz, J. (2005). Impaired "affective theory of mind" is associated with right ventromedial prefrontal damage. *Cognitive and Behavioral Neurology, 18*(1), 55–67.

Tassy, S., Oullier, O., Duclos, Y., Coulon, O., Mancini, J., Deruelle, C., ... Wicker, B. (2011). Disrupting the right prefrontal cortex alters moral judgement. *Social Cognitive and Affective Neuroscience, 7*(3), 282–288.

Taylor, D. M., & Doria, J. R. (1981). Self-serving and group-serving bias in attribution. *The Journal of Social Psychology, 113*(2), 201–211.

Tettamanti, M., Rognoni, E., Cafiero, R., Costa, T., Galati, D., & Perani, D. (2012). Distinct pathways of neural coupling for different basic emotions. *Neuroimage, 59*(2), 1804–1817. https://doi.org/10.1016/j.neuroimage.2011.08.018

Thomson, J. J. (1984). The trolley problem. *Yale LJ, 94*, 1395.

Van Overwalle, F., D'aes, T., & Mariën, P. (2015). Social cognition and the cerebellum: A meta-analytic connectivity analysis. *Human Brain Mapping, 36*(12), 5137–5154.

Wager, T. D., Davidson, M. L., Hughes, B. L., Lindquist, M. A., & Ochsner, K. N. (2008). Prefrontal-subcortical pathways mediating successful emotion regulation. *Neuron, 59*(6), 1037–1050.

Wang, G., Li, J., Yin, X., Li, S., & Wei, M. (2016). Modulating activity in the orbitofrontal cortex changes trustees' cooperation: A transcranial direct current stimulation study. *Behavioural Brain Research, 303*, 71–75.

Wexler, A. (2017). Recurrent themes in the history of the home use of electrical stimulation: Transcranial direct current stimulation (tDCS) and the medical battery (1870–1920). *Brain Stimulation, 10*(2), 187–195.

Willis, M. L., Murphy, J. M., Ridley, N. J., & Vercammen, A. (2015). Anodal tDCS targeting the right orbitofrontal cortex enhances facial expression recognition. *Social Cognitive and Affective Neuroscience, 10*(12), 1677–1683.

Woods, A. J., Antal, A., Bikson, M., Boggio, P. S., Brunoni, A. R., Celnik, P., … Nitsche, M. A. (2016). A technical guide to tDCS, and related non-invasive brain stimulation tools. *Clinical Neurophysiology, 127*(2), 1031–1048.

Yuan, H., Tabarak, S., Su, W., Liu, Y., Yu, J., & Lei, X. (2017). Transcranial direct current stimulation of the medial prefrontal cortex affects judgments of moral violations. *Frontiers in Psychology, 8*, 1812.

Zheng, H., Lu, X., & Huang, D. (2018). tDCS over DLPFC leads to less utilitarian response in moral-personal judgment. *Frontiers in Neuroscience, 12*, 193.

Chapter 16
What Our Eyes Can Tell Us About Our Social and Affective Brain?

Paulo Guirro Laurence, Katerina Lukasova, Marcus Vinicius C. Alves (ID)**, and Elizeu Coutinho de Macedo**

> *The only true voyage of discovery, (… would be) to possess*
> *other eyes, to behold the universe through the eyes of another,*
> *of a hundred others, to behold the hundred universes that each*
> *of them beholds, that each of them is. Marcel Proust,*
> *Remembrance of Things Past (or In Search of Lost Time)*

Abstract The eyes are windows to the soul. This phrase present in the common sense popularly expresses that it is possible to deeply understand people's minds just by how their eyes behave. This assumption is not that far from reality. Analyzing the eyes of subjects, researchers have answered questions of how people think, remember, pay attention, recognize each other, and many other theoretical and empirical ones. Recently, with the advancement of research in social and affective neuroscience, researchers are starting to look at human interactions and how the individuals' eyes can relate to their behaviors and cognitive functions in social contexts. To measure individuals' gaze, a machinery specialized in recording eye movements and pupillary diameter changes is used: a device known as an eye tracker.

Keywords Eye tracking · Pupillometry · Cognitive ethology

P. G. Laurence · E. C. de Macedo (✉)
Social and Cognitive Neuroscience Laboratory and Developmental Disorders Program,
Center for Health and Biological Sciences, Mackenzie Presbyterian University,
São Paulo, Brazil

K. Lukasova
Postgraduate Program in Neuroscience and Cognition – PPGNC, Federal University
of ABC – UFABC, São Bernardo, Brazil

M. V. C. Alves
Faculty of Health Sciences of Trairi, Universidade Federal do Rio Grande do Norte,
Santa Cruz, Brazil

271

Introduction

The eyes are windows to the soul. This phrase present in the common sense popularly expresses that it is possible to deeply understand people's minds just by how their eyes behave. This assumption is not that far from reality. Analyzing the eyes of subjects, researchers have answered questions of how people think, remember, pay attention, recognize each other, and many other theoretical and empirical ones. Recently, with the advancement of research in social and affective neuroscience, researchers are starting to look at human interactions and how the individuals' eyes can relate to their behaviors and cognitive functions in social contexts. To measure individuals' gaze, a machinery specialized in recording eye movements and pupillary diameter changes is used: a device known as an eye tracker.

Eye tracking as a research tool is more accessible than ever, and since it allows different inferences about mental functioning at a less expense of researcher grants, its popularity has grown exponentially in psychology and cognitive neuroscience laboratories. The eye-tracking device is a nonintrusive machine that normally emits infrared/near-infrared light to create a reflection in the cornea of the subject. This cornea reflection corresponds to the first Purkinje image (P1) obtained from the reflection of eye structures and is commonly known as a "glint." This reflection and the center of the pupil are used to track eye movements. The corneal reflection is captured by a camera in the eye-tracking device, and it is possible to calculate a vector formed by the angle between the corneal reflection and the pupil. Those features enable the software to calculate the gaze direction. For the software to be fully capable of capturing eye movements, a calibration procedure is required, consisting of a presentation of dots on the screen which the subject should normally follow while the device registers the position of the eye (with the reflection) in order calculate references of where the person is looking (Hansen & Ji, 2010).

The eye-tracking equipment is able to record some helpful measures. Eye-tracking measurements can be divided into four large groups, being (1) movement measures (how the eyes move through space and the properties of these movements), (2) position measures (dealing with where a participant has or has not been looking and its properties), (3) numerosity measures (proportion or rate of any countable eye movement event), and (4) latency measures (how long these events take to start and finish). Thus, depending on the question asked by the researcher, it is possible to answer with several possible measures, for example, if the study shows different emotional faces to the participant, it is possible to verify how much and in which places the participant's eyes move around to process those faces, the time it takes to do these scanpaths, the parts of the face that the participant looks at, the number of times he checks essential points of the faces (i.e., mouth, eyes), and even the time he keeps processing any of these points. Each of these indices will be able to answer different questions and may be also integrated so that we can make inferences about underlying cognitive processes.

In relation to movements, eye-tracking devices can record saccades and fixations. Saccades are the movements that the eyes do when they are searching for stimuli in the environment, while the fixations are brief moments when the eyes stop to look at something more carefully. During fixations, the visual resolution is optimal, and the visual system receives information about retinal input that is the moment when we process information and plan the next saccade to the objects of interest. In other words, the eyes are always in movement; even when fixation takes place, the eyes perform very small jitter, but for classification purposes, the eye movements are divided into those two categories (Liversedge et al., 2011). The measure recorded by eye-tracking devices is the pupil dilation, which is calculated from pupil diameter changes during the task execution (Sirois & Brisson, 2014).

Fixations are a great way to study emotion recognition based on facial expressions. For example, when a person is visually scanning a human face in order to recognize an emotion, they fixate approximately 88% of the time on facial regions including the eyes, nasion, nose, or upper lip (see Fig. 16.1). Emotions such as fear, anger, sadness, and shame have fixations predominantly on the region of eyes, while other emotions, such as joy and disgust, draw more attention toward the upper lip. This fixation pattern is related to optimizing the visual search for cues that are important for emotion identification. For example, the deformation of the upper lip characteristic during a smile is an important feature of joy. Moreover, the lower part of the nose seems to be a key region to differentiate between emotional faces. Those

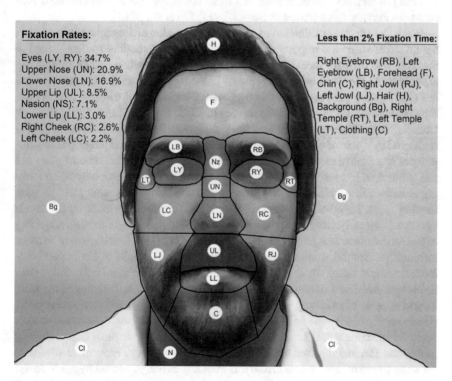

Fixation Rates:

Eyes (LY, RY): 34.7%
Upper Nose (UN): 20.9%
Lower Nose (LN): 16.9%
Upper Lip (UL): 8.5%
Nasion (NS): 7.1%
Lower Lip (LL): 3.0%
Right Cheek (RC): 2.6%
Left Cheek (LC): 2.2%

Less than 2% Fixation Time:

Right Eyebrow (RB), Left Eyebrow (LB), Forehead (F), Chin (C), Right Jowl (RJ), Left Jowl (LJ), Hair (H), Background (Bg), Right Temple (RT), Left Temple (LT), Clothing (C)

Fig. 16.1 Regions of interest in the face for emotion recognition. (This image and the regions of interest were based on the manuscript of Schurgin et al. (2014))

results suggest that there are certain diagnostic regions in the face for emotion processing (Schurgin et al., 2014).

Besides looking at the emotion expressed by the face, another important aspect is face recognition per se. One important finding has shown that people are generally better at face recognition of their own race, and this process is called own-race bias (ORB). Then trying to recognize faces of own race than faces of another race, participants had a shorter response time. Studies with eye tracking helped to understand this phenomenon. In a study with Caucasians trying to remember if they saw other Caucasian faces or Asiatic faces, it was possible to understand that a more complex scanning happened when Caucasians looked in their own-race faces. In own-race faces, they performed more saccades and more fixations. Additionally, these fixations were shorter than in other-race faces. The distance of saccades was not different when trying to recognize faces of own race or other races (Wu et al., 2012).

When trying to recognize a face, a person looks more than 70% of the time on the eyes, nose, and mouth. The participants spent more time looking in the region of eyes and forehead, while less time is spent looking at the nose when the face is an own-race face. This gaze pattern points to a different strategy of visual processing when trying to recognize faces of own race compared to other races. The visual scanning of own-race faces is done in a more automatic, quick, and effortless process than in other faces (Alves & Bueno, 2017; Wu et al., 2012).

Since the effective evaluation of the facial expression and a correct inference of the affective states are important for people's social interaction, studies looked at the strategies used for face processing. They showed that when people are looking at static faces, they tend to direct their gaze to the right side of the face, the so-called left (hemispace) gaze bias, and this preferential looking is already present in children (Gilbert & Bakan, 1973; Sackeim et al., 1978; Heller & Levy, 1981; Hisao & Cottrel, 2008; Chiang et al., 2000; Taylor et al., 2012). Balas and Moulson (2011) registered eye gaze of children 5–10 years old while looking and judging face similarity of proof and a target face. They confirmed left-side bias in children 5 years old and showed an increase for left-side preference with age, however only when looking at human faces. No effect was found when children were looking and judging monkey faces. Together with other studies, the findings indicate that over the developmental trajectory, people improve their looking strategies together with acquiring expertise in human face judgment. Indeed, looking to the left hemiface may be more informative. Several studies examined composite photographs of human and chimera faces and asked whether the left-left composites were more informative than right-right hemiface composites. In most cases, the left-left photographs were judged as more emotionally expressive (Moreno et al., 1990) and more trustworthy (Okubo et al., 2013) and had more muscle movements (Dimberg & Peterson, 2000). Nicholls and colleagues (2002) found the left-gaze bias also in faces turned slightly to the side 15°, and it raised a question whether the same eye movement strategies are to be found in faces viewed from different angles and in natural dynamic setup.

Pupillometry

Pupillometry is a measure of pupil diameter variance (i.e., pupil dilation) in the course of time. In one of the earliest studies with pupil diameters, scientists took pictures of the participants when performing tasks and then compared them with a baseline, that is, during a period when no task was done (Hess & Polt, 1960, 1964; Kahneman & Beatty, 1966). Since then, with the development of video-based eye trackers, the scientific interest in pupillometry has been growing.

Pupils' diameter changes in order to allow more light to enter the eye and reach the retina, increasing our vision in dim light conditions. However, the pupil diameter also increases in response to cognitive processing, such as performing a test or contemplating a photograph with strong emotional content. Numerous studies that used pupillometry as a complementary measure in the execution of cognitive tasks demonstrated that the magnitude of change is directly related to the tasks' cognitive demands. The change in pupil dilation related to the use of cognitive resources is minimal if compared to the change due to the change in luminosity, and while the former tends to vary by less than 1 millimeter, the latter may imply changes of up to 8 millimeters. This small, but conspicuous, difference is used to infer the way participants are allocating mental resources to perform demanding tasks. It is well known that change in pupil diameter is an effective indicator of a person's mental activity (Hess & Polt, 1964; Kahneman & Peavler, 1969). Pupillometric studies provide evidence that pupil dilation is related not only to processing emotional states but also to increasing mental effort that is undertaken on a task (Eckstein et al., 2016; Hess & Polt, 1964; Kahneman & Peavler, 1969; Mathôt, 2018; Wierda et al., 2012).

Another two indices useful to the mental effort-related hypothesis are the pupil dilation peak and the eye blink rate. The peak of dilation – arguably as reliable as pupil dilation – can be related with the peak of effort during a task, since stabilization of the dilation can happen after the beginning of tasks (Beatty & Lucero-Wagoner, 2000; Hershaw & Ettenhofer, 2018), while eye blink rate is a complementary measure that can reflect cognitive engagement, usually, before a high-demanding task begins (Siegle et al., 2008; Van Bochove et al., 2013). In view of that, more blinks represent more preparation for doing a hard task and, with pupil dilation, can be used to indicate an effortful task (Fukuda et al., 2005; Ichikawa & Ohira, 2004).

Cognitive Ethology: From the Real World to the Lab and from the Lab to Virtual Reality

A prominent research approach to eye tracking is called cognitive ethology, mostly studying everyday attention and social interactions (Kingstone, 2009; Smilek et al., 2006). The goal is to first begin one's research approach at the level of natural performance before moving it into the lab where it can be recreated, controlled, and manipulated. Cognitive ethology ends up being an alternative way of studying attentional processes when related to social interactions. By starting at the real-world level, the main focus is on what people really do in real life, and hence, one can determine what behaviors are, and are not, specific to the laboratory environment (Kingstone, 2009).

People have strong tendencies to follow gaze cues. With the help of an eye-tracking device, MacDonald and Tatler (2013) investigated whether social perceptions of a collaborator affects how people look at them and follow their gaze. Namely, they aimed to understand how social context can affect our gaze behavior during social interaction. With an experiment in which two participants worked together to perform a task (in their case, cooking), they found results showing that social context can affect gaze behavior, that is, the social context influenced the way the participants interacted with their eyes, focusing their attention depending on the action of the other. This result points out the use of eye tracking in social research and attempts to carry out an experiment in naturalistic environments to show how social attention works in natural social contexts.

Besides eye tracking, another technology that may help investigate social neuroscience is virtual reality (VR). VR is interesting to social neuroscience because it allows the creation of ecologically valid experiments that can be fully interactive and three-dimensional (Parsons et al., 2017). One study proposed to create the trolley dilemma in VR. This is a well-known series of experiments on moral decision-making on whether to sacrifice one person to save a larger number of people by making a certain action, such as diverting the incoming trolley on a sidetrack. In the VR version of the task, participants had to choose killing either ten victims or one victim. They created three conditions for the experiment environment, the first one with randomized women and men as possible victims, the second one with possible victims of different ethnicity, and the third one with a possible victim facing toward them and a possible victim facing away from them. Results from eye tracking pointed that the participant spent more gazing time on the chosen victim, which was an unexpected result since it was expected that they would avoid looking at the victim (Skulmowski et al., 2014).

In this experiment, the variation of the pupillary diameter was also verified. In relation to pupil variation, in all conditions, the pupil presented an increased diameter after the moment of decision, indicating that the participants had an increased cognitive load in the moment of decision. In the different ethnicity condition, participants presented a higher pupil dilatation, suggesting that the participants had a higher cognitive load in an extreme social decision situation due to the controversial topic (Skulmowski et al., 2014). In the previous study on the faces of different or equal races, the pupillary diameter was also recorded, indicating that a person will have a bigger pupil variation when trying to recognize other race faces. This is consistent with the gaze pattern that was already described above, indicating that a person will have more cognitive effort to recognize the face of another race, while own-race faces will be more automatic.

Eye Tracking in Clinical Populations

Another interesting way to use eye tracking is with studies in clinical populations with impaired social interaction, such as study with individuals diagnosed with schizophrenia, autism, or social anxiety disorder. Since eye tracking can demonstrate underlying cognitive patterns of a person, it can be a good tool to understand how different clinical populations understand and process different stimuli. For example, eye tracking can help us understand which part of a stimulus (e.g., faces) a person more fixates on, indicating where is the part of the face that a person applies most attention to. Thus, it is possible to infer different cognitive processes of a clinical population when comparing their eye gaze with a typical population.

In relation to schizophrenia, there have been a large number of studies with eye tracking that goes beyond the scope of this work. The findings point out to different aspects of eye movement impairments in persons with schizophrenia, one of them being an impaired smooth pursuit. Smooth pursuit happens when eyes follow a moving object. In persons with schizophrenia, the smooth pursuit lags behind the moving object, and thus a series of saccades is made to catch up the target (O'Driscoll & Callahan, 2008). It has long been known that this population presents a worse performance in anti-saccade tasks (Fukushima et al., 1988) during which the participant must avoid looking at a suddenly appearing target and is supposed to look in the opposite direction. Recently, new experiments revealed a worse performance in the fixation task (Benson et al., 2012) assessed by a study that asked the participants to visually fixate on the point ignoring a cue appearing in the peripheral area. Furthermore, on free-viewing tasks, participants with schizophrenia tend to focus their gaze on a smaller area, if compared with typical participants (Sprenger et al., 2013).

Since people with schizophrenia present a different eye movement pattern, compared with typical persons, there are some discussions regarding the use of the eye gaze as a biomarker for schizophrenia. This is possible because eye movements are underlaid by different neurological mechanisms that can be altered in persons with schizophrenia. In this regard, Morita et al. (2020) made a review describing the findings in this area. The results suggest that eye movements can be used to discriminate between persons with schizophrenia and typical subjects at a rate of ~75–90% (Morita et al., 2020).

The eye movements of persons with autism spectrum disorder (ASD) also seem to be different from typical persons. These regions may not be identified by persons with some developmental type of disorders. One meta-analysis reviewed studies on face processing and showed that children with ASD have significantly reduced the number of fixations in the region of the eyes. Furthermore, diminished attention on eyes negatively impacts social interaction because not looking at social cues may lead to worse interaction and emotion recognition (Papagiannopoulou et al., 2014). The same meta-analysis demonstrated that there were no significant differences in mouth region fixations for children with or without ASD. Another meta-analysis of 38 studies revealed that individuals with ASD present reduced social attention if compared with typical individuals and that the social attention in persons with ASD is influenced by social contents (Chita-Tegmark, 2016). However, a comparison of the gaze pattern in different regions of interest, this time in a meta-analysis involving 122 studies, found differences of small and medium magnitudes (Frazier et al., 2017). In special, participants with ASD presented a higher difficulty in selecting socially relevant or nonrelevant stimuli. The biggest difference was again found in the eyes and whole face regions of interest (Frazier et al., 2017).

Very promising results come from studies on social attention in toddlers (18–35 months old) with and without ASD. Specific signs of ASD may be indicated by subtle variation in the way the child follows another person's look to the target of interest, an ability called joint attention. There are two principal kinds of joint attention: the *response joint attention* that requires to spot the change of the other person's look and follow it to the new destiny and the *initiation joint attention* that requires the child to look at a moving object and by her/his own gaze indicate this fact to another person. While at 24 months of age the eye-tracking pattern, especially in initiating joining attention, was different in ASD toddlers compared to typically developing children, by 6 months later, this difference disappeared. Due to the natural maturation, the ASD improved their ability to disengage from the face stimuli and explore the global aspect of the scene approaching their eye moving pattern to the performance of typically developing children (Muratori et al., 2019).

Lastly, persons with social anxiety disorder (SAD) also present peculiarities in their eye gaze. A meta-analysis containing 13 studies demonstrated that participants with SAD presented a hypervigilance-avoidance effect in their eye gaze when looking into faces, compared to typical participants (Claudino et al., 2019). This eye gaze effect can be understood by a big number of fixations in the face at the first moment and then less fixation in the stimulus at a second moment. Claudino et al. (2019) also found that this effect was more prominent in faces presenting negative emotions, such as anger.

Conclusion

The measurement of eye movement and pupil dilation is a valid undertaking for studies in cognitive, social, and affective neurosciences. With this technique, it is possible to carry out an ecological evaluation, which is cheaper and answers several important experimental questions. Using typically developing or clinical populations of different age groups and even allowing constant social interactions during the experiment, the device allows a series of inferences on cognitive processing based on objective, simple, and noninvasive physiological measures. The use of eye tracking by different behavioral disciplines depends more on the limit of what the researcher is willing to investigate than on the technique per se. To sum up, the possibilities of research questions that can be answered by participants' eyes go much further than expected or, rather, go beyond what the eyes can see.

References

Alves, M. V. C., & Bueno, O. F. A. (2017). Retroactive interference: Forgetting as an interruption of memory consolidation. *Trends Psychology, 25*(3), 1043–1054.

Andreassi, J. L. (2000). *Pupillary response and behavior. Em: Psychophysiology: Human behavior & physiological response* (pp. 289–307). Lawrence Erlbaum Assoc.

Balas, B., & Moulson, M. C. (2011). Developing a side bias for conspecific faces during childhood. *Developmental Psychology, 47*(5), 1472–1478. https://doi.org/10.1037/a0024494

Beatty, J., & Kahneman, D. (1966). Pupillary changes in two memory tasks. *Psychonomic Science, 5*(10), 371–372. https://doi.org/10.3758/BF03328444

Beatty, J., & Lucero-Wagoner, B. (2000) The Pupillary System. In: John T. Cacioppo, Louis G. Tassinary, & Gary G. Bernston (Eds.), Handbook of Psychophysiology (2 ed.), USA: Cambridge University Press, p. 142–161.

Benson, P. J., Beedie, S. A., Shephard, E., Giegling, I., Rujescu, D., & Clair, D. S. (2012). Simple viewing tests can detect eye movement abnormalities that distinguish schizophrenia cases from controls with exceptional accuracy. *Biological Psychiatry, 72*(9), 716–724.

Chiang, C. H., Ballantyne, A. O., & Trauner, D. A. (2000). Development of perceptual asymmetry for free viewing of chimeric stimuli. *Brain and Cognition, 44*(3), 415–424. https://doi.org/10.1006/brcg.1999.1202

Chita-Tegmark, M. (2016). Social attention in ASD: A review and meta-analysis of eye-tracking studies. *Research in Developmental Disabilities, 48*, 79–93.

Claudino, R. G., Lima, L. K. S. D., Assis, E. D. B. D., & Torro, N. (2019). Facial expressions and eye tracking in individuals with social anxiety disorder: A systematic review. *Psicologia: Reflexão e Crítica, 32*.

Dimberg, U., & Petterson, M. (2000). Facial reactions to happy and angry facial expressions: Evidence for right hemisphere dominance. *Psychophysiology, 37*(5), 693–696.

Eckstein, M. K., Guerra-Carrillo, B., Singley, A. T. M., & Bunge, S. A. (2016). Beyond eye gaze: What else can eyetracking reveal about cognition and cognitive development? *Developmental Cognitive Neuroscience*.

Frazier, T. W., Strauss, M., Klingemier, E. W., Zetzer, E. E., Hardan, A. Y., Eng, C., & Youngstrom, E. A. (2017). A meta-analysis of gaze differences to social and nonsocial information between individuals with and without autism. *Journal of the American Academy of Child & Adolescent Psychiatry, 56*(7), 546–555.

Fukuda, K., Stern, J. A., Brown, T. B., & Russo, M. B. (2005). Cognition, blinks, eye-movements, and pupillary movements during performance of a running memory task. *Aviation, Space, and Environmental Medicine 76*(7 Suppl):C75–85.

Fukushima, J., Fukushima, K., Chiba, T., Tanaka, S., Yamashita, I., & Kato, M. (1988). Disturbances of voluntary control of saccadic eye movements in schizophrenic patients. *Biological Psychiatry, 23*(7), 670–677.

Gilbert, C., & Bakan, P. (1973). Visual asymmetry in perception of faces. *Neuropsychologia, 11*(3), 355–362.

Hansen, D. W., & Ji, Q. (2010). In the eye of the beholder: A survey of models for eyes and gaze. *IEEE Transactions on Pattern Analysis and Machine Intelligence, 32*(3), 478–500. https://doi.org/10.1109/TPAMI.2009.30

Heller, W., & Levy, J. (1981). Perception and expression of emotion in right-handers and left-handers. *Neuropsychologia, 19*(2), 263–272.

Hershaw, J. N., & Ettenhofer, M. L. (2018). Insights into cognitive pupillometry: Evaluation of the utility of pupillary metrics for assessing cognitive load in normative and clinical samples. *International Journal of Psychophysiology, 134*, 62–78. https://doi.org/10.1016/j.ijpsycho.2018.10.008.

Hess, E. H., & Polt, J. M. (1960). Pupil size as related to interest value of visual stimuli. *Science, 132*, 349–350. https://doi.org/10.1126/science.132.3423

Hess, E. H., & Polt, J. M. (1964). Pupil size in relation to mental activity during simple problem-solving. *Science, 143*, 1190–1192. https://doi.org/10.1126/science.143.3611.1190

Hisao, J. H. W., & Cottrell, G. (2008). Two fixations suffice in face recognition. *Psychological Science, 19*(10), 998–1006.

Ichikawa, N., & Ohira, H. (2004). Eyeblink activity as an index of cognitive processing: Temporal distribution of eyeblinks as an indicator of expectancy in semantic priming. *Perceptual and Motor Skills, 98*(1), 131–140.

Kahneman, D., & Peavler, W. S. (1969). Incentive effects and pupillary changes in association learning. *Journal of Experimental Psychology, 7*(2), 312–318.

Kingstone, A. (2009). Taking a real look at social attention. *Current Opinion in Neurobiology, 19*, 52–56.

Liversedge, S. P., Gilchrist, I. D., & Everling, S. (2011). *The Oxford handbook of eye movements.* Oxford University Press.

Mathôt, S. (2018). Pupillometry: Psychology, Physiology, and Function. *Journal of Cognition, 1*(1), 1–23. https://doi.org/10.5334/joc.18.

Macdonald, R. G., & Tatler, B. W. (2013). Do as eye say: Gaze cueing and language in a real-world social interaction. *Journal of Vision, 13*(4), 6.

Moreno, C. R., Borod, J., Welkowitz, J., & Alpert, M. (1990). Lateralization for the perception and expression of facial expression as a function of age. *Neuropsychologia, 28*, 199–209. https://doi.org/10.1016/0028-3932(90)90101-S

Morita, K., Miura, K., Kasai, K., & Hashimoto, R. (2020). Eye movement characteristics in schizophrenia: A recent update with clinical implications. *Neuropsychopharmacology Reports, 40*, 2–9.

Muratori, F., Billeci, L., Calderoni, S., Boncoddo, M., Lattarulo, C., Costanzo, V., Turi, M., Colombi, C., & Narzisi, A. (2019). How attention to faces and objects changes over time in toddlers with autism spectrum disorders: Preliminary evidence from an eye tracking. *Study Brain Science, 9*, 344. https://doi.org/10.3390/brainsci9120344

O'Driscoll, G. A., & Callahan, B. L. (2008). Smooth pursuit in schizophrenia: a meta- analytic review of research since 1993. *Brain and Cognition, 68*(3), 359–370.

Okubo, M., Ishikawa, K., & Kobayashi, A. (2013). No trust on the left side: Hemifacial asymmetries of trustworthiness an emotional expressions. *Brain and Cognition, 82*, 181–186. https://doi.org/10.1016/j.bandc.2013.04.004

Papagiannopoulou, E. A., Chitty, K. M., Hermens, D. F., Hickie, I. B., & Lagopoulos, J. (2014). A systematic review and meta-analysis of eye-tracking studies in children with autism spectrum disorders. *Social Neuroscience, 9*(6), 610–632. https://doi.org/10.1080/17470919.2014.934966

Parsons, T. D., Gaggioli, A., & Riva, G. (2017). Virtual Reality for Research in Social Neuroscience. *Brain Sciences, 7*(4), 42. https://doi.org/10.3390/brainsci7040042

Sackeim, H. A., Gur, R. C., & Saucy, M. C. (1978). Emotions are expressed more intensely on the left side of the face. *Science, 202*, 434–436.

Schurgin, M. W., Nelson, J., Iida, S., Ohira, H., Chiao, J. Y., & Franconeri, S. L. (2014). Eye movements during emotion recognition in faces. *Journal of Vision, 14*(13), 1–16. https://doi.org/10.1167/14.13.14

Siegle, G. J., Ichikawa, N., & Steinhauer S. (2008). Blink before and after you think: Blinks occur prior to and following cognitive load indexed by pupillary responses. *Psychophysiology, 45*, 679–687.

Sirois, S., & Brisson, J. (2014). Pupillometry. *Wiley Interdisciplinary Reviews: Cognitive Science, 5*(6), 679–692. https://doi.org/10.1002/wcs.1323

Skulmoski, A., Bunge, A., Kaspar, K., & Pipa, G. (2014). Forced-choice decision- making in modified trolley dilemma situations: A virtual reality and eye tracking study. *Frontiers in Behavioral Neuroscience, 8*. https://doi.org/10.3389/fnbeh.2014.00426

Smilek, D., Birmingham, E., Cameron, D., Bischof, W., & Kingstone, A. (2006). Cognitive Ethology and exploring attention in real-world scenes. *Brain Research, 1080*(1), 101–119. https://doi.org/10.1016/j.brainres.2005.12.090

Sprenger, A., Friedrich, M., Nagel, M., Schmidt, C. S., Moritz, S., & Lencer, R. (2013). Advanced analysis of free visual exploration patterns in schizophrenia. *Frontiers in Psychology, 4*, 737.

Taylor, S., Workman, L., & Yeomans, H. (2012). Abnormal patterns of cerebral lateralization as revealed by the universal chimeric faces task in individuals with autistic disorder. *Laterality, 17*, 428–437. https://doi.org/10.1080/1357650X.2010.521751

van Bochove, M. E., Van der Haegen, L., Notebaert, W., & Verguts, T. (2013). Blinking predicts enhanced cognitive control. *Cognitive, Affective, & Behavioral Neuroscience, 13*(2), 346–354.

Wierda, S. M., van Rijin, H., Taatgen, N. A., & Martens, S. (2012) Pupil dilation deconvolution reveals the dynamics of attention at high temporal resolution. *PNAS, 109*(22), 8456–8460.

Wu, E. X., Laeng, B., & Magnussen, S. (2012). Through the eyes of the own-race bias: Eye-tracking and pupillometry during face recognition. *Social Neuroscience, 7*(2), 202–216. https://doi.org/10.1080/17470919.2011.596946

Chapter 17
Facial EMG – Investigating the Interplay of Facial Muscles and Emotions

Tanja S. H. Wingenbach (iD)

Abstract This chapter provides information about facial electromyography (EMG) as a method of investigating emotions and affect, including examples of application and methods for analysis. This chapter begins with a short introduction to emotion theory followed by an operationalisation of facial emotional expressions as an underlying requirement for their study using facial EMG. This chapter ends by providing practical information on the use of facial EMG.

Keywords Electromyography · Facial EMG · Facial emotional expressions · Facial muscles

Introduction

This chapter provides information about facial electromyography (EMG) as a method of investigating emotions and affect, including examples of application and methods for analysis. This chapter begins with a short introduction to emotion theory followed by an operationalisation of facial emotional expressions as an underlying requirement for their study using facial EMG. This chapter ends by providing practical information on the use of facial EMG.

Theory: From Emotional States to Their Expression

Darwin (1872/1965) studied emotions and their expression across species and argued that emotion phenomena were the products of natural selection. According to this evolutionary perspective, emotions constitute an interrelated suite of

T. S. H. Wingenbach (✉)
School of Human Sciences, Faculty of Education, Health, and Human Sciences,
University of Greenwich, Greenwich, London, UK
e-mail: tanja.wingenbach@bath.edu

© The Author(s) 2023

P. S. Boggio et al. (eds.), *Social and Affective Neuroscience of Everyday Human Interaction*, https://doi.org/10.1007/978-3-031-08651-9_17

physiological and behavioural systems that have guided adaptive action over evolutionary time. According to Tomkins (1962), specific response patterns related to emotion experience are elicited automatically by certain events. For example, a threat or danger in the perceived environment should elicit fear. Emotion responses are characterised by coordinated patterns of activity that can include physiological changes, signalling behaviours in the voice and face, subjective experience, and relevant action. For example, a fearful response includes changes in brain activity in the amygdala (a region in the brain associated with emotion processing in general but specifically with fear) (Janak & Tye, 2015). Changes in physiology during a fearful episode can manifest as an associated facial expression (i.e. wide opened eyes, eyebrows pulled upwards and drawn together, and the corners of the mouth pulled outwards), the face turning pale, sweating, and a vocal expression (e.g. fear scream). Fear can direct our attention to the dangerous situation and facilitate adaptive action, such as fleeing. Emotions allow us to navigate life's challenges, and each emotion is governed by its own adaptive logic.

Several theories consider emotions as distinct entities and as biologically innate (e.g. Ekman et al., 1982; Izard, 1977; Plutchik, 1980; Tomkins, 1984). A very prominent theory is the 'basic emotion theory' (Ekman, 1992a, b), according to which some emotions are considered universal, meaning they occur in humans across all cultures. Most theorists agree on at least six basic emotion categories: anger, disgust, fear, sadness, surprise, and happiness (Ortony & Turner, 1990). According to Tomkins (1962), each emotion has its unique affect programme such as the example outlined above in the case of fear. Ultimately, these categories map onto distinct patterns of activity shaped by evolutionary processes to solve different kinds of adaptive problems faced by our highly social hominin ancestors. Research has provided evidence for distinct patterns in physiology on the basis of heart rate, temperature, and electrodermal activity for the six basic emotions, and these varying physiological patterns can be linked to functions of emotions on a behavioural level (as proposed by Darwin). In a state of anger, a preparation for fighting occurs by increasing the blood flow to the hands (Levenson et al., 1990). In a state of fear, the blood flow to large skeletal muscles increases which prepares for a flight reaction (Levenson et al., 1990). A state of disgust will lead to a rejection of the eliciting stimulus by restricting airflow to olfactory receptors and triggering a gag reflex (Koerner & Antony, 2010). A state of sadness results in a loss of muscle tone (Oberman et al., 2007), slowing us down, allowing us to focus on the issue that induced the sadness (Wolpert, 2008). A state of happiness leads to an increase in the energy available to the organism by releasing respective transmitters (Uvnäs-Moberg, 1998). A surprised state results in air being quickly inhaled which increases the ability to react fast (Ekman & Friesen, 1975), as it interrupts ongoing processes (Tomkins, 1962). Even participants' subjective understanding (i.e. conceptualisation) of emotion reflects distinct patterns for each of the six basic emotions. When asking participants to colour in the body parts they perceive to be affected by either an increase or decrease in sensations when being in a state of each of the six basic emotions, the obtained results were in line with associated physiological changes as outlined above (see Nummenmaa et al., 2014). Neuroscientific research has shown that the distinctiveness of emotions is also evident in brain activity patterns. Vytal and Hamann (2010) conducted a neuroimaging meta-analysis and found distinct patterns of neural correlates for anger, disgust, fear, happiness, and sadness. The evidence presented here supports the assumption that there are distinct response patterns of emotions at least for the basic emotions.

One alternative view is that emotions can be characterised as the integration of at least two fundamental dimensions: *valence* and *arousal* (Russell, 1980). Russell (1994) views the dimensions of valence and arousal as universal to emotions but questions the universality of distinct emotion categories. The valence dimension spans from negative (i.e. unpleasant) to positive (i.e. pleasant). The arousal dimension ranges from low (i.e. deactivated) to high (i.e. activated). Any affective state can be represented as a combination of these two dimensions. Multidimensional scaling thus reveals similarities and dissimilarities between affective states. For example, sadness is an emotion considered as negative in valence and low in arousal, whereas anger is considered also as negative in valence but high in arousal. As such, the dimensional conceptualisation of affect and the categorisation of emotions are not mutually exclusive and can actually complement each other (see Harmon-Jones et al., 2017). However, it should be noted that not all affective states are emotions, while emotions always are affective states. For example, the longer-lasting affective states are called 'moods', and emotions are rather short-lasting, while other affective states overlap with cognitive states, e.g. confusion and boredom.

Facial Emotional Expressions

The changes occurring throughout the body in an emotional state such as a face turning pale in a state of fear are visible to an observer and provide information about the affective state. Moreover, some physiological changes during the experience of emotion result in movement. For example, the activation of facial muscles leads to facial movement manifesting as facial expressions. Unlike skeletal muscles in the human body that are generally attached to bones, facial muscles also attach to each other or to the skin of the face. This anatomical set-up allows even slight contractions of facial muscles to pull the facial skin and create a facial expression visible to others. The general number of facial muscles in humans is 43, although this number can vary between people (Waller et al., 2008). This large concentration of muscles in a narrowly defined space (i.e. the face) allows for the execution of many different facial movements and results in various expressions. The *Facial Action Coding System* (*FACS*; Ekman & Friesen, 1978; new edition: Ekman et al., 2002) is an anatomical catalogue describing all movement-related facial actions (i.e. action units (AUs)) possible in humans. As a result, *FACS* has become a widely used tool in facial emotion research.

For emotional facial expressions to send an interpretable signal and serve as a means of communication, the emotion needs to be expressed in a certain way for it to be clearly attributable to a specific emotion. Ekman et al. (2002) provided suggestions for AU combinations that align with basic emotional expressions. For example, the activations of AU 9 (nose wrinkle), AU 10 (upper lip raise), and AU25 (lips parted) together result in a facial expression displaying disgust. Since facial actions as outlined by the AUs are the result of facial muscle activations, facial muscles can be linked to specific AUs. Sticking with the example of disgust, the activation of the levator labii muscle leads to a wrinkling of the nose and a raised upper lip. The connection between facial action and muscles also provides the association with specific emotions. Table 17.1 shows the six basic emotions with associated AUs and facial muscles. The facial expressions resulting from AU activations per emotion category are considered prototypical and align with the universality assumption of basic emotions as proposed by Ekman (Ekman & Friesen, 1971). When participants are shown images (or videos) displaying these prototypes, attributions of the respective emotion label are generally high. Most facial emotion recognition research utilises prototypes of basic facial emotional expressions, and many stimulus sets including these prototypes have been developed for these purposes (e.g. Ekman & Friesen, 1976; Krumhuber et al., 2013; Matsumoto & Ekman, 1988; Tottenham et al., 2009; Van Der Schalk et al., 2011; Wingenbach et al., 2016; Young et al., 2002).

As mentioned above, there are inter-individual differences in humans regarding their number of facial muscles. This variability raises the question of how

Table 17.1 Basic emotions with associated AUs and facial muscles

Emotion	AU	Description	Facial muscle
Anger	AU 4	Eyebrow lowered	M. corrugator supercilii
	AU 5	Upper lid raised	M. levator palpebrae superioris
	AU 7	Eyelid tightened	M. orbicularis oculi
	AU 23	Lips tightened	M. orbicularis oris
Disgust	AU 9	Nose wrinkled	M. levator labii
	AU 10	Upper lip raised	M. levator labii
	AU 25	Lips parted	M. depressor labii
Sadness	AU 1	Inner eyebrow raised	M. frontalis medialis
	AU 4	Eyebrow lowered	M. corrugator supercilii
	AU 15	Lip corners depressed	M. depressor anguli oris
Fear	AU 1	Inner eyebrow raised	M. frontalis medialis
	AU 2	Outer eyebrow raised	M. frontalis lateralis
	AU 4	Eyebrow lowered	M. corrugator supercilii
	AU 5	Upper lid raised	M. levator palpebrae superioris
	AU 20	Lips stretched	M. risorius
Surprise	AU 1	Inner eyebrow raised	M. frontalis medialis
	AU 2	Outer eyebrow raised	M. frontalis lateralis
	AU 5	Upper lid raised	M. levator palpebrae superioris
	AU 26	Jaw dropped	M. masseter
Happiness	AU 6	Cheeks raised	M. orbicularis oculi
	AU 12	Lip corners pulled up	M. zygomaticus major

prototypical displays of facial emotion are possible. Would it not require a standard set of facial muscles to produce expressions specific to basic emotions (as presented in Table 17.1)? To address this question, Waller et al. (2008) investigated whether the facial muscles underlying facial movements associated with facial emotional expressions of basic emotions are affected by inter-individual variability. These researchers dissected recent human cadavers and documented whether specific facial muscles were absent or present and whether this was the case for both sides of the face. The facial muscles investigated were the frontalis, orbicularis oculi, zygomaticus major, depressor anguli oris, orbicularis oris, procerus, corrugator supercilii, zygomaticus minor, buccinator, mentalis, depressor labii inferioris, risorius, levator labii superioris, levator labii superioris alaeque nasi, nasalis, and depressor septi. The first five facial muscles of this list were considered essential for the production of facial emotional expressions associated with the expression of basic emotions by Waller et al. (2008). Their results showed that the facial muscles assumed to be necessary to produce basic facial emotional expressions were present, mostly bilaterally, in all of the dissected cadavers. In addition, muscles commonly associated with the expression of basic emotions (as outlined in Table 17.1) were, although not always bilaterally, present in all cadavers, i.e. the corrugator, mentalis, depressor labii inferioris, and both levator labii muscles. The other facial muscles investigated were not present in all cadavers, and many were only present unilaterally. These findings support the universality assumption of basic emotions, at least in terms of facial expressions.

Investigating Facial Emotional Expressions Using Facial EMG

In some instances, participants' facial expressions are video-recorded while they are undergoing an experiment, and the recorded facial expressions are subjected to analyses. The *FACS* (Ekman & Friesen, 1978) can be used to code the presence of specific facial AUs, and a combination of certain facial AUs can be indicative of the presence of a specific facial emotion. For example, the co-presence of the AU6 (raising the cheek) and AU12 (pulling lip corners outwards) would indicate the presence of a facial expression of happiness. Applying this method requires *FACS* training and is subject to inter-individual perceptual differences. For these reasons, automated facial action coding software has been developed based on *FACS* (e.g. FaceReader). When using the FaceReader software, video recordings of faces can be imported, and the software output provides coded AUs as well as timings for the six basic emotions, valence, and arousal values. However, good video quality and clearly visible faces are necessary for automatic detection of AUs/emotions, and thus, trained human decoders can outperform the software.

Whether AUs are coded by humans or by software, visible movements are required for an AU to be coded. An alternative method for investigating facial emotional expressions is using facial electromyography (EMG). A great advantage of facial EMG is that it is a highly sensitive method able to ascertain the slightest contractions in facial muscles. Since fatty tissue and skin are covering the muscles in the face, very slight muscle contractions are not necessarily visible to the naked eye but do occur nonetheless during the processing of emotion-related stimuli or the presence of emotion. It should be noted that emotional states are not always expressed, as the expression thereof often has communicative or signalling function (Fridlund, 1994) that does not apply to all emotion-inducing situations. However, facial muscle contractions non-visible to observers are measurable using facial EMG (Cacioppo et al., 1986). Consequently, facial EMG can also detect facial muscle activity congruent with the affective state even when participants are instructed to suppress their emotional expression (Cacioppo et al., 1992).

So, how does facial EMG work? Whenever muscles are contracted, electricity is generated through the combined action potentials of an active motor with the measurement unit being either millivolt (mV) or microvolt (μV). These action potentials are the result of depolarisation and repolarisation at the muscle fibre membrane. When a motor nerve is excited, transmitters are released in the motor endplates, and a potential is formed in the muscle fibre (Nazmi et al., 2016). Even during a resting state when muscles are not contracted, a muscle tonus is present which can be measured with EMG. The presence of this muscle tonus is the reason why baseline measures often need to be taken, i.e. to be able to evaluate the reaction to a stimulus relative to the baseline activity; the fast nature of facial expressions makes using a prestimulus baseline necessary. Two detecting electrodes are needed to assess the electricity in one muscle, one negative electrode (VIN-) and one positive electrode (VIN+). An additional electrode is used as a reference point, i.e. ground electrode. There are two different kinds of electrodes for EMG. Needle electrodes are more commonly used within medical settings, and surface electrodes (which are non-invasive) are generally used in psychological studies. This is because surface electrodes do not require medical training and do not risk infection and discomfort. It should be noted though that surface electrodes are not necessarily muscle-specific, as they can pick up muscle activity from a greater area than the confined area around the needle insertion point. Thus, it is advised to speak of facial muscle *sites* instead of specific muscles when measuring facial EMG. Guidelines on using facial EMG were published by Fridlund and Cacioppo (1986), which are still considered the gold standard today.

Investigating Affect and Emotion Using Facial EMG

Affective states are associated with physiological responses across the body as described earlier, so one obvious use of facial EMG within emotion research is to investigate the presence of these affective states. Physiological measures such as electrocardiogram and galvanic skin response have long been applied when examining affect or specifically affective arousal (Alexander & Adlerstein, 1958; Block, 1957; Dimascio et al., 1957; Goldstein et al., 1965; Kaiser & Roessler, 1970; Oken, 1962; Vogel et al., 1958). Whereas most physiological measures are useful tools to measure affective arousal, they do not allow one to easily identify the valence of the experienced affective state. But in the 1970s, researchers started to use facial EMG and demonstrated its usefulness for differentiating affective states based on valence. For example, Schwartz et al. (1976) instructed participants to imagine happy, sad, and angry situations. The researchers distinguished between sad and happy states based on measurements from the corrugator and zygomaticus muscle sites. Cacioppo et al. (1986) demonstrated that based on measurements of the corrugator and zygomaticus facial muscle sites, mildly and moderately experienced affect can be differentiated according to its valence and also intensity. It should be noted that the resulting facial muscle activity in Cacioppo et al. (1986) was mainly covert (i.e. not visible), again highlighting the sensitivity of facial EMG. Such research findings underpin the association between the corrugator muscle site activity and negative affect (i.e. frowning) and the zygomaticus site activity with positive affect (i.e. smiling).

Published research thus far has most often investigated the facial muscle sites of corrugator and zygomaticus despite there being at least five muscles that are considered essential for the facial expression of basic emotions (see Waller et al., 2008). A reason for the preference of investigating the corrugator and zygomaticus facial muscle sites could be that a rudimentary differentiation of stimuli as either positive or negative is considered the first occurring process when faced with affective stimuli (Zajonc, 1980) and allows for investigation including a variety of stimuli of positive or negative valence without having to categorise the stimuli in distinct emotion categories. The categorisation and interpretation of an affective stimulus in specific emotion categories is often difficult. For example, a visual stimulus such as a static picture or a movie scene is often complex and can elicit a range of emotions. For instance, a scene of a bully physically attacking a person (from the film *My Bodyguard*) can elicit disgust and contempt for the bully and anger (and/or sadness) about the situation (see Gross & Levenson, 1995). The general responsiveness of the corrugator and zygomaticus muscle sites to negative and positive valence stimuli, respectively, overcomes this difficulty and makes them the standard choice within facial EMG research related to affect and emotion.

Another potential reason for not generally including multiple facial muscle sites in facial EMG research can be the issue of 'crosstalk'. That is, when neighbouring facial muscle sites are investigated, electrode pairs are necessarily placed close to one another. It is possible that an electrode pair of a non-activated muscle site records some of the activity from an adjacent activated muscle site, thus confounding results (Farina et al., 2004). Challenges like this might constitute one reason researchers generally measure fewer facial muscle sites that are not in close proximity. Corrugator and zygomaticus facial muscle sites are of sufficient distance from one another to not create crosstalk but also do not tend to activate simultaneously. Technological advances, however, have led to the recent development of smaller electrodes (i.e. with an outer diameter of <1 cm) which when placed carefully can potentially increase the number of electrode pairs used while still minimising possible crosstalk.

Corrugator and zygomaticus muscle sites are standard in facial EMG research, but there are many studies that included more facial muscles sites. For example, Vrana (1993) investigated multiple facial muscle sites to discriminate varying emotion experiences based on facial EMG. This researcher employed an imagery technique to have participants experience disgust, anger, pleasure, and joy while facial muscle activity was measured from the levator labii, corrugator, and zygomaticus sites. Results showed (1) higher activity in the levator site during disgust imagery than during anger imagery, (2) greater corrugator site activity during disgust and anger imagery compared to pleasure and joy imagery, and (3) increased zygomaticus site activity during joy imagery compared to anger, disgust, and pleasure imagery. This approach of comparing various emotion categories to each other based on the facial EMG activity at one muscle site is very common in facial EMG research. The approach is based on the assumption that specific facial action activation is indicative of a specific emotion such as a wrinkled nose resulting from the levator labii activation during the expression of disgust. However, facial emotional expressions generally include more than one facial feature activation, and some emotion categories share facial features. For example, corrugator activation is associated with facial expressions of anger, sadness, and fear (see Table 17.1) based on the overlapping facial feature of eyebrows pulled together. Such overlaps can make it difficult to draw precise conclusions about specific emotions based on individual muscle sites.

An alternative to investigating one facial muscle site per emotion category is to examine co-activations across several facial muscle sites for each emotion category. According to basic emotion theory, patterns of facial muscle activity should distinguish well between emotion categories. Fridlund et al. (1984) instructed participants to imagine situations related to feeling happiness, fear, anger, and sadness but also to pose the respective expressions while facial muscle activity was measured using EMG from the zygomaticus, corrugator, orbicularis oris, and orbicularis oculi sites. Their results showed that these emotion categories were differentiated from each other in valence based on facial EMG patterns across muscles for some, but not all, participants. But multiple emotions can be experienced during imagery, and there is significant inter-individual variability in displaying posed emotional expressions, both of which pose important limitations for this methodological approach.

Studies presented thus far involve participants imagining emotional situations and measuring aspects of their resultant emotional experience. However, facial reactions can also be measured as a participant's affective response to visual or auditory affective stimuli. For example, Larsen et al. (2003) presented participants with pictures, sounds, and words of positive and negative affective content and measured the zygomaticus and corrugator facial muscle sites while participants reported their affective states. A relationship was found between self-reported valence ratings and facial EMG activity. Positive valence ratings were associated with activity in the zygomaticus muscle site and negative valence ratings with corrugator site activity. Facial reactions to emotional stimuli can also be assessed using EMG. Dimberg (1988) presented happy and angry facial expressions to participants and measured corrugator and zygomaticus site activity as well as heart rate.

Increased corrugator site activity, heart rate deceleration, and more subjective experiences of fear were found in response to angry stimuli compared to happy stimuli. Conversely, increased zygomatic site activity and more subjective experiences of happiness were found in response to happy stimuli. A wide range of stimuli types with varying intensities can be used in research on the experiences and expression of affect and emotion and responses measured with facial EMG.

Investigating Emotion-Related Processes Using Facial EMG

The sensitivity of facial EMG in detecting facial muscle activity is of particular importance when examining phenomena that are difficult to observe with other approaches. For example, consider the investigation of covert facial mimicry. When we see a facial emotion expression, it is very likely that the muscles in our own face will become subtly activated in a manner that matches the observed expression. This phenomenon is commonly termed 'facial mimicry' and was first reported by Dimberg (1982). He investigated facial EMG from the zygomaticus and corrugator muscle sites while participants observed pictures of facial emotional expressions of anger and happiness. The results showed greater zygomaticus site activity in response to happiness than anger expressions and greater corrugator site activity in response to anger than happiness expressions. This phenomenon has since been replicated numerous times from the zygomaticus and corrugator muscle sites (for a review, see Hess & Fischer, 2013). These authors also list facial EMG studies where additional muscles were investigated in facial mimicry. For example, the levator labii muscle site has been reported to respond to observing facial expressions of disgust (Lundqvist, 1995; Lundqvist & Dimberg, 1995; Murata et al., 2016; Oberman et al., 2007; Rymarczyk et al., 2016) and the lateralis frontalis muscle site to expressions of fear (Lundqvist, 1995; Rymarczyk et al., 2016) and surprise (Lundqvist, 1995; Lundqvist & Dimberg, 1995; Murata et al., 2016). Nonetheless, the evidence is rather limited for matched facial muscle activation in observers for muscle sites other than the zygomaticus or corrugator.

Generally, studies on facial mimicry listed above investigated emotion-specific facial muscle activation in individual facial muscle sites for multiple emotion categories. As described earlier, some facial muscles are involved in the expression of various emotions (see Table 17.1). The corrugator muscle constitutes a prime example—it is involved in many expressions of negative affect and emotion. Thus, a different approach to showing differential facial muscle activation related to facial mimicry would be to investigate facial EMG across several muscles and consider the emerging activation patterns per emotion category, similar to the approach taken by Fridlund et al. (1984). Wingenbach et al. (2020) measured facial EMG from the corrugator, zygomaticus, depressor, levator, and frontalis facial muscle sites while participants watched dynamic facial expressions of the six basic emotions as well as the more complex emotions of contempt, pride, embarrassment, and neutral facial expressions (i.e. blank stares). The expected activation per muscle site based on previous work on facial emotional expressions was prespecified (as contrast coefficients) for each emotion category and treated as patterns (see https://www.nature.com/articles/s41598-020-61563-5/tables/1 Table 1 in Wingenbach et al., 2020). The measured EMG data across facial muscle sites per emotion category were compared to the theory-based expected patterns to investigate facial mimicry per emotion category. The measured EMG pattern of each emotion category with its expected pattern was also contrasted to expected patterns of emotion categories of the same valence category (positive, neutral, and negative) to test for distinctiveness. Results showed that the measured EMG data matched the expected patterns for most tested emotions. Additionally, the measured EMG patterns for individual emotion categories were distinct within their own valence category for most tested emotions (see Figure 3 in Wingenbach et al., 2020). That is, the measured EMG data better fit the expected patterns of the target emotions than the expected patterns of non-target emotions of the same valence. These findings suggest that facial mimicry is a categorical mirroring of the observed facial emotional expression.

As many studies have now demonstrated, facial EMG can be a useful tool for emotion-specific investigations. Moreover, facial EMG can also be used to investigate variations in facial expressions within an emotion category. For example, research has shown that subtle variations in kinds of smiles are mimicked by observers (Korb et al., 2014; Krumhuber et al., 2014). These researchers recorded facial muscle activity from the corrugator, orbicularis oculi, and zygomaticus sites while participants viewed dynamic displays of various smiles operationalised as variations of AU combinations. These variations are possible because facial expressions of emotion can be posed volitionally, and such posed expressions often differ from spontaneous felt expressions in terms of included AUs. Moreover, judges can reliably discriminate between posed and felt facial expressions (e.g. McLellan et al., 2010). Results from Korb et al. (2014) showed that the recorded EMG activity corresponded with the AUs displayed in the stimuli, demonstrating feature-specific mimicry, similar to the results by Wingenbach et al. (2020). Such findings of specificity in facial muscle activation, in line with the observed stimulus, hint at facial mimicry being a mirroring of the stimulus content rather than an affective reaction to the stimulus, although more research is needed to examine this issue.

Facial EMG can also be used to differentiate between participants' felt and posed facial expressions of emotion. This differentiation is based on divergent temporal characteristics in posed and spontaneous facial expressions (Ekman & Friesen, 1982). For example, spontaneous smiles have a longer duration than posed smiles (Schmidt et al., 2006). Hess et al. (1988) instructed participants to pose or feel happiness and measured facial muscle activation across the zygomaticus, depressor anguli oris, corrugator, and masseter muscle sites. Temporal aspects of the facial EMG measurements (i.e. time mean, time variance, time skewness, and time kurtosis; Cacioppo et al., 1983) distinguished between posed and felt smiles. Such research findings demonstrate that facial EMG is a useful tool in assessing not only participants' different expressions across elicitation conditions but also their defining characteristics.

Based on EMG's high temporal frequency, it is further possible to identify the onset and offset of an expression and to illustrate the development of an expression (e.g. identifying the peak). Achaibou et al. (2008) segmented the recorded signal of the facial muscle activity in the zygomaticus and the corrugator in response to observing expressions of happiness and anger (i.e. a facial mimicry paradigm) in 100 ms epochs. Facial muscle response onsets were defined by comparing the mean facial muscle activity per epoch in response to happy and angry facial emotional expressions to one another (per muscle). The onset of corrugator activity in response to observing angry facial expressions was found at 200 ms after stimulus onset and 500 ms after stimulus onset for happy facial expressions in the zygomaticus. These findings suggest that the corrugator is activated more quickly. Angry expressions might be processed more rapidly than happy expressions which could serve an evolutionary adaptive function. It is further possible that the corrugator is involved in the (stimulus-unspecific) orienting response (Dimberg, 1982) preceding the mimicry response. Moreover, since morphed dynamic stimuli were used in this study, which create artificial facial movements, it remains to be seen whether these timing differences also occur when participants view video-recorded facial emotional expressions including the natural temporal characteristics of the facial emotional expressions. The investigation of the onsets of facial muscle activity when participants observe static facial emotional expressions has not shown differing onsets in the EMG signal in response to the stimuli (Dimberg & Thunberg, 1998).

We have now seen application possibilities of facial EMG to assess the experience of affect and emotion, posed expressions of emotion, and responses related to the processing of stimuli of emotional content (e.g. facial expressions, words, sounds). Another application possibility is using facial EMG as a manipulation check. Some investigations include the manipulation of facial muscle activation in participants, and facial EMG can demonstrate the success of the manipulation. Examples of the manipulation of facial muscle activation are biting on a pen or holding a pen with the lips (e.g. Oberman et al., 2007; Wingenbach et al., 2018) or imitating observed facial expressions (e.g. Wingenbach et al., 2018). In Wingenbach et al. (2018), participants solved a facial emotion recognition task across two conditions with manipulated facial muscle activation, i.e. explicit imitation and pen in the mouth, next to a control condition with no manipulation, while five different facial muscle sites were measured across the face. Participants showed increased activity (compared to the control condition) in all five facial muscle sites in the explicit imitation condition. The pen-holding condition showed the highest activity in the electrodes placed below the left mouth corner (see Figure 2 in Wingenbach et al., 2018). The measured facial muscle activity thus showed a pattern as was intended by the manipulations, and the facial EMG results served to verify the method. The study further showed that an incongruence between visual input (facial emotional expression in the stimuli) and motor action (activity induced under the mouth corner from pen-holding) hampered the recognition of facial emotional expressions with feature saliency in the lower part of the face/mouth region (here, disgust, happiness, embarrassment, contempt, and pride) based on accuracy rates. Emotional expressions with feature saliency in the lower part of the face all include lip movement either outwards or upwards, which is inhibited by the pressing of the lips induced by the pen-holding. Judges' lowered recognition rates might be due to a conflict between facial muscle movement observed in the stimuli and muscular feedback to the brain, which might also be part of a representation of the observed emotion (for more information on embodiment on emotion, see Niedenthal, 2007). Thus, not only can facial EMG serve as a means to verify applied facial muscle manipulations, facial EMG results can also inform interpretation of obtained behavioural results (e.g. recognition rates), and new theoretical insights might be gained.

Challenges of Using Facial EMG and How to Overcome Them

In summary, this chapter highlighted the many strengths of facial EMG and some possible applications in research. Its most notable advantages are (1) increased objectivity relative to self-reports, (2) high sensitivity in detecting small muscle activations, and (3) high temporal frequency allowing for the assessment of rapidly changing activations characteristic of facial expressions. Nonetheless, facial EMG also comes with challenges. It is well-known that awareness about the purpose of a measure or the hypotheses of a study can alter participants' behaviour. To avoid potential influences on the obtained EMG data, it is custom to keep participants blind to the true purpose of the electrodes. This can be achieved by using a cover story in the instructions provided to participants, such as the electrodes measure temperature in various parts of the face. It is also possible that participants alter their natural facial behaviour simply because they have electrodes attached to their face. Some participants report during attachment that they are afraid the electrodes would come off, and others report that they feel restricted in their movements. These challenges can be overcome by ensuring proper electrode attachment (e.g. thorough cleaning of the skin) and asking participants to make grimaces to demonstrate secure electrode attachment. Generally, participants habituate to the electrodes quickly and do not actively feel them anymore. Acceptance of having electrodes attached in the face is generally high in participants, as participants do not perceive the electrodes as disturbing or restricting (Wingenbach, 2010).

Facial muscles are rather small, and the guidelines for electrode placement must thus be carefully followed. When misplacing an electrode by just 1 cm, it is already likely that non-targeted muscles are being recorded. While assessing facial muscle activity from multiple facial sites has numerous advantages, one should be aware of the potential for crosstalk between EMG sites. Researchers should make sure to have sufficient distance between electrode pairs; the smaller the electrodes, the better. Since facial EMG electrodes measure electricity, they are affected by ambient electromagnetic fields creating noise in the data, which can be minimised by collecting data within Faraday cages. It is further recommended to use shielded electrodes, keep electrical devices in the laboratory to a bare minimum, and use a notch filter on the recorded signal. Moreover, further filtering of the EMG data is necessary (e.g. high pass, low pass, moving average), and spike artefacts should be eliminated (e.g. see Wingenbach et al., 2020). Movement artefacts are common during EMG recordings, including sneezing, coughing, scratching, and yawning. Since these artefacts cannot easily be separated from the rest of the signal based on visual inspection, it is recommended to observe participants via camera, take notes including exact timing, and exclude those segments from data analysis. Every face is anatomically different which also includes variations in fatty tissue and muscle size. As a consequence, the recorded strength of the EMG signal has high inter-individual variability in addition to variability in responsiveness per se. To tackle this challenge, normalisation of the EMG data per participant is recommended (e.g. see Wingenbach et al., 2020).

Many investigations using facial EMG opt to z-standardise each participant's data before entering it into analyses. This is then done for each measured facial muscle site across all experimental conditions but individually per participant. While this is indeed a legitimate approach to make the data comparable between participants, researchers are urged to consider the implications of z-standardisation for their results and whether the posed research question can be answered with z-standardised data. For example, should researchers wish to investigate whether there was an increase in facial muscle activity in response to a stimulus, then z-standardisation should not be done. Z-standardisation scales the mean activity from one channel (i.e. facial muscle site) across all trials to zero. Resulting positive z-values are thus to be interpreted as higher than average in response to a specific stimulus and negative z-values as lower than average. Care must thus be taken when interpreting these kinds of results. This problem is exemplified in a recent study by Wingenbach et al. (2020). The corrugator facial muscle site did not show an increase in activity in response to anger facial expression stimuli after a prestimulus baseline correction based on the non-standardised data. However, after z-standardisation of the corrugator site, positive z-values were obtained in response to anger facial expression stimuli (compare the third to fourth column in Fig. 17.1). That is, the corrugator site showed higher than average activity in response to anger facial expressions than to other stimulus categories included in the task. But the resulting positive and negative z-values did not represent an increase or decrease in activity, respectively, as was demonstrated by the non-standardised data, which in fact showed a decrease in activity in response to anger facial expressions. An alternative to z-standardisation is provided by range correction, which does not alter the interpretation of the results. That is, after prestimulus baseline correction, positive values represent an increase in activity in response to a stimulus, and negative values represent a decrease.

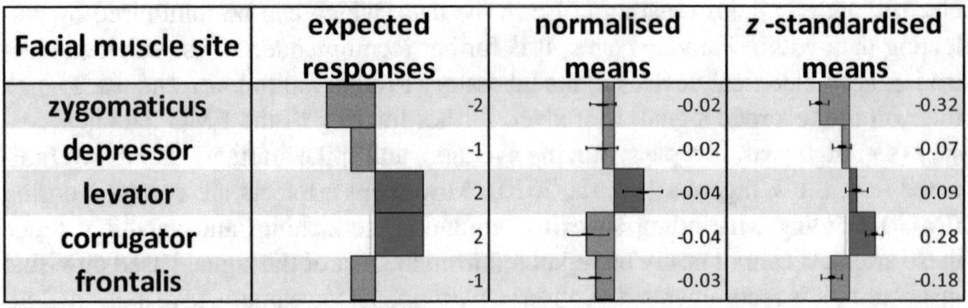

Facial muscle site	expected responses	normalised means	z-standardised means
zygomaticus	-2	-0.02	-0.32
depressor	-1	-0.02	-0.07
levator	2	-0.04	0.09
corrugator	2	-0.04	0.28
frontalis	-1	-0.03	-0.18

Fig. 17.1 Facial muscle responses to facial expression of anger
Note. This figure is a composite of Figures 2 and 4 in Wingenbach et al. (2020). The first column shows the five measured facial muscle sites and the second column the expected facial muscle responses when participants viewed angry facial expressions. Blue bars indicate an (expected) increase compared to a prestimulus baseline, and gold bars indicate an (expected) decrease. The third column shows the measured facial muscle responses to angry faces; the EMG data were range-corrected, and no increase in activity occurred in the corrugator. The fourth column shows the z-standardised means with positive z-values for the corrugator, which are in fact based on a decrease in corrugator activity in response to angry faces

Conclusion

Facial EMG is a sophisticated measurement tool that allows researchers to uncover subtle emotional components and thus deepen our understanding of emotion-related phenomena that occur in face-to-face social interaction (e.g. facial mimicry). It can also add to our knowledge of the experience and expression of emotions through faces, such as fine-grained temporal characteristics of facial emotional signalling. Based on sociocultural norms, people sometimes suppress their emotional feelings and experiences, which can include suppressing the associated facial expression. Otherwise, not easily observable facial EMG can provide information about the presence of a suppressed emotion. This provides researchers with a nice alternative to self-report, which are subjective in nature and require introspective abilities that vary across individuals, and is subject to a host of biases and normative constraints. Facial EMG further allows us to differentiate authentically felt emotion from posed affect/emotion and can uncover phenomena that we would not otherwise be aware of (e.g. facial mimicry). Overall, facial EMG is a valuable tool that is expanding our current knowledge on phenomena and processes associated with and underlying affect and emotion.

Acknowledgement I would like to thank Greg Bryant for procrastinating on his work by reviewing and proofing this chapter.

References

Achaibou, A., Pourtois, G., Schwartz, S., & Vuilleumier, P. (2008). Simultaneous recording of EEG and facial muscle reactions during spontaneous emotional mimicry. *Neuropsychologia, 46*(4), 1104–1113. https://doi.org/10.1016/j.neuropsychologia.2007.10.019

Alexander, I. E., & Adlerstein, A. M. (1958). Affective responses to the concept of death in a population of children and early Adolescents. *The Journal of Genetic Psychology, 93*(2), 167–177. https://doi.org/10.1080/00221325.1958.10532416

Block, J. (1957). A study of affective responsiveness in a lie-detection situation. *The Journal of Abnormal and Social Psychology, 55*(1), 11–15. https://doi.org/10.1037/h0046624

Cacioppo, J. T., Marshall-Goodell, B., & Dorfman, D. D. (1983). Skeletal muscular patterning: Topographical analysis of the integrated electromyogram. *Psychophysiology, 20*(3), 269–283.

Cacioppo, J. T., Petty, R. E., Losch, M. E., & Kim, H. S. (1986). Electromyographic activity over facial muscle regions can differentiate the valence and intensity of affective reactions. *Journal of Personality and Social Psychology, 50*(2), 260–268. https://doi.org/10.1037/0022-3514.50.2.260

Cacioppo, J. T., Bush, L. K., & Tassinary, L. G. (1992). Microexpressive facial actions as a function of affective stimuli: Replication and extension. *Personality and Social Psychology Bulletin, 18*(5), 515–526. https://doi.org/10.1177/0146167292185001

Darwin, C. (1872). *The expression of the emotions in man and animals.* John Murray. https://doi.org/10.1037/10001-000

Dimascio, A., Boyd, R. W., & Greenblatt, M. (1957). Physiological correlates of tension and antagonism during psychotherapy. *Psychosomatic Medicine, 19*(2), 99–104. https://doi.org/10.1097/00006842-195703000-00002

Dimberg, U. (1982). Facial reactions to facial expressions. *Psychophysiology, 19*(6), 643–647. https://doi.org/10.1111/j.1469-8986.1982.tb02516.x

Dimberg, U. (1988). Facial electromyography and the experience of emotion. *Journal of Psychophysiology, 2*(4), 277–282.

Dimberg, U., & Thunberg, M. (1998). Rapid facial reactions to emotional facial expressions. *Scandinavian Journal of Psychology, 39*(1), 39–45. https://doi.org/10.1111/1467-9450.00054

Ekman, P. (1992a). An argument for basic emotions. *Cognition & Emotion, 6*(3), 169–200. https://doi.org/10.1080/02699939208411068

Ekman, P. (1992b). Are there basic emotions? *Psychological Review, 99*(3), 550–553. https://doi.org/10.1037/0033-295X.99.3.550

Ekman, P., & Friesen, W. V. (1971). Constants across cultures in the face and emotion. *Journal of Personality and Social Psychology, 17*(2), 124–129. https://doi.org/10.1037/h0030377

Ekman, P., & Friesen, W. V. (1975). *Unmasking the face: A guide to recognizing emotions from facial clues.* Prentice-Hall.

Ekman, P., & Friesen, W. (1976). *Pictures of facial affect.* Consulting Psychologists Press.

Ekman, P., & Friesen, W. (1978). *Facial Action Coding System: A technique for the measurement of facial movements.* Consulting Psychologists Press.

Ekman, P., & Friesen, W. V. (1982). Felt, false, and miserable smiles. *Journal of Nonverbal Behavior, 6*(4), 238–258. https://doi.org/10.1007/BF00987191

Ekman, P., Friesen, W. V., & Ellsworth, P. (1982). What emotion categories or dimensions can observers judge from facial behaviour? In P. Ekman & I. P. Ekman (Eds.), *Emotion in the Human Face* (2nd ed., pp. 39–55). Cambridge University Press.

Ekman, P., Friesen, W. V., & Hager, J. C. (2002). *Facial action coding system.* Research Nexus.

Farina, D., Merletti, R., Indino, B., & Graven-Nielsen, T. (2004). Surface EMG crosstalk evaluated from experimental recordings and simulated signals. Reflections on crosstalk interpretation, quantification and reduction. *Methods of Information in Medicine, 43*(1), 30–35. Retrieved from http://www.ncbi.nlm.nih.gov/pubmed/15026832

Fridlund, A. J. (1994). *Human facial expression: An evolutionary view* (Vol. xiv). Academic, 369 p.

Fridlund, A. J., & Cacioppo, J. T. (1986). Guidelines for human electromyographic research. *Psychophysiology, 23*(5), 567–589. https://doi.org/10.1111/j.1469-8986.1986.tb00676.x

Fridlund, A. J., Schwartz, G. E., & Fowler, S. C. (1984). Pattern recognition of self-reported emotional state from multiple-site facial EMG activity during affective imagery. *Psychophysiology, 21*(6), 622–637. https://doi.org/10.1111/j.1469-8986.1984.tb00249.x

Goldstein, M. J., Jones, R. B., Clemens, T. L., Flagg, G. W., & Alexander, F. G. (1965). Coping style as a factor in psychophysiological response to a tension-arousing film. *Journal of Personality and Social Psychology, 1*(4), 290–302. https://doi.org/10.1037/h0021917

Gross, J. J., & Levenson, R. W. (1995). Emotion elicitation using films. *Cognition and Emotion, 9*(1), 87–108. https://doi.org/10.1080/02699939508408966

Harmon-Jones, E., Harmon-Jones, C., & Summerell, E. (2017). On the importance of both dimensional and discrete models of emotion. *Behavioral Sciences (Basel, Switzerland), 7*(4). https://doi.org/10.3390/bs7040066

Hess, U., & Fischer, A. (2013). Emotional mimicry as social regulation. *Personality and Social Psychology Review, 17*(2), 142–157. https://doi.org/10.1177/1088868312472607

Hess, U., Kappas, A., McHugo, G. J., Kleck, R. E., & Lanzetta, J. T. (1988). An analysis of the encoding and decoding of spontaneous and posed smiles: The use of facial electromyography. *Journal of Nonverbal Behavior, 13*(2), 121–137. https://doi.org/10.1007/BF00990794

Izard, C. E. (1977). *Human emotions*. Plenum Publishing Corp.

Janak, P. H., & Tye, K. M. (2015). From circuits to behaviour in the amygdala. *Nature, 517*(7534), 284–292. https://doi.org/10.1038/nature14188

Kaiser, C., & Roessler, R. (1970). Galvanic skin responses to motion pictures. *Perceptual and Motor Skills, 30*(2), 371–374. https://doi.org/10.2466/pms.1970.30.2.371

Koerner, N., & Antony, M. M. (2010). Special series on disgust and phobic avoidance: A commentary. *International Journal of Cognitive Therapy, 3*(1), 52–63. https://doi.org/10.1521/ijct.2010.3.1.52

Korb, S., With, S., Niedenthal, P., Kaiser, S., & Grandjean, D. (2014). The perception and mimicry of facial movements predict judgments of smile authenticity. *PLoS One, 9*(6), e99194. https://doi.org/10.1371/journal.pone.0099194

Krumhuber, E. G., Kappas, A., & Manstead, A. S. R. (2013). Effects of dynamic aspects of facial expressions: A review. *Emotion Review, 5*(1), 41–46. https://doi.org/10.1177/1754073912451349

Krumhuber, E. G., Likowski, K. U., & Weyers, P. (2014). Facial Mimicry of spontaneous and deliberate Duchenne and non-Duchenne smiles. *Journal of Nonverbal Behavior, 38*(1), 1–11. https://doi.org/10.1007/s10919-013-0167-8

Larsen, J. T., Norris, C. J., & Cacioppo, J. T. (2003). Effects of positive and negative affect on electromyographic activity over zygomaticus major and corrugator supercilii. *Psychophysiology, 40*(5), 776–785. Retrieved from http://www.ncbi.nlm.nih.gov/pubmed/14696731

Levenson, R. W., Ekman, P., & Friesen, W. V. (1990). Voluntary facial action generates emotion-specific autonomic nervous system activity. *Psychophysiology, 27*(4), 363–384. https://doi.org/10.1111/j.1469-8986.1990.tb02330.x

Lundqvist, L.-O. (1995). Facial EMG reactions to facial expressions: A case of facial emotional contagion? *Scandinavian Journal of Psychology, 36*(2), 130–141. https://doi.org/10.1111/j.1467-9450.1995.tb00974.x

Lundqvist, L.-O., & Dimberg, U. (1995). Facial expressions are contagious. *Journal of Psychophysiology, 9*(3), 203–211.

Matsumoto, D., & Ekman, P. (1988). *Japanese and Caucasian facial expressions of emotion and neutral faces (JACFEE and JACNeuF)* (p. 401). Human Interaction Laboratory, University of California.

McLellan, T., Johnston, L., Dalrymple-Alford, J., & Porter, R. (2010). Sensitivity to genuine versus posed emotion specified in facial displays. *Cognition and Emotion, 24*(8), 1277–1292.

Murata, A., Saito, H., Schug, J., Ogawa, K., & Kameda, T. (2016). Spontaneous facial mimicry is enhanced by the goal of inferring emotional states: Evidence for moderation of "automatic" mimicry by higher cognitive processes. *PLOS ONE, 11*(4), e0153128. https://doi.org/10.1371/journal.pone.0153128

Nazmi, N., Abdul Rahman, M., Yamamoto, S.-I., Ahmad, S., Zamzuri, H., Mazlan, S., ... Mazlan, S. A. (2016). A review of classification techniques of EMG signals during isotonic and isometric contractions. *Sensors, 16*(8), 1304. https://doi.org/10.3390/s16081304

Niedenthal, P. M. (2007). Embodying emotion. *Science, 316*, 1002–1005. https://doi.org/10.1126/science.1136930

Nummenmaa, L., Glerean, E., Hari, R., & Hietanen, J. K. (2014). Bodily maps of emotions. *Proceedings of the National Academy of Sciences, 111*(2), 646–651. https://doi.org/10.1073/pnas.1321664111

Oberman, L. M., Winkielman, P., & Ramachandran, V. S. (2007). Face to face: Blocking facial mimicry can selectively impair recognition of emotional expressions. *Social Neuroscience, 2*(3–4), 167–178. https://doi.org/10.1080/17470910701391943

Oken, D. (1962). Relation of physiological response to affect expression. *Archives of General Psychiatry, 6*(5), 336. https://doi.org/10.1001/archpsyc.1962.01710230004002

Ortony, A., & Turner, T. J. (1990). What's basic about basic emotions? *Psychological Review, 97*(3), 315–331. https://doi.org/10.1037/0033-295X.97.3.315

Plutchik, R. (1980). *A general psychoevolutionary theory of emotion. Theories of emotion* (Vol. 1). Academic Press.

Russell, J. A. (1980). A circumplex model of affect. *Journal of Personality and Social Psychology, 39*(6), 1161–1178. https://doi.org/10.1037/h0077714

Russell, J. A. (1994). Is there universal recognition of emotion from facial expression? A review of the cross-cultural studies. *Psychological Bulletin, 115*(1), 102–141. Retrieved from http://www.ncbi.nlm.nih.gov/pubmed/8202574

Rymarczyk, K., Żurawski, Ł., Jankowiak-Siuda, K., & Szatkowska, I. (2016). Emotional empathy and facial mimicry for static and dynamic facial expressions of fear and disgust. *Frontiers in Psychology, 7*, 1853. https://doi.org/10.3389/fpsyg.2016.01853

Schmidt, K. L., Ambadar, Z., Cohn, J. F., & Reed, L. I. (2006). Movement differences between deliberate and spontaneous facial expressions: Zygomaticus major action in smiling. *Journal of Nonverbal Behavior, 30*(1), 37–52. https://doi.org/10.1007/s10919-005-0003-x

Schwartz, G. E., Fair, P. L., Salt, P., Mandel, M. R., & Klerman, G. L. (1976). Facial expression and imagery in depression: An electromyographic study. *Psychosomatic Medicine, 38*(5), 337–347. Retrieved from https://journals.lww.com/psychosomaticmedicine/Fulltext/1976/09000/Facial_Expression_and_Imagery_in_Depression__An.6.aspx

Tomkins, S. S. (1962). *Affect, imagery, consciousness: Vol. I. The positive affects.* Springer.

Tomkins, S. S. (1984). Affect theory. In K. R. Scherer & P. Ekman (Eds.), *Approaches to emotion* (Vol. 163, p. 195). Psychology Press.

Tottenham, N., Tanaka, J. W., Leon, A. C., McCarry, T., Nurse, M., Hare, T. A., ... Nelson, C. (2009). The NimStim set of facial expressions: Judgments from untrained research participants. *Psychiatry Research, 168*(3), 242–249. https://doi.org/10.1016/j.psychres.2008.05.006

Uvnäs-Moberg, K. (1998). Oxytocin may mediate the benefits of positive social interaction and emotions. *Psychoneuroendocrinology, 23*(8), 819–835. https://doi.org/10.1016/S0306-4530(98)00056-0

Van Der Schalk, J., Hawk, S. T., Fischer, A. H., & Doosje, B. (2011). Moving faces, looking places: Validation of the Amsterdam dynamic facial expression set (ADFES). *Emotion, 11*(4), 907.

Vogel, W., Baker, R. W., & Lazarus, R. S. (1958). The role of motivation in psychological stress. *The Journal of Abnormal and Social Psychology, 56*(1), 105–112. https://doi.org/10.1037/h0040719

Vrana, S. R. (1993). The psychophysiology of disgust: Differentiating negative emotional contexts with facial EMG. *Psychophysiology, 30*(3), 279–286. https://doi.org/10.1111/j.1469-8986.1993.tb03354.x

Vytal, K., & Hamann, S. (2010). Neuroimaging support for discrete neural correlates of basic emotions: A voxel-based meta-analysis. *Journal of Cognitive Neuroscience, 22*(12), 2864–2885. https://doi.org/10.1162/jocn.2009.21366

Waller, B. M., Cray, J. J., & Burrows, A. M. (2008). Selection for universal facial emotion. *Emotion, 8*(3), 435–439. https://doi.org/10.1037/1528-3542.8.3.435

Wingenbach, T. S. H. (2010). *Feasibility and acceptability of a morphed faces emotion recognition paradigm – A pilot study*. University of Basel.

Wingenbach, T. S. H., Ashwin, C., & Brosnan, M. (2016). Validation of the Amsterdam dynamic facial expression set – Bath intensity variations (ADFES-BIV): A set of videos expressing low, intermediate, and high intensity emotions. *PLoS One, 11*(1), e0147112. https://doi.org/10.1371/journal.pone.0147112

Wingenbach, T. S. H., Brosnan, M., Pfaltz, M. C., Plichta, M. M., & Ashwin, C. (2018). Incongruence between observers' and observed facial muscle activation reduces recognition of emotional facial expressions from video stimuli. *Frontiers in psychology, 9*, 864.

Wingenbach, T. S. H., Brosnan, M., Pfaltz, M., Peyk, P., & Ashwin, C. (2020). Perception of discrete emotions in others: Evidence for distinct facial mimicry Patterns. *Scientific Reports, 10*, 4692. https://doi.org/10.1038/s41598-020-61563-5

Wolpert, L. (2008). Depression in an evolutionary context. *Philosophy, Ethics, and Humanities in Medicine, 3*(1), 8. https://doi.org/10.1186/1747-5341-3-8

Young, A. W., Perrett, D. I., Calder, A. J., Sprengelmeyer, R., & Ekman, P. (2002). *Facial expressions of emotion: Stimuli and tests*. Thames Valley Test Company.

Zajonc, R. B. (1980). Feeling and thinking: Preferences need no inferences. *American Psychologist, 35*(2), 151–175. https://doi.org/10.1037/0003-066X.35.2.151

Index

A

Active inference, 122, 130, 131, 136
Adolescence and childhood, 162–165,
    167, 170
Aesthetic appreciation, 53, 57, 59
Aesthetic stimuli, 54–56, 58, 59
Amygdala, 9, 12–14, 24, 26, 29–34, 56, 66,
    93, 97, 99, 100, 110, 127, 223–225,
    239, 265, 284
Audio visuomotor neurons, 75

B

Brain imaging methods, 214, 218

C

Cognitive ethology, 275, 276

D

Default-mode network, 58, 147
Dopamine system, 7, 8, 10, 12, 15

E

EEG and ERP, 5, 24, 31–33, 44–47, 66–68,
    70–73, 76, 80, 87, 89, 94, 97–99, 101,
    149, 154, 166, 186, 195, 196, 198–200,
    202, 204–207, 209, 214, 215, 224, 226,
    234–236, 247, 262
Electromyography, 44, 57, 283, 287
Electrophysiology of language, 196, 197

Embodiment, 37–48, 294
Emergence, 101, 122, 136, 215
Emotion/affective decoding, 223–224
Emotional processing, 4, 13, 14, 23, 24,
    26–29, 31–34, 56, 108–111, 224, 265
Emotional recognition, 162, 167
Emotional states, 11, 34, 44, 47, 54, 55, 70,
    88, 90, 171, 238, 275, 285, 288
Emotion embodiment, 38, 43–48, 294
Empathy, 40, 42, 43, 47, 65, 70, 82, 86, 92,
    98–102, 113, 132–134, 162, 163, 170,
    238–240, 245, 258, 261
Eye tracking, 246, 247, 272, 274–277

F

Face *pareidolia*, 98
Facial EMG, 283, 287–296
Facial emotional expressions, 44, 46, 283,
    286, 287, 290–293
Facial expressions, 14, 24–27, 29, 31–33, 40,
    43–46, 48, 70, 86, 88–96, 98, 102, 163,
    166, 262–264, 273, 274, 284–296
Facial muscles, 44, 46, 56, 285–296
Functional magnetic resonance imaging
    (fMRI), 4, 5, 9, 12–15, 24, 28–31, 33,
    43, 56–58, 65–67, 69, 74, 76, 77, 79,
    90, 97, 147, 154, 178, 179, 181, 186,
    207, 214, 215, 224, 225, 232–237, 239,
    240, 242–243, 246, 247
Functional near infrared spectroscopy
    (fNIRS), 186, 214, 215, 224, 226,
    232–240, 244–247

H
Halo effect, 58
Hemispheric asymmetries, 68, 69, 87
Homeostasis, 3, 38, 121, 122, 127, 128, 132
Human emotions, 10, 14, 224
Hyperscanning, 184, 186, 187, 225, 232,
        235–240, 242–244, 246, 247

M
Machine learning, 216
Mental improvisation, 146, 148, 149, 151, 152
Mental navigation, 146, 148–152
Mind wandering, 58, 146–154
Mirror neuron system, 39, 40, 65, 67, 96,
        238–240, 245
Molecular imaging, 6, 11, 13
Moral psychology, 108
Morality, 107–109, 111–113, 167

N
Neurofeedback, 214, 224–226, 246
Neuropsychiatry, 179, 184
Neuroscience machine learning, 214–227

O
Opioid system, 9, 11–14, 182
Orthographic analysis, 197–199

P
Parental response, 86, 90, 93, 94
Perceptual decoupling, 146, 148–150, 152
Psychiatric disorders, 162, 169, 178–180,
        182–187, 214, 223
Pupillometry, 275

R
Racial bias, 42, 43, 47, 210

S
Second-person neuroscience, 183, 184
Serotonin system, 14, 15
Sex hormones, 100–102
Social brain, 86, 184, 226, 232, 238, 239, 243,
        244, 246, 247
Social embodiment, 40–43, 48
Social interaction, 12, 34, 42, 86, 102, 120,
        121, 129, 130, 134–136, 162–164, 168,
        169, 171, 178, 179, 183–187, 204, 225,
        226, 238, 240, 242, 243, 245–247,
        274–278, 296
Stereotypes and prejudices, 43, 48, 131, 200,
        204–210, 258–259, 266
Subcortical visual pathway, 24, 26, 31, 34

T
Theory of mind, 66, 69, 152, 153, 162, 164,
        165, 204, 238–240
Transcranial Direct Current Stimulation
        (tDCS), 256–259, 261–266
Transcranial Magnetic Stimulation (TMS), 79,
        226, 256–260, 262, 263, 265, 266
Trust, 92, 102, 120–137, 259

U
Unconscious emotional responses,
        26–29, 33

W
Whole brain coverage, 247

Printed in the United States
by Baker & Taylor Publisher Services